D0375119

*The twentieth century has seen biology come of age as a conceptual and quantitative science. Biochemistry, cytology, and genetics have been unified into a common framework at the molecular level. However, cellular activity and development are regulated not by the interplay of molecules alone, but by interactions of molecules organized in complex arrays, subunits, and organelles. Emphasis on organization is, therefore, of increasing importance.*

*So it is too, at the other end of the scale. Organismic and population biology are developing new rigor in such established and emerging disciplines as ecology, evolution, and ethology, but again the accent is on interactions between individuals, populations, and societies. Advances in comparative biochemistry and physiology have given new impetus to studies of animal and plant diversity. Microbiology has matured, with the world of viruses and procaryotes assuming a major position. New connections are being forged with other disciplines outside biology—chemistry, physics, mathematics, geology, anthropology, and psychology provide us with new theories and experimental tools while at the same time are themselves being enriched by the biologists' new insights into the world of life. The need to preserve a habitable environment for future generations should encourage increasing collaboration between diverse disciplines.*

*The purpose of the Modern Biology Series is to introduce the college biology student—as well as the gifted secondary student and all interested readers—both to the concepts unifying the fields within biology and to the diversity that makes each field unique.*

*Since the series is open-ended, it will provide a greater number and variety of topics than can be accommodated in many introductory courses. It remains the responsibility of the instructor to make his selection, to arrange it in a logical order, and to develop a framework into which the individual units can best be fitted.*

*New titles will be added to the present list as new fields emerge, existing fields advance, and new authors of ability and talent appear. Only thus, we feel, can we keep pace with the explosion of knowledge in Modern Biology.*

*James D. Ebert*
*Ariel G. Loewy*
*Richard S. Miller*
*Howard A. Schneiderman*

**Modern Biology Series**

## Consulting Editors



## Published Titles

*Delevoryas* **Plant Diversification**
*Ehrenfeld* **Biological Conservation**
*Fingerman* **Animal Diversity**
*Griffin-Novick* **Animal Structure and Function,** *second edition*
*Levine* **Genetics,** *second edition*
*Savage* **Evolution,** *second edition*
*Sistrom* **Microbial Life,** *second edition*
*Van der Kloot* **Behavior**

## Forthcoming Titles

*Burnett-Eisner* **Animal Adaptation,** *second edition*
*Gillespie* **Cell Function**
*Novikoff-Holtzman* **Cells and Organelles**
*Odum* **Ecology,** *second edition*
*Ray* **The Living Plant,** *second edition*

# Interacting Systems in Development

**second edition**

RDCRR

**James D. Ebert**
*Carnegie Institution of Washington*

**Ian M. Sussex**
*Yale University*

*Holt, Rinehart and Winston, Inc.*
*New York Chicago San Francisco*
*Atlanta Dallas Montreal*
*Toronto London Sydney*

*Library of Congress Catalog Card Number: 78–100552*
ISBN: 0-03-081306-9
*Printed in the United States of America*
234 006 98765



*Text and cover design by Margaret O. Tsao*
*Illustrations by Gaetano di Palma*

# *Preface*

The purpose of this second edition, like that of its predecessor, is to instill in students who are beginning their study an understanding of the concepts of developmental biology and the key experiments and observations on which they are based. We have tried to convey not only the fundamental facts but an awareness of how they were obtained in the past, and are being obtained today.

In stating our objectives, we can hardly improve on the introduction to this Series as a whole which reads, in part, "... cellular activity and development are regulated not by the interplay of molecules alone, but by interactions of molecules organized in complex arrays, subunits and organelles. Emphasis on organization is, therefore, of increasing importance." Thus our title, *Interacting Systems in Development*, accurately expresses the essence of our approach.

The worldwide response to the first edition, in the original English as well as in the several other languages in which it has appeared, has been gratifying. The reception accorded the book—the evidence that it has played at least a small part in the reshaping of the teaching of developmental biology—was one of two major forces in our decision to accept our publisher's invitation to prepare this new edition. The second was the belief, based on our own experience and the constructive criticism of students and teachers, that the book could be substantially improved. Thus, although the plan of the book remains the same, we believe it has been executed more effectively in this second edition.

Although our phrase "executed more effectively" means many things—better writing, better definitions, improved illustrations (thanks to Gaetano di Palma), and of course updating to insure timeliness—it means especially that the book is now more truly representative of developmental biology. It now contains more than token integration of the essentials of plant development.

The word integration is used advisedly. We have not attempted to assimilate examples drawn from plant development into a textual framework based solely on what has been learned about the development of animals, but we have given due recognition to the contributions that each of these two groups of organisms has made to the understanding of developmental processes.

At the present time there seems to be little difficulty in discussing the molecular and cellular aspects of animal and plant development jointly, and we have done this. However, morphogenetic cell movements have no counterpart in plant development; neither does postembryonic organ initiation play a significant role in animal development. These are, therefore, treated in separate chapters. Should future research show that cell movements and meristem growth are simply different developmental tactics, and that in the larger strategy of organismal development the resulting interactions are of a fundamentally similar nature, we shall then be able to unify our treatment further.

Developmental biology is a focal field of research: its implications are far-reaching and its concepts are profoundly influencing today's intellectual framework of the biological and behavioral sciences. In planning our book for beginning students we have been fully aware that it may be used by a variety of "beginners": by some in general biology courses whose teachers emphasize the cellular and developmental approach; by others in core programs, in which development may be coupled, here with cellular and genetic biology, there with anatomy and physiology; and, of course, by students in courses in embryology and developmental biology. Our principal concern has not been the "package" or framework within which our readers will be receiving their training in developmental biology. That responsibility rests with the instructor, who knows the background and experience and the motivations of his students. Rather, we have tried to convey the ideas of our subject, its spirit, and the sense of excitement surrounding it—to effect a synthesis that accurately depicts the quality of the field.

The preparation of this edition was aided by the constructive criticism of many colleagues. In some sections we drew heavily on the contributions of the National Academy of Sciences Panel on The Biology of Development. A number of suggestions emerged during the book's translation into other languages; the comments of Giovanni Giudice, G. M. Ignatieva, Tokindo and Eiko Okada, and A. K. Tarkowski especially should be mentioned. A. W. Blackler, D. Branton, T. Delevoryas, R. J. Goss, F. Kafatos, W. M. Laetsch, G. E. Lesh, J. B. Phillips, D. Rudnick, and F. Wilt were frank and helpful critics. To them, and to numerous others who helped along the way, we are deeply grateful.

*Baltimore, Maryland* *J.D.E.*
*New Haven, Connecticut* *I.M.S.*
*February 1970*

# Contents

# *Interacting Systems in Development*

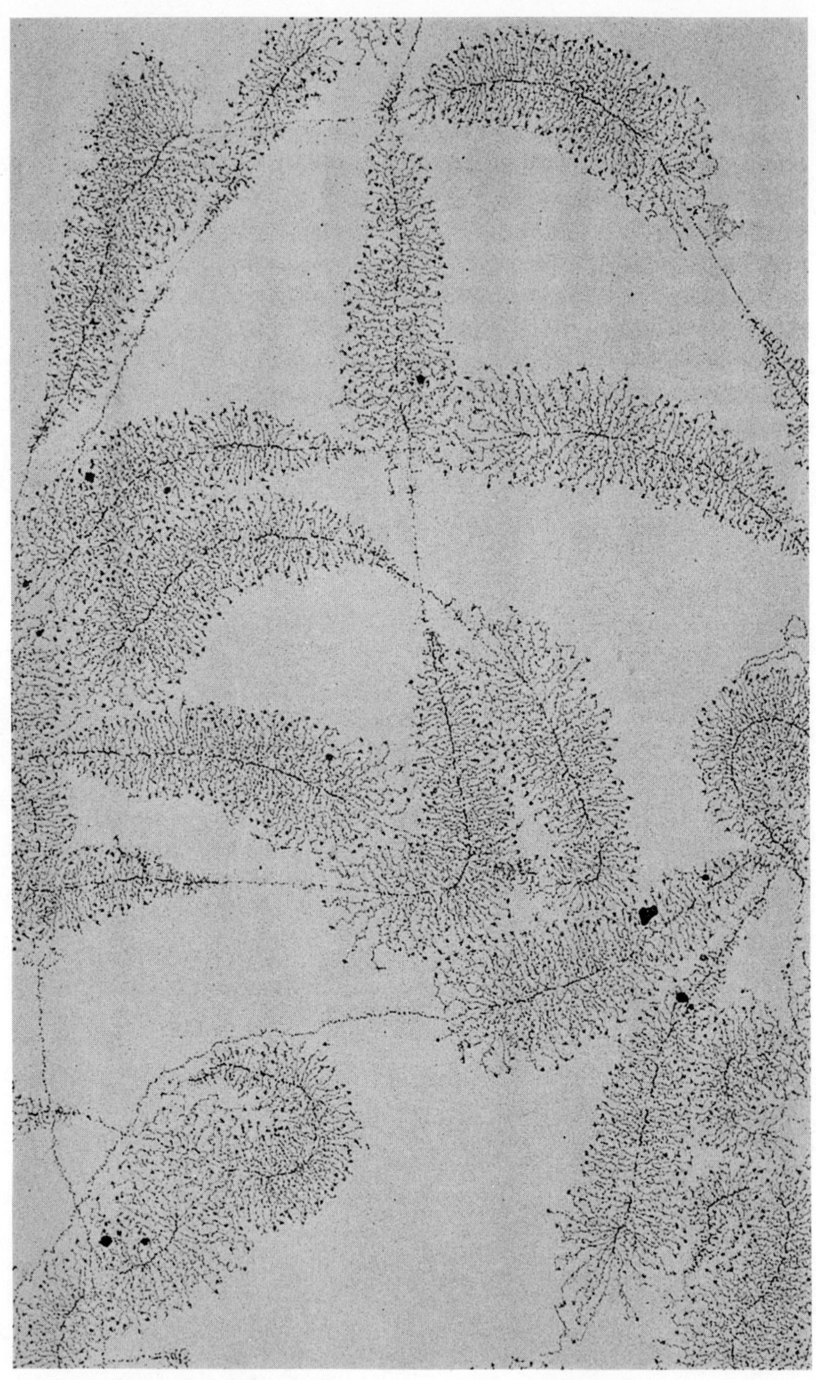

*Nucleolar genes from an amphibian oöcyte. The presence of extrachromosomal nucleoli in amphibian oöcytes permits isolation of the genes coding for ribosomal RNA precursor molecules, which appear as a gradient of fibrils in progressive stages of completion. Active genes are separated by "spacers," stretches of DNA that are apparently inactive at the time of ribosomal RNA synthesis. Electron micrograph (× 22,200) by O. L. Miller, Jr. and Barbara R. Beatty, Biology Division, Oak Ridge National Laboratory.*

*chapter* **1**

# *Interacting Systems in Development*

*The body of a worm and the face of a man alike have to be taken as chemical responses. The alchemists dreamed of old that it might be so. Their dream however supposed a magic chemistry. There they were wrong. The chemistry is plain everyday chemistry. But it is complex. Further, the chemical brew, in preparation for it, Time has been stirring unceasingly throughout some millions of years in the service of a final cause. The brew is a selected brew.*

SHERRINGTON, 1951

*Life is a relationship among molecules and not a property of any one molecule.*

PAULING, 1960

Thus, each in his own way, with clarity and grace, Charles Sherrington and Linus Pauling have written the prologue of our introduction to the mechanisms of development.

The body of a worm and the face of a man alike are composed of cells, organized into tissues. Yet the cells of each tissue are not generalized cells, prefabricated "modules" from which tissues are constructed. They are specialized cells adapted for specific functions in bone and muscle, in blood and nerve. Nor have they, nor the fertilized egg from which they were first derived by cell division, ever been truly generalized, archetypal cells.

They have been embryonic cells, constantly changing in a changing chemical environment. From the very beginning, each cell has its characteristic specificity, which can often be traced back to a specific region of the cytoplasm in the egg in which development began. To be sure, a motor neuron has not always been a motor neuron, nor a root hair cell always a root hair cell. But what are their "birth dates"? Their development is gradual (Figs. 1-1 through 1-6). How long before they extend their processes—axons and root hairs, respectively—are these cells irrevocably committed to become neurons and hair cells?

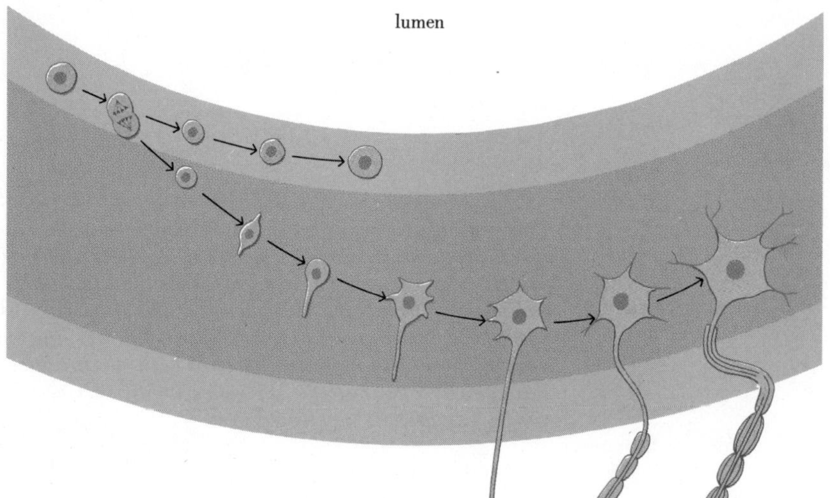


***Fig. 1-1*** *Stages in the development of a multipolar neuron.*

The daughter cells derived from the fertilized egg follow increasingly diverse and specialized pathways resulting in the heterogeneity of cell types characteristic of the adult. Yet we know that during cell division each daughter cell receives an identical set of chromosomes, hence should be genetically identical. In this and in our earlier phrase, "a specific region of the cytoplasm," are the major clues that mark the beginning of our trail. There is no compelling reason to doubt the validity of our underlying assumption that the genes of developing cells of any given species of animal or plant—of muscle and kidney, of root and shoot—are identical; hence we must look beyond the gene, to regional differences in the cytoplasm of the egg. We must look to interrelations between the genes and cytoplasm, to the cell's inner controls.

But in the embryo the kidney cells, muscle cells, and neurons do not develop, or function in isolation. They develop as part of the whole.

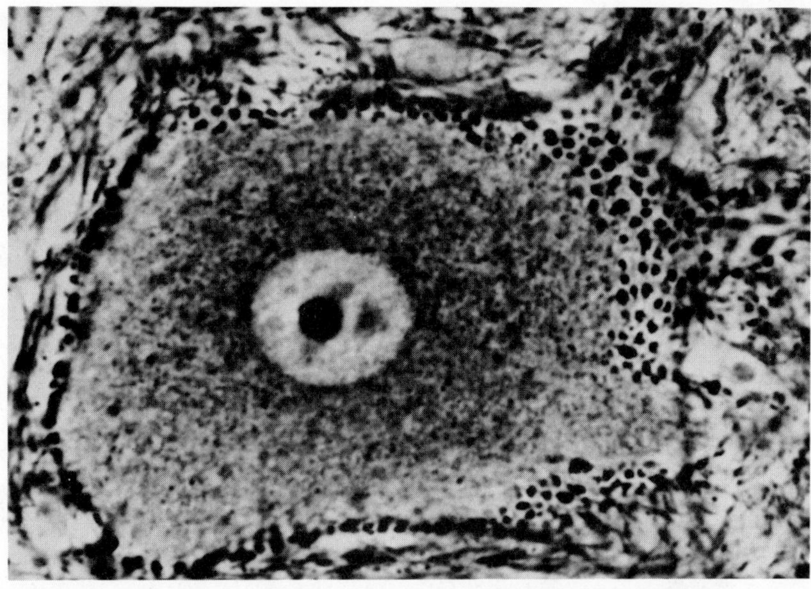

*Fig. 1-2 Photomicrograph of the cell body of a motor neuron with many synapses on its surface. Short lengths of presynaptic axons are stained black. (From G. L. Rasmussen,* New Research Techniques of Neuroanatomy, *by courtesy of Charles C Thomas, Publisher.)*

What properties of the many types of cells of which it is composed determine the shape and size of a worm—or a man? Clearly we must examine not only the cell's inner controls, but the manner in which a cell interacts with its neighbors and its environment. How do cells impinge upon and influence each other?

We have now enumerated two levels of developmental control, the first at the level of the cell and the second at the intercellular level. However, there is yet a third: external environmental factors may exert a profound influence on the developmental response. A specific illustration of environmental control is the photoperiodic control of flowering—the fact that, in many plants, the length of the daily exposure to light determines whether flowers can develop, and thus, in effect, controls sexual reproduction. Nor is the importance of environmental factors limited to plant development. Sexuality in birds is subject to photoperiodic control, and changes in photoperiod and temperature may exert profound effects on the development of insects.

We next wish to bring into sharp focus two concepts, which underlie all developmental biology and will permeate our discussion.

First, all development rests ultimately on the genes. The fabrication of a macromolecule and the final form of an organ must involve a

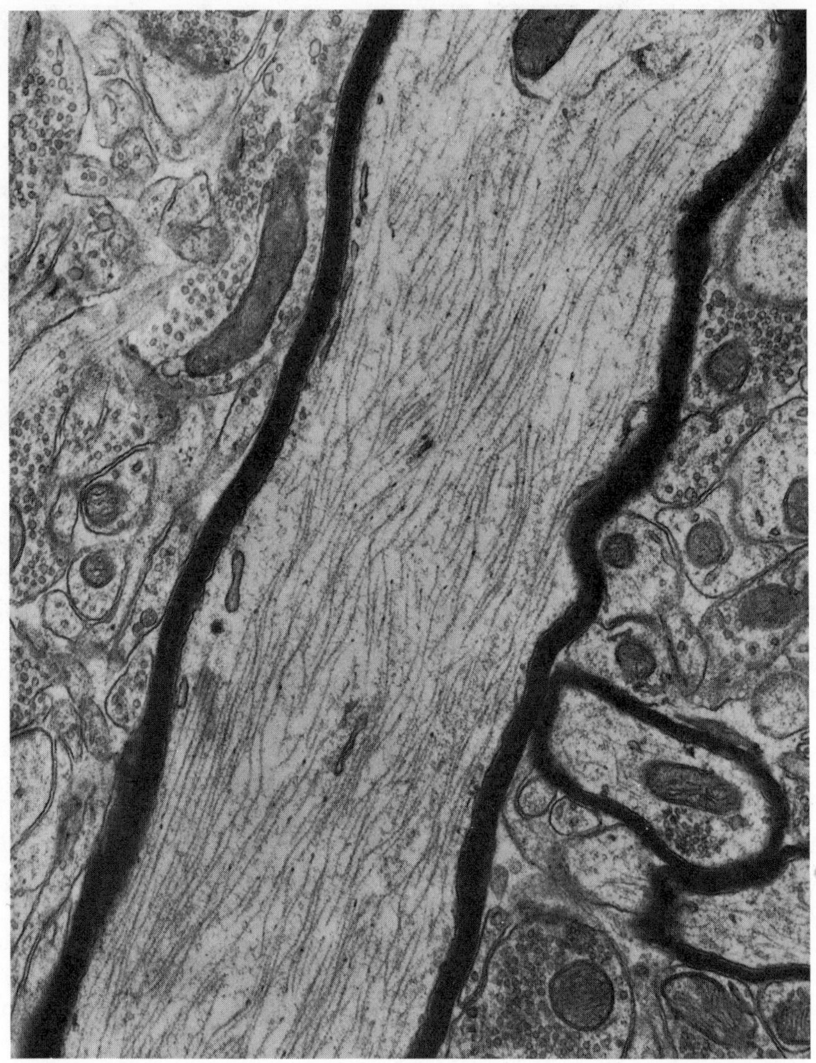

***Fig. 1-3*** *A longitudinal section through a myelinated nerve fiber in the rat cerebral cortex. The axoplasm, filled with numerous long filaments, is bounded by the dense myelin sheath. (From K. R. Porter, in* The Nature of Biological Diversity, *by courtesy of the author and McGraw-Hill Book Company.)*

series of interactions beyond the gene, beyond the individual cell, even with the external environment but eventually it will be necessary to trace the origin of these interactions to the structure of deoxyribonucleic acid (DNA) and the control of its function.

Second, we would stress the properties of embryos and other developing systems as distinct from properties of adults. Too often in studying development we think of an embryo as an adult in miniature, failing to realize that the characteristic properties of an embryo or an embryonic cell change with time. Although the metabolic requirements of a neuron in the adult brain may differ from those of a neuron in the adult peripheral nervous system, the requirements for the *formation* of either during embryonic life may differ in yet other ways. *The requirements for making a neuron may differ drastically from those for maintaining it.* The importance of recognizing the special properties of embryos is underlined if we recall that agents like the German measles virus and the tranquilizing drug, thalidomide, which have only minor effects later in life, produce drastic malformations in early human embryos.

There are developing systems other than the emergence of an individual from the fertilized egg; later we will consider some of them.

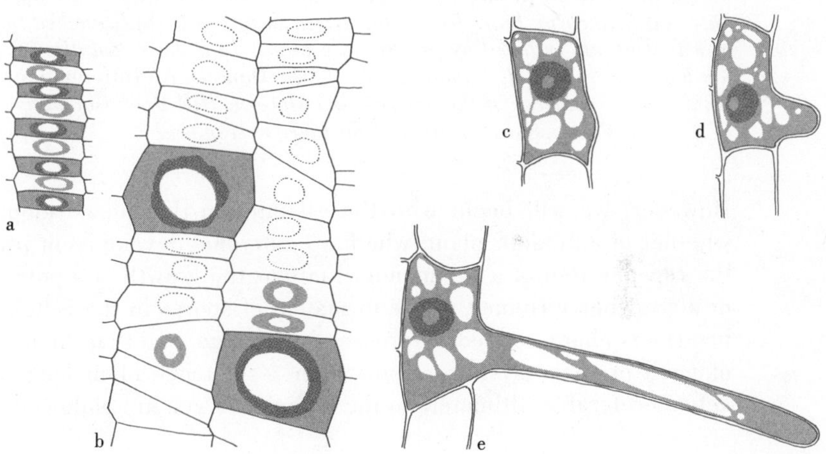


***Fig. 1-4*** *Differentiation of root hairs.* (a) *Epidermal cells that will develop root hairs are first identifiable by the increased density of their cytoplasm and nucleus.* (b) *As the root cells enlarge those that will form root hairs become much larger than adjacent nonroot-hair-forming cells. They retain their characteristic dense cytoplasm and the nucleus undergoes considerable enlargement.* (c) *Outgrowth of part of the cell to form the root hair is a local event and does not involve the entire outer part of the wall. The position of the outgrowth is a site of cytoplasmic accumulation.* (d *and* e) *Later stages of root hair growth showing the polarized nature of the cytoplasm in the growing tip and the nongrowing basal part of the hair. (From E. Bünning,* Entwicklungs- und Bewegungsphysiologie der Pflanze, *by courtesy of the author and Springer-Verlag.)*

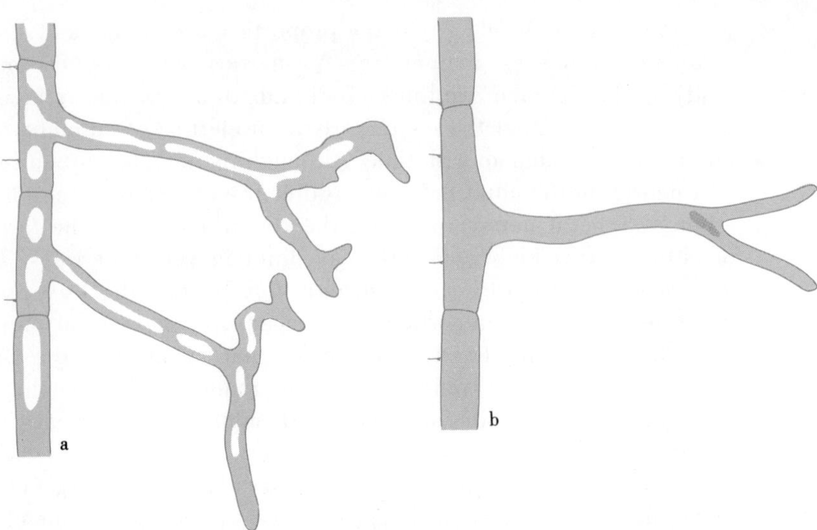


**Fig. 1-5** *Branched root hairs. Developmental control of root hair growth usually results in a single outgrowth. The root hairs in this figure are branched, indicating that the outgrowth is not committed to forming a single unbranched hair.* (a) *Branching hairs from roots grown in dry soil.* (b) *Branched root hairs of wheat plants induced by growing the plants in a dilute solution of phenylurethane. (a from A. J. Eames and L. H. MacDaniels,* An Introduction to Plant Anatomy, *by courtesy of the authors and McGraw-Hill Book Company.* b *from G. Friesen,* Planta 8, *by courtesy of Springer-Verlag.)*

However, we will begin with the assumption that all development—whether of animal or plant, whether embryonic development proper or the regeneration of a salamander limb, or the growth of a potato tuber or a strawberry runner, or the formation of spores in the cellular slime molds—is characterized by progressive change, and that the properties of each of these systems result from a sequence of molecular interactions referable ultimately to the organism's genetic endowment.

## THE COMPONENT PROCESSES OF DEVELOPMENT

Thus we have defined development. Or have we? We have both directly and indirectly stated the cardinal criterion of development: progressive and cumulative change—evident at all levels of biological organization, molecules, cells, tissues, and organs—during an organism's life history. Beyond this, an unfailing definition cannot be given. Certainly no one would doubt that the birth of an infant, or the emergence of a flower, is the culmination of developmental change, or that the formation of a fruiting body in a

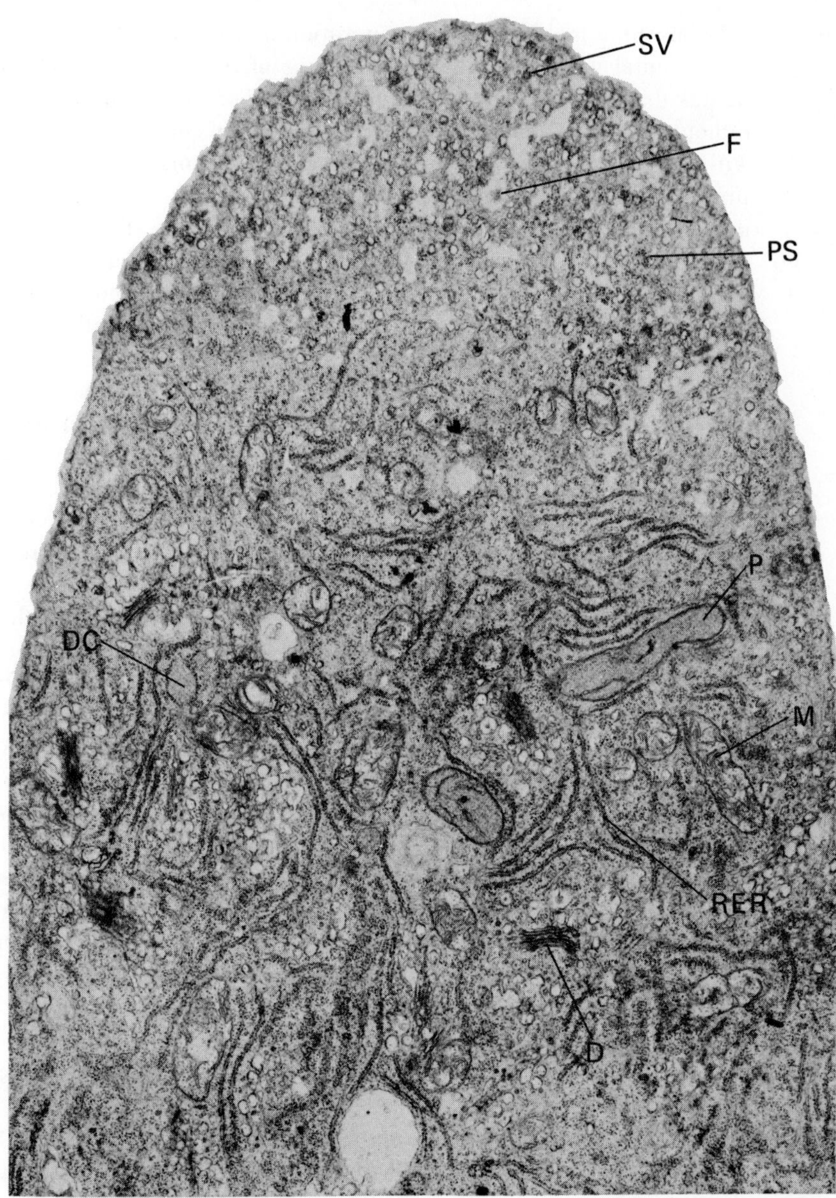


*Fig. 1-6 Electron micrograph of the apex of a growing root hair. The extreme tip contains cytoplasm with only filamentous inclusions* (F), *polysomes* (PS), *and smooth surfaced vesicles* (SV). *The cytoplasmic organization behind the tip is different, and in this region are found plastids* (P), *mitochondria* (M), *dictyosomes* (D), *and rough surfaced endoplasmic reticulum* (RER) *in which are occasional dilated regions* (DC). × *15,000. (From H. T. Bonnett, Jr., and E. H. Newcomb,* Protoplasma, *62, by courtesy of the authors and Springer-Verlag.)*

cellular slime mold and the regeneration of a head in *Planaria* are developmental processes; the consequence of the stepwise events is unmistakable. But what of the individual steps? Here our task is more difficult, because a given process, if taken out of context and considered without reference to whether its impact on the organism has been to produce a recognizable change in state, cannot be clearly defined as developmental or physiological or pathological; the distinction is not meaningful. In the shaping of a limb, a key step is the death of certain cells, a step that is highly ordered in space and time. Knowing the final product, several steps removed, we may speak of cell death as a developmental event—but only in that context.

In the characterization of development, two features stand out: (1) Developmental processes lead to more or less permanent structural change, whereas nondevelopmental processes (such as muscle contraction or nerve impulse conduction) involve only transitory and reversible changes in structure. (2) Developmental phenomena are typically sizeable segments of the life history, whereas other physiological processes occur over relatively brief periods.

Yet we cannot draw a sharp line between these two kinds of change. We see continuity between development and rapid processes in the area of chemical synthesis, since the long-term control of what, and how much, material is synthesized is central to developmental change. Yet we see at the other extreme a continuity between development and the slow course of evolution. In the relative pace of its mechanisms, therefore, development grades into both the rapid processes of biochemistry and the slow processes of evolution. Just as the continuous presence of DNA in successive generations of a species forms the basis of quantitative study of cause and effect in evolution, so does the presence of the same genetic material in all the cells of a developing organism form the starting point for a consideration of developmental mechanisms.

Development is a unified process, which includes the arising of orderly, recognizable patterns as a consequence of (or at least accompanied by) the formation of new constituents, the fabrication of these constituents into larger units, and their rearrangement in space. When we examine a complex process there is some advantage in breaking it down into its component parts, if we realize that the separation is only conceptual.

It is meaningful to consider at least four such component processes: determination, differentiation, growth, and morphogenesis. Although the word *differentiation* is often used in the general sense to mean the full sequence of changes involved in the progressive diversification of cell structure and function that is the hallmark of development, we prefer at the outset to recognize two distinct components. *Determination* is the process by which a cell or a part of an embryo becomes restricted to a given pathway; *differentiation* is the actual appearance of new properties, whether defined in biochemical or structural terms—for example, the appearance of the contractile protein, myosin, in the muscle cell, identified by its chemical or immunological properties or by its characteristic structure when analyzed by electron microscopy. It is important, therefore, that we ask: *differentiation of what?* Of hemoglobin or myosin? Of a single cell? Of a tissue or an embryo?

Do we need to define *growth*, or is the meaning self-evident? Growth means permanent enlargement—that is, developmental increase in total mass. Our definition must exclude physiological fluctuations; on the other hand it must be broad enough to include plant cell enlargement, which involves the entry of water into the cell and ultimately into cytoplasmic vacuoles. Eventually, although the original rigid cell wall is capable of some expansion, a new wall must be synthesized and deposited. Later on we shall discuss the measurement of growth, for protoplasmic increase can be described by counting the number of cells in a population (or a measured sample from it) or by determining the change in a constituent, say protein, by measuring nitrogen content. In fact, usually we want to know how much protein nitrogen has increased per cell, or per unit of DNA, since cell numbers and constituents do not invariably increase together. But before we consider these questions, there are others of a more general nature to be raised. For example, what are the relations between cell enlargement and cell division? Between cell division and cell differentiation? Moreover, as biological systems grow, they often change their form. How are the two phenomena causally related? Does growth always produce a change in form?

Another group of questions is related to the regulation of growth. What initiates growth? Does the newly fertilized egg grow, or do other steps have to be completed before growth begins? What factors determine the rate of growth? We know that different parts of the animal and plant grow at different rates, and that their proportions gradually change. And not least in importance, why does growth stop? Certainly there are no more intriguing observations than the following: a given number—5, 10, 15—of segments are removed from the tail end of a segmented worm; the worm regenerates a new tail of 5, 10, or 15 segments, usually no more or no less. Regeneration stops when the total number of segments, old plus new, approximates the number characteristic of the species (Fig. 1-7). As we shall learn, in higher animals the capacity for regenerating lost parts is gradually restricted, but some regeneration goes on throughout life, even in man. Some clues as to possible mechanisms of growth regulation have come from studies of the regenerating kidney in rats and other animals. If part of a kidney or, better still, one and one half kidneys are removed from a rat, the cells

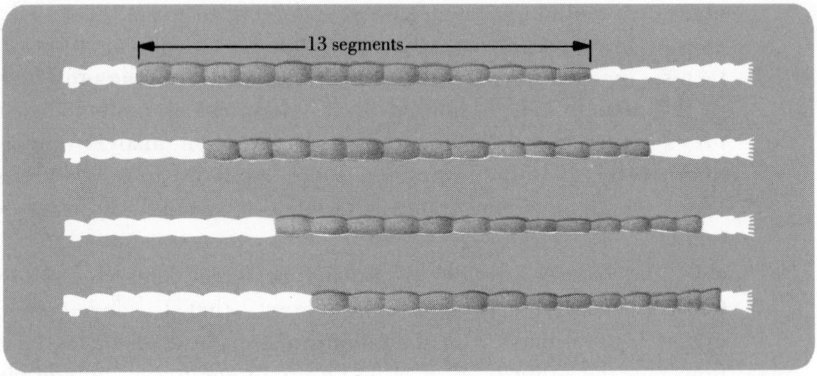

*Fig. 1-7 Regeneration in the segmented marine worm,* Clymenella torquata. *Adults of this species have exactly 22 segments. When anterior and posterior ends are severed at various levels, leaving pieces 13 segments long, simultaneous anterior and posterior regeneration restores the normal number of segments. (From G. B. Moment, in* Journal of Experimental Zoology, *117, by courtesy of the Wistar Institute.)*

in the remaining stump begin to proliferate. If we couple two rats surgically so that their blood circulations are connected (the technique is known technically as *parabiosis*) and remove three kidneys from the two animals, the growth of the remaining kidney proceeds even more vigorously. Does this experiment suggest to you that the kidney is constantly putting into the circulation some specific factor or factors that regulate kidney growth? Or does the organ "monitor" the amount of body to be served and adjust its size accordingly? These ideas have occurred to many biologists, but proof has been elusive and, thus far, the hypothetical factors remain just that.

The term *morphogenesis* means simply the generation of form, the assumption of a new shape. We have implied that it may be a consequence of differentiation and growth—that is to say, the synthesis and aggregation of macromolecules. We have also stated that selective cell death is a morphogenetic mechanism. Although morphogenesis encompasses both of these mechanisms, it is most commonly associated with two others: in animals with morphogenetic movements, and in plants with differential growth. Changing form in animal embryos embraces the organized movements of cells and groups of cells—in other words, rearrangements involving the shifting of cells from one place to another, bringing a given cell into a new environment in which it can be used in the ensuing steps of development. In this sense, it is a process of redistribution.

In plant embryos, because each cell is surrounded by a rigid wall secreted by the cell itself, cells cannot move. Therefore, shape changes result not from cellular migration but from cell growth.

This is but one of the ways in which development differs in animals and plants. However we do not wish to emphasize the contrast between animals and plants or the exclusivity of processes. We wish instead to stress the fact that different processes are used to different extents by different organisms. For example, most animal embryos contain essentially all the organs and tissues that the adults will have; but insect larvae and plant embryos contain only rudimentary primordia of some of the organs, with most of development taking place in the postlarval and postembryonic periods.

These differences are reflected in the organization of this book. Fundamental biochemical pathways are the same in cells of plants, worms, and men. The basic laws of heredity operate in lower forms of life just as they do in man. Some aspects of the behavior of cells—their mode of division and death—are also similar in the many forms studied. Thus in considering some questions we have attempted a synthesis of the information on animals and plants. On the other hand, some questions must be considered in one kingdom, with only passing reference to the other. It is important to emphasize that not all developmental phenomena are readily observable and analyzable in all animal or plant forms. Certain organisms lend themselves particularly to the investigation of cell division. In others, cell migration may be most easily studied; still others are particularly suited for the study of gene action. The investigator therefore must choose the organism most likely to yield information on a particular phenomenon. Thus it is of strategic advantage not to restrict ourselves to the study of mice and men. We must exploit such diverse forms as sea urchins and green plants, frogs, and fungi.

Once it was fashionable to stress the separability of differentiation, growth, and morphogenesis; much was made of experiments in which differentiation and growth, or differentiation and morphogenesis, were separated. There is still much to be gained from such experiments, but in approaching our subject, although we shall study individual events, we shall stress not the dissociability of these processes, but their interactions. Hence our title, *Interacting Systems in Development.*

## FURTHER READING

Balinsky, B. I., *An Introduction to Embryology*, 2d ed. Philadelphia: Saunders, 1965.

Barth, L. J., *Development: Selected Topics.* Reading, Mass.: Addison-Wesley, 1964.

Bodemer, C. W., *Modern Embryology.* New York: Holt, Rinehart and Winston, 1968.

Bonner, J. T., *Morphogenesis* (reprint), New York: Atheneum, 1963.

Corner, G. W., *Ourselves Unborn.* New Haven, Conn.: Yale University Press, 1944.

DeHaan, R. L., and H. Ursprung, *Organogenesis.* New York: Holt, Rinehart and Winston, 1965.

Harris, M., *Cell Culture and Somatic Variation.* New York: Holt, Rinehart and Winston, 1964.

Levine, R. P., *Genetics*, 2d ed. New York: Holt, Rinehart and Winston, 1968.

Loewy, A., and P. Siekevitz, *Cell Structure and Function*, 2d ed. New York: Holt, Rinehart and Winston, 1969.

Novikoff, A., and E. Holtzman, *Cells and Organelles.* New York: Holt, Rinehart and Winston, 1970.

Oppenheimer, J. M., *Essays in the History of Embryology and Biology.* Cambridge, Mass.: M.I.T. Press, 1967.

Pauling, L., "The molecular basis of genetic defects," in *Congenital Defects*, M. Fishbein, ed. Philadelphia: Lippincott, 1963, p. 15.

Saxén, L., and J. Rapola, *Congenital Defects.* New York: Holt, Rinehart and Winston, 1969.

Sherrington, C. S., *Man on His Nature.* New York: Cambridge, 1951, p. 104.

Sinnott, E. W., *Plant Morphogenesis.* New York: McGraw-Hill, 1960.

Steward, F. C., *Growth and Organization in Plants.* Reading, Mass.: Addison-Wesley, 1968.

Taussig, H. B., "The Thalidomide Syndrome," *Scientific American*, August 1962, p. 29.

Torrey, T. W., *Morphogenesis of the Vertebrates*, 2d ed. New York: Wiley, 1967.

Wardlaw, C. W., *Morphogenesis in Plants.* London: Methuen, 1968.

Weiss, P., *Principles of Development.* New York: Holt, Rinehart and Winston, 1939.

Willier, B. H., and J. M. Oppenheimer, *Foundations of Experimental Embryology.* Englewood Cliffs, N.J.: Prentice-Hall, 1964.

Wilson, E. B., *The Cell in Development and Heredity.* New York: Macmillan, 1925.

chapter 2

# Fertilization: Interactions of Eggs and Sperm

When does an individual first deserve this name? In embryonic development, from the very moment the egg is activated. Usually eggs are activated by sperm; however, fertilization by sperm is not necessary to initiate development of many eggs. There are a number of examples of development without fertilization, or *natural parthenogenesis*. In honeybees the males (but not the females), develop from eggs without action of sperm. In the water flea, *Daphnia*, every other day for a lifetime of several weeks a female may produce a hundred live, fatherless young. Within the rotifers, or wheel animalcules as they are sometimes called, at least three different patterns have evolved. In one subclass there is no parthenogenesis;

males are well-developed and all reproduction is bisexual. In a second subclass, bisexual reproduction occurs regularly, but is interspersed with periods of asexual parthenogenesis; in yet a third, no males have ever been identified. In the vertebrates, natural parthenogenesis is very rare. However, it has been shown convincingly that turkey eggs can develop parthenogenetically.

Most eggs can be activated artificially under laboratory conditions. Such *artificial parthenogenesis* may lead to complete or partial development. Studies of parthenogenesis have contributed importantly to our understanding of the mechanisms of activation. They prove that eggs are endowed with the machinery needed for the initiation of development, requiring only to be "triggered."

Thus far we have considered only one of the fundamental aspects of fertilization, activation of the egg. There is a second basic feature: the union of the sperm nucleus with that of the egg, thus bringing together in the primary nucleus of the *zygote*, or fertilized egg, the genetic information contained in the paternal and maternal gametes.

We shall consider the stepwise events of fertilization and their consequences. Our attention shall be confined almost entirely to animals. Occasionally we will be able to compare discrete events in the process in animals and plants, but on the whole, the study of fertilization in plants has hardly begun.

## THE PRIMORDIAL GERM CELLS IN ANIMALS

Where do the gametes originate, and how are they fitted for their respective roles? The functional gametes that mature in the *gonads*—the *ovary* and *testis*, respectively—are eggs and sperm. They are highly differentiated, specialized cells. From studies of parthenogenesis we know that the egg itself contains all the essentials for development. It is specialized for the function of development, just as a muscle is differentiated for contraction or a neuron for transmitting impulses. In short, as higher animals have evolved, the capacity for duplicating the entire organism has been restricted to the egg, a cell specialized for the function of development. And as we have said, and will document shortly, the sperm is highly differentiated for its own subsidiary role.

The cells destined to become the functional gametes do not begin their development in the gonads. The diploid *primordial germ cells*, as the cells which give rise to eggs and sperm are called, are set aside very early in development. By "set aside" we mean that these cells can be identified in early embryos even before the gonads themselves have developed, the specific sites varying with the species. In many species of both invertebrates and vertebrates, the gonads begin to develop independently of the germ cells. Ultimately, the germ cells migrate to the gonad, to settle down and embark on their own specialized course. We shall defer our discussion of the evidence for the precise origin of these cells and their migratory paths until we know more about the terrain of the embryo. For the present we will pick up the trail in the ovary and testis, respectively.

## The Egg

It was an embryologist who wrote, "The avian egg is an architectural marvel." And so it is. The familiar chicken's egg—with its nutrient yolk, nutrient and bacteriostatic albumen, shell membrane, and shell—is a remarkable feat of engineering. But these trappings are only a small part of our story, as are all the protective and supporting membranes that vary so greatly from species to species. Even casual observation reveals that organisms have adopted particular characteristics that enable their young to survive under the special environmental conditions in which they are found. There is far more to tell than the number of eggs in a clutch or their mode of dispersal, or the way in which the embryo is nourished. These facts are important; many are discussed in *Animal Adaptation*, in this series. We shall cite only a few key facts, sufficient to achieve an understanding of their importance, but we are concerned less with compiling of differences than with general principles and ideas.

***Oögenesis*** Once in the ovary, the germ cells pass through three principal phases: (1) proliferation; (2) growth; and (3) maturation. We can say little about the proliferation phase. The germ cells, now called *oögonia*, divide by mitosis, and thus their number is increased. We know little about the factors limiting these divisions, except that as a consequence of them, a number of cells begin to grow and are now recognized as primary oöcytes. The appearance of primary oöcytes is a further expression of germ cell differentiation, as distinct from their earlier determination.

In the frog, for example, this growth begins at about the time the adult emerges at metamorphosis; over a three-year span the oöcyte may increase in diameter from 50 to 1500–2000 $\mu$. Each year a new batch of oöcytes is produced by division of oögonia. Since full growth is not attained for three years, an ovary may contain three "generations" of oöcytes at one time.

The rate of growth of oöcytes varies widely with the species. Contrast the frog with the chicken and mouse: in the laying hen the oöcyte grows rapidly during the two weeks immediately preceding ovulation, whereas in the mouse the proliferative phase is confined to the period of intrauterine life. All of the mature eggs to be ovulated by a mouse during her lifetime are derived from oöcytes laid down in embryonic life. They are present at birth and persist throughout adulthood. This pattern of oögenesis is probably not unique for the mouse; it seems to be a universal feature in mammals, including the human species. Mammalian oöcytes grow rapidly a few (or one) at a time.

In the growth phase we can identify three principal activities, and relate them to the future functional activities of the cell.

1. Large amounts of ribonucleic acids (RNAs) are made. In Chapters 7 and 8 we shall discuss the synthesis of the different classes of RNA, especially "messenger" RNA and ribosomal RNA during oögenesis. Generalizing at this point, however, we can say that oöcytes are highly *basophilic*—that is, they stain with basic dyes which have an affinity for RNA—the specificity being confirmed by the fact that this basophilia is removed by the enzyme ribonuclease. The relative amounts of the different classes of RNA vary with the species; as we shall see in some species, such as frogs and toads, there is an intense synthesis of ribosomal RNA during oögenesis. In short, *during oögenesis the "machinery" for operating the cell during early stages of development is formed.*

2. Since there is an extreme demand for ribosomal RNA during oögenesis, the oöcyte faces a unique problem: its single nucleus must supply ribosomal RNA to a mass of cytoplasm which in other tissues would contain several thousand nuclei. Thus at the beginning of the growth phase, the intense synthesis of ribosomal RNA is preceded by the synthesis of the DNA that codes for it. There is a remarkable *amplification of the genes for ribosomal RNA.*

3. The third activity is the accumulation of proteins. Do oöcytes synthesize the proteins used in their growth? The answer may be given simply: many proteins (and other molecular species as well) are produced outside the ovary and brought to it via the blood stream. To appreciate the evidence, however, it is necessary to look into the relations between the oöcytes and their surroundings. In the ovaries of chordates and many invertebrates, oöcytes are surrounded by *follicle cells* (Fig. 2-1). The membranes of the follicle cells and oöcytes are closely apposed; each sends out fingerlike projections, increasing the opportunity for exchange (Figs. 2-2 and 2-3). In other forms, for example in some of the insects, special *nurse cells* are closely associated with the oöcyte. In the fruit fly, nurse cells arise by divisions of an oögonium. In four divisions, 16 cells are formed: one oocyte and 15 nurse cells. The contributions to the oöcyte by nurse cells are even more extensive than are the follicular contributions in the vertebrates. The entire cytoplasm may pour into the oöcyte.

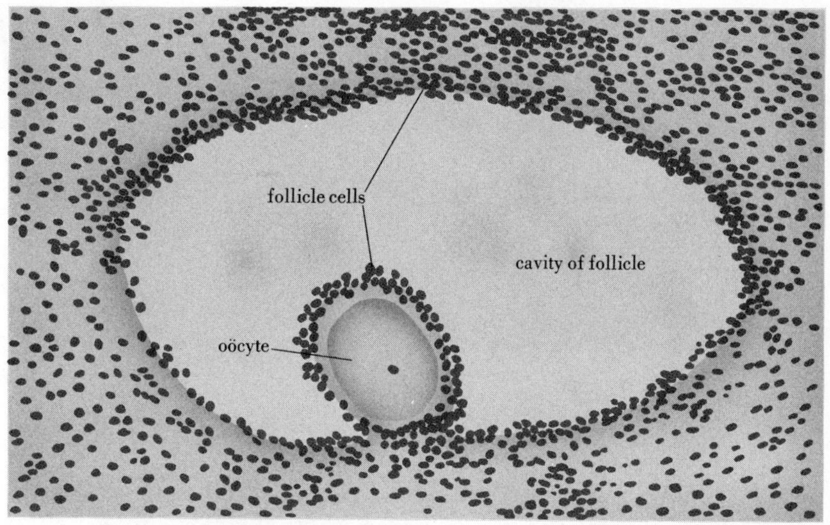


*Fig. 2-1 Diagram of a mammalian oöcyte surrounded by follicle cells.*

Several specialized mechanisms exist for the transfer of materials into the oöcyte. The evidence for the passage of "ready-made" materials is compelling; much of it is derived from experiments using labeled molecules. In the silk moth *Cecropia*, it was suspected that large quantities of a circulating protein, found only in females, were deposited in the growing oöcyte. How might it be demonstrated? The blood proteins of different species of silk moths are themselves specific, that is, they can be distinguished by immunological techniques. As antigens, they engender specific antibodies in rabbits. These antibodies may then be used as diagnostic tools, much as blood-typing antibodies are used. If female blood of one species is transferred into a female of a second species, the donor's blood proteins are eventually found in large quantities in the oocyte. Moreover, when these antibodies are labeled, by use of a fluorescent dye, it is possible to see that the blood proteins are concentrated in spaces between the follicle cells and within the oöcyte, suggesting that blood proteins are incorporated into the oöcyte. Electron microscopic observations of another insect, the mosquito, suggest that after protein in the extracellular space is absorbed onto the cell membrane, the membrane invaginates, or "pinches in" as small vesicles within the cytoplasm. These vesicles coalesce and eventually form yolk droplets. In the mosquito it appears (but it has not been proved) that these yolk proteins are formed in the midgut (Fig. 2-4). Similar evidence of transfer exists for the chicken, frog, and mouse.

It remains to be determined, however, not only just what fraction of the egg proteins is made in the egg and what fraction is "imported,"

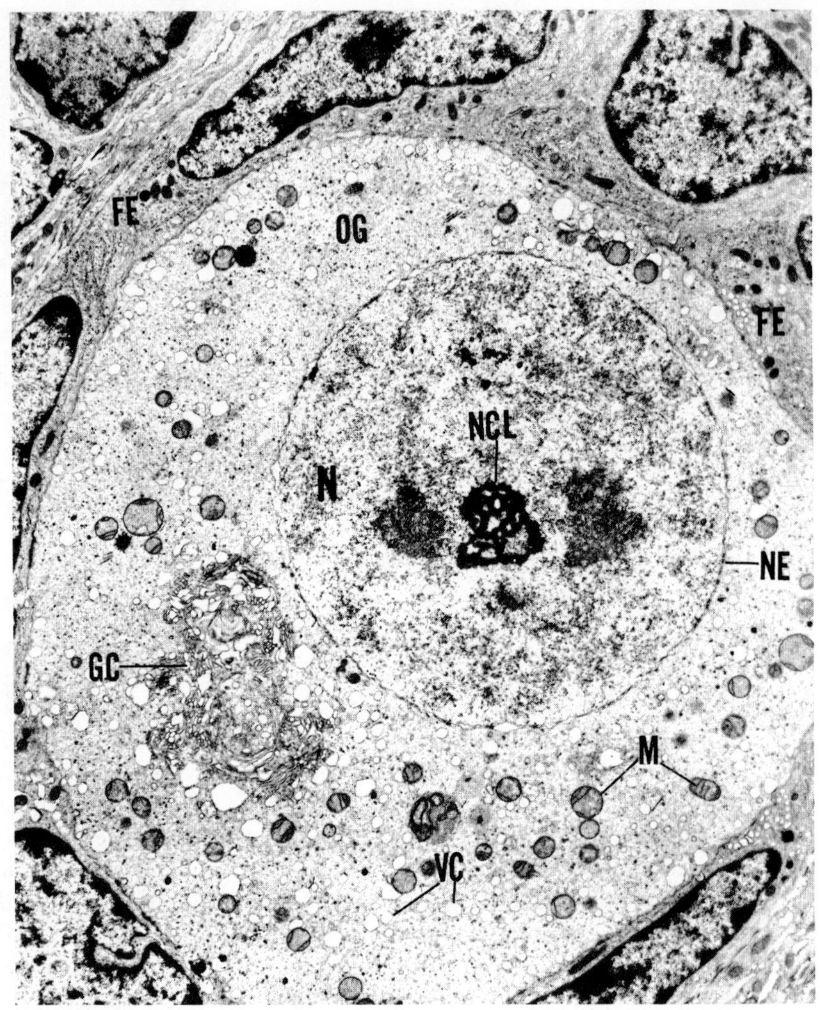


*Fig. 2-2* *Electron micrograph of a part of a guinea pig ovary showing a primary follicle containing an oögonium* (OG) *encompassed by follicular epithelial cells* (FE). *The oögonium contains a spherical nucleus* (N) *circumscribed by a double membrane envelope* (NE). *The nucleolus* (NCL) *in the granular nucleoplasm is composed of dense particles arranged in a reticular pattern. The cytoplasm contains ribosomes, a large Golgi complex* (GC), *mitochondria* (M), *and a vesicular component* (VC). × *8000 (By courtesy of Everett Anderson.)*

but also whether these two classes of molecules may be separated according to their distribution or functions. For example, we have observed that in some species there is an intense synthesis of ribosomal RNAs during oögenesis. However, in addition to the RNAs, ribosomes

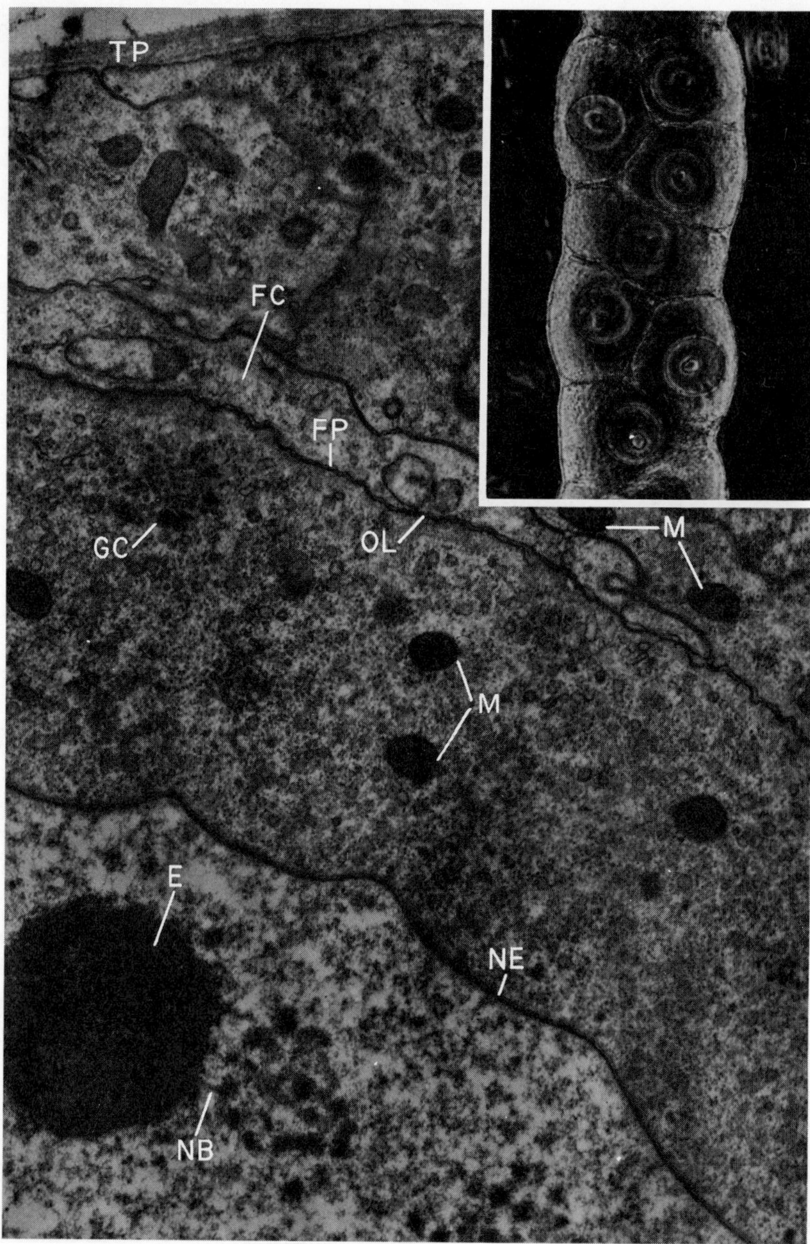


**Fig. 2-3** *(Inset, upper right) Phase contrast photomicrograph of an ovariole (part of ovary) of the cockroach, showing oöcytes.* × *200.* (Lower) *Electron micrograph of small portion of an oöcyte and associated follicle cells* (FC). TP: *tunica propria;* FP: *plasma membrane of follicle cells;* OL: *membrane of oöcytes;* M: *mitochondria;* GC: *Golgi complex;* NE: *nuclear envelope;* E: *nucleolar emission;* NB: *particles attached to nucleolar emission.* × *23,000. (Inset by courtesy of Everett Anderson; electron micrograph also from Anderson, in* Journal of Cell Biology, *20, by courtesy of the author and the Rockefeller University Press.)*

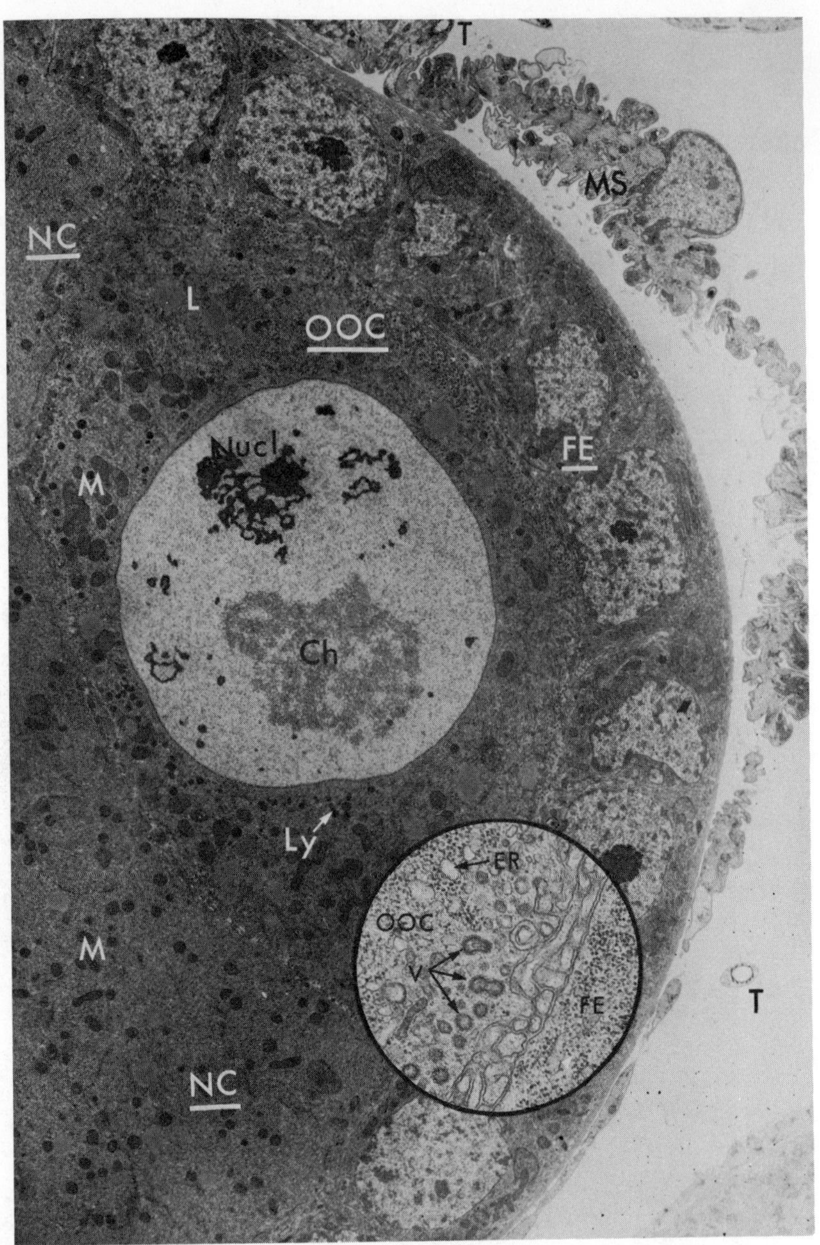


*Fig. 2-4 Electron micrograph showing part of an ovarian follicle in the mosquito, with a portion of two adjacent nurse cells* (NC) *and an oöcyte* (OOC) *enclosed by a follicular epithelial layer* (FE). L: *Lipid droplets;* LY: *dense droplets;* Ch: *clumped chromatin;* Nucl: *nucleolus;* T: *trachea and tracheoles;* MS: *muscular sheath.* × *3700.* (Inset) *Higher magnification of a small segment of the vesiculated interface between the oöcyte and follicular epithelium. The oöcyte cortex contains numerous bristle-coated vesicles* (V) *and elements of endoplasmic reticulum.* × *20,000. (From T. F. Roth and K. R. Porter, in* Journal of Cell Biology, *20, by courtesy of the authors and the Rockefeller University Press.)*

are composed of a number of proteins. In species in which the ribosomal RNAs are made in the oöcyte, where are the several ribosomal proteins formed? At the moment we have no answer. Moreover generalizations are difficult, for in some insects, ribosomes themselves are apparently contributed to oöcytes by nurse cells.

We have been discussing the synthesis of proteins and other materials. We have not, however, taken up the question of their organization. During the growth phase, the structure of the oöcyte changes gradually until the definitive form of the egg is attained.

Poised and ready for activation, an egg possesses many of the accoutrements of any cell: a large premeiotic nucleus, or *germinal vesicle*; mitochondria; and ribosomes. Although the pattern varies from species to species, the general tendency is for a progressive change during oögenesis from a predominance of free ribosomes toward the state in which they are found in more orderly arrangements of free polyribosomes or polysomes. There is relatively little endoplasmic reticulum even in mature eggs.

Although we list here the usual constituents of cells, they do not help us recognize a "typical egg" any more than we can recognize a "typical cell." They are put together in different ways, the sum total being an egg characteristic of the species. Other features that add distinction to eggs include the yolk, which serves as the food supply of the embryo and is organized into fat droplets, lipid, and protein and lipoprotein granules; pigment granules; the membranes of the egg proper, frequently containing polysaccharides; plus the outer, protective envelopes, such as the jelly coat of polysaccharide and protein in frogs, toads, and salamanders and the shell membrane and inorganic shell in birds and reptiles. In the latter, of course, the outer layer must be added *after* fertilization.

The problems of the emergence of organization are exemplified by consideration of the formation of yolk granules. Yolk is not a single molecular species, but a mixture of proteins, phospholipids, and neutral fats. In the *yolk platelets* of amphibians, for example, there are two principal proteins, phosvitin and lipovitellin, the latter containing bound lipid. A single unit contains two molecules of phosvitin and one of lipovitellin. These molecules are synthesized not in the oöcyte, but in the liver. Once in the oöcyte they must somehow be converted from the soluble into the insoluble form and deposited in the pattern characteristic of the species.

Even before fertilization, then, eggs are not alike; many exhibit polarity, and bilateral symmetry is recognizable in some. Among the characteristics that provide an index of polarity we might list the position of the polar bodies (Fig. 2-5); the distribution of yolk, of pigments, and of other cytoplasmic inclusions; the position of the germinal vesicle; and indentations of the egg surface.

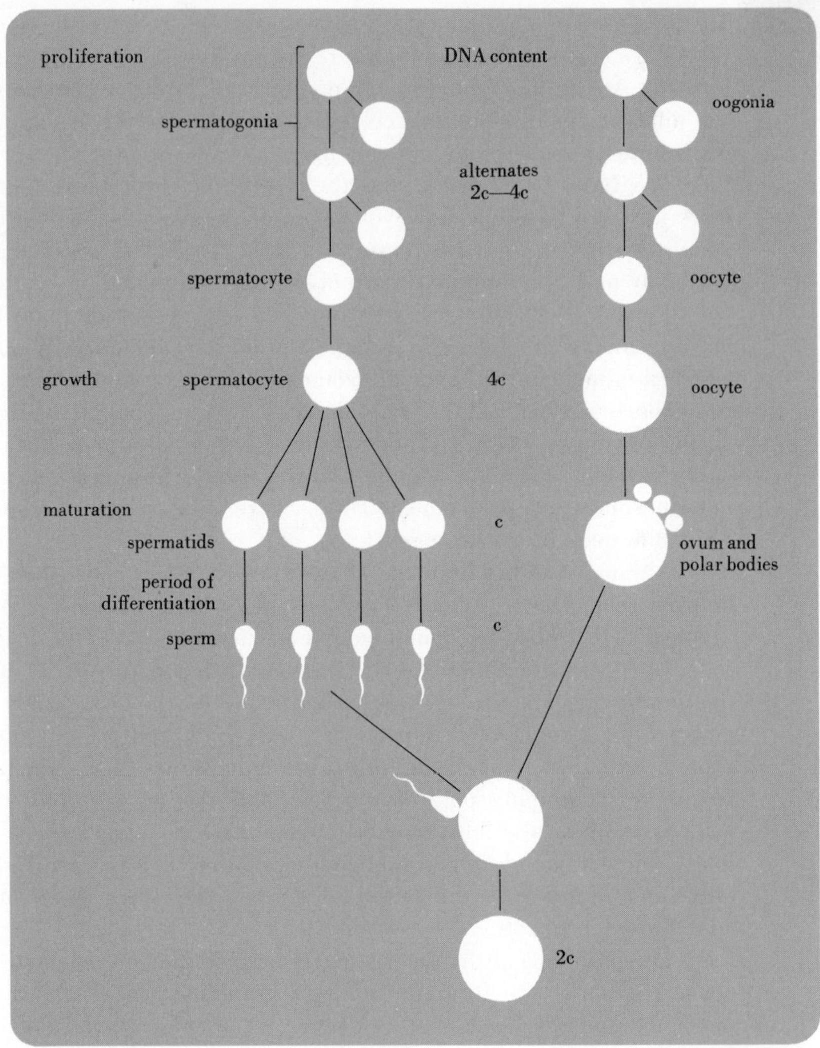


*Fig. 2-5 Diagram of the formation of sperm and eggs. (From C. W. Bodemer,* Modern Embryology, *by courtesy of Holt, Rinehart and Winston, Inc.)*

Almost nothing is known about the origin of the polarity of the egg. Numerous ideas have been advanced, but all have been flawed. Thus, for example, there is no clear relation between the course of blood vessels in the ovary and the ultimate position of the animal (upper) and vegetal (lower) poles of the egg; nor is there a consistent pattern in the relation of the vegetal pole to the oöcyte surface through which nutrients pass. It is true that in mollusks, for example, the vegetal pole arises at the end of the oöcyte that is attached to the ovarian wall, but this relationship has not been shown to be a general one. It is not very helpful to say that there is evidence of innate polarity even in the young oöcyte. In discussing innate polarity, we are in the difficult position of believing it to be significant, for we can hardly dismiss the importance of preexisting structure in the delineation of future structure, yet of knowing next to nothing about its origin. In Chapter 5 we shall consider the brown alga, *Fucus*, in which polarity may be influenced by environmental stimuli.

Before concluding our inventory of eggs and their constituents, we should at least recognize that the amount and distribution of yolk has an important consequence. The eggs of species in which the embryo or larva begins active feeding at an early stage contain little yolk; it is usually evenly distributed. At the other end of the scale, the eggs of birds contain vast amounts of yolk relative to the size of the embryo. The mammals have specialized in still another direction: Few food reserves are required, since the embryo receives its nourishment through the mother; for that purpose, mammals have evolved a specific structure, the *placenta*.

The important consequence is this: since many of the key steps in the early development of animals involve morphogenetic movements (the orderly movements of cells and tissues), the amount and distribution of the inert yolk must be an important factor in determining the type of movements possible. Thus, although yolk has a primary function that is nutritional, it has by virtue of its very presence an influence on the shape of things to come.

We have to consider, finally, the third major phase of oögenesis: the maturation phase. During the proliferative phase, which involves continuous passage through the cell cycle, oögonia are alternately diploid (2c) or tetraploid (4c) in DNA content. During the growth phase, oöcyte nuclei are tetraploid in DNA content (4c). To reach the haploid state characteristic of functional germ cells, they undergo meiosis. We speak of this phase as *maturation*. The process quite properly is treated fully in *Genetics*, another book in this series. But since unfailingly this topic seems to engender confusion, a glance at Fig. 2-5 recalls that in meiosis two successive cell divisions result in a reduction in DNA content to the "c" level.

We shall find, repeatedly, that phases of a process or stages of development are merely arbitrary designations enabling us to discuss phenomena in convenient "packages." So it is with the phases of oögenesis; there can be no sharp separation between growth and maturation. As the oöcyte grows, its nucleus enters into the prophase of the first meiotic division. Not only does the nucleus increase in size, but photometric determinations reveal that it has four times the haploid content of DNA (4c). The subsequent events of meiosis are postponed until the end of the growth period, when the reduction divisions begin.

As a consequence of this process, the haploid egg is formed. The divisions are unequal in the sense that finally one large daughter cell, containing most of the cytoplasm and yolk, is produced along with three tiny cells, the *polar bodies*, which are destined to degenerate.

In some species the steps of the maturation process follow rapidly one after the other until it is completed; in others, maturation stops at one of the intermediate stages, most commonly after the first polar body is extruded, until the egg is activated, as a consequence of which maturation is completed.

***The Sperm*** A monograph on *Spermatozoon Motility* was published recently; it encompassed over 300 pages. What are the cardinal facts that must be presented in about 1/100 of that volume? First, the sperm has two principal functions to perform: It activates the egg to initiate development, and it supplies a haploid nucleus, contributing half of the chromosomal complement of the embryo. Its contribution of cytoplasm appears to be insignificant, and unlike the egg, it does not supply food reserves. It is specialized in another direction: by virtue of its morphology and its metabolism, it is highly motile—albeit for short periods.

***Spermatogenesis*** The process of spermatogenesis is like oögenesis in many fundamental respects. The primordial germ cells originate ouside the testis; once within it they undergo proliferation. Spermatogenesis occurs in the *seminiferous tubules* of the testis (Fig. 2-6). In the mammal, for example, the proliferating *spermatogonia* are located at the outer surface of the tubule. Since spermatogenesis is a continuous process, some spermatogonia retain their capacity to divide and thus represent a continuing source of new cells. Others are moved toward the lumen of the tubule and enter the next phase, that of growth. In this stage they are called *primary spermatocytes*.

In contrast to oöcytes, spermatocytes grow very little. They do become perceptibly larger, but there is no striking accumulation of reserves, no dramatic increase in size.

In the maturation phase, the end result is the formation of four haploid *spermatids*, equal in size. However, the spermatids are not yet functional gametes. They must undergo a further specialization (in a process called *spermiogenesis*) into characteristic sperm (Fig. 2-7).

Before continuing with our study of sperm, let us ask how a field of inquiry develops. Can one cite for any given field the major forces that have shaped its dimensions? In large measure it is possible to do just that for the field of fertilization. Clearly, three major forces have

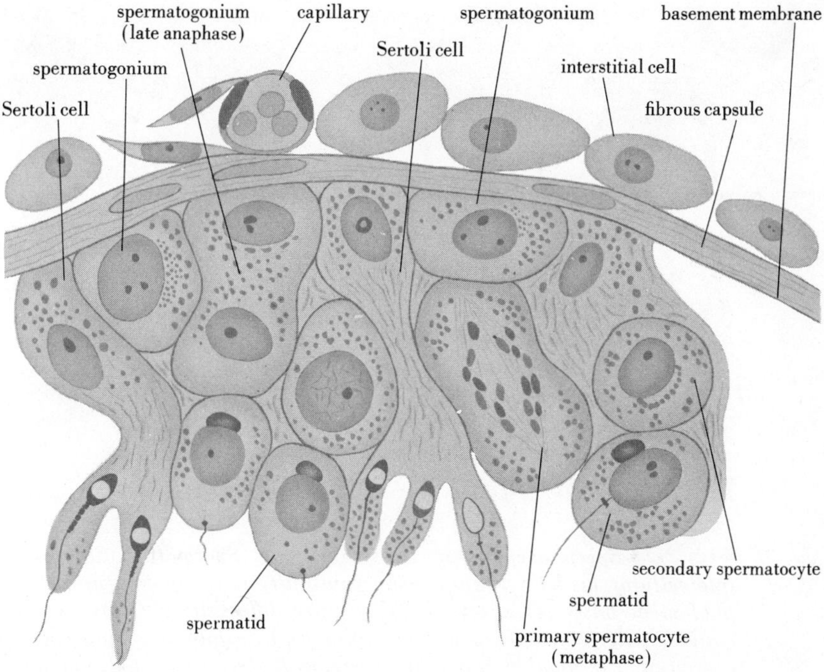

*Fig. 2-6 Section of part of human seminiferous tubule, showing spermatozoa in all stages of maturation. (Redrawn from Hamilton, Boyd, and Mossman,* Human Embryology, *2d ed., by courtesy of W. Heffer & Sons, Ltd.)*

shaped this field. First, technical advances in microscopy have been largely responsible for our understanding of the structure of sperm. Beginning with van Leeuwenhoek's description of sperm, published in 1678, each new step forward in microscopy—light microscopy, phase contrast microscopy, electron microscopy—has clarified sperm structure and function. Second, perhaps more than in most aspects of biology, pressures from, and advances in, practical aspects of science have helped to shape the course of basic research. Often we are told that such forces operate only in the reverse direction. Here we refer to the widespread need for, and development of, techniques of handling and preserving sperm for artificial insemination, both in animal husbandry and medical practice. Much of what we know, and an idea of what we need to know, about metabolism of sperm is derived from such programs.

The third major directive force is the far-reaching hypothesis. Not all hypotheses are correct; moreover, large ideas that turn out to be wrong, or partly wrong, may sometimes have as great an impact as small ideas that prove to be correct. As we shall see, such a sweeping idea was Frank R. Lillie's *fertilizin–antifertilizin* hypothesis of fer-

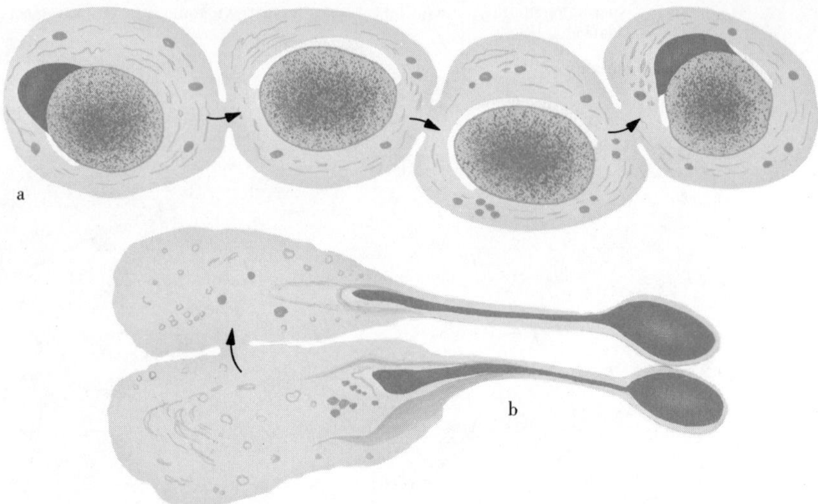


**Fig. 2-7** *Spermiogenesis in the guinea pig. Spermatids* (a) *are joined by intercellular bridges. Protoplasmic continuity is thought to provide the morphological basis for synchrony of further differentiation.* (b) *Two late spermatids, still joined by a bridge, although transformation into the sperm is advanced. Presumably two other spermatids were connected, but not seen in this plane of section. (After D. W. Fawcett, "Changes in the fine structure of the cytoplasmic organelles during differentiation," in* Developmental Cytology, *D. Rudnick, ed. The Ronald Press Company, 1959.)*

tilization, which we shall consider because it throws light on the methods of science.

***Structure and composition of sperm*** Because of the great variation in their shape and size, animal sperm have been a favored object of study by students of comparative biology. Sperm range in size from sea urchin sperm, some 40 $\mu$ long, to the sperm of hemipterans, which may reach a length of 12 mm. Again, there is no "typical sperm," but there are certain general features common to most of them (Fig. 2-8). It is on these key points that we shall focus our attention.

At the risk of oversimplifying, we may begin by observing that the three main divisions of the sperm—(1) head, (2) middle piece, and (3) tail—correspond roughly to (1) activating and genetic, (2) metabolic, and (3) motile functions (Fig. 2-9). Let us amplify this statement, basing our discussion largely on studies using the electron microscope, the need for this approach being emphasized when we consider that most sperm are measured in microns. Another index of size is weight: a single bull sperm weighs about $2.86 \times 10^{-8}$ mg.

The *genetic* functions of the sperm are embodied in the nucleus, which may extend the entire length of the *head* or be restricted to its

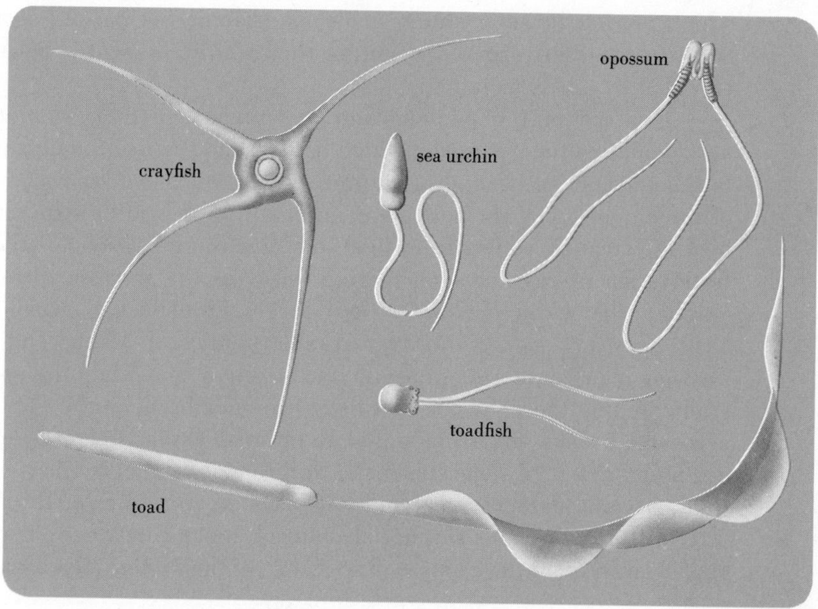


*Fig. 2-8 Variations in sperm morphology, showing sperm of crayfish, sea urchin, toadfish, toad, and opossum. (From original drawings by D. W. Bishop; not drawn exactly to scale.)*

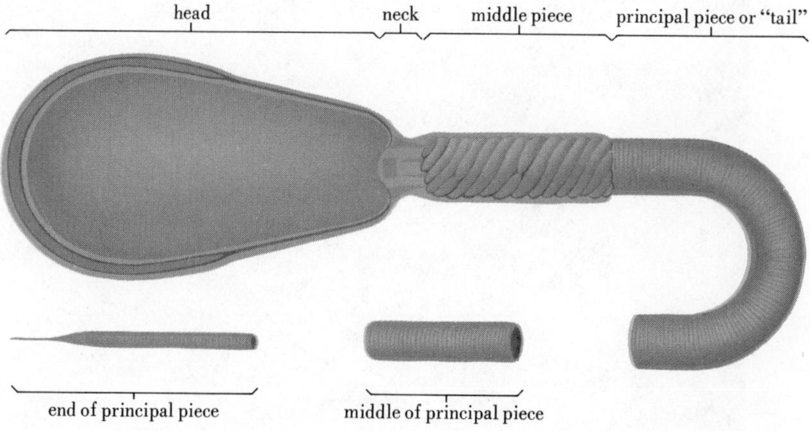


*Fig. 2-9 Diagram of bull sperm. (From an original drawing by D. W. Bishop.)*

posterior portion. In the mature nucleus the chromatin is condensed into a homogeneous, electron-dense mass, which characteristically absorbs strongly in the ultraviolet and has a marked affinity for basic dyes. The chromosomes are not usually seen clearly, even in the electron microscope. The chromatin material—largely DNA and basic protein, which is rich in such amino acids as arginine—is packed so tightly that any precisely ordered structure that may be present is not readily revealed by current methods.

Forward of the nucleus and just under the thin, membranous *head cap*, lies the *acrosome*, which as we shall see is intimately related to the entrance of the sperm into the egg. Although we know something of the chemistry of the acrosome, and many details of its structure, our best information comes from light and electron microscopic studies of fertilization of eggs of such marine forms as the starfish, the sea cucumber, the worm *Hydroides*, and the enteropneust or acorn worm, *Saccoglossus*. Based on the remarkable studies of Arthur and Laura Colwin, Jean Clark Dan, and others, we know that when the sperm is stimulated by the presence of an egg, the membrane at the apex of the acrosome breaks down, leading to an intimate association of sperm and egg membranes. In some species an acrosomal filament is released. This rigid, straight filament may be as long as 50 $\mu$, depending on the species (Fig. 2-10). It provides the initial point of contact between sperm and egg. Bear in mind the relative dimensions of the system; the area of the

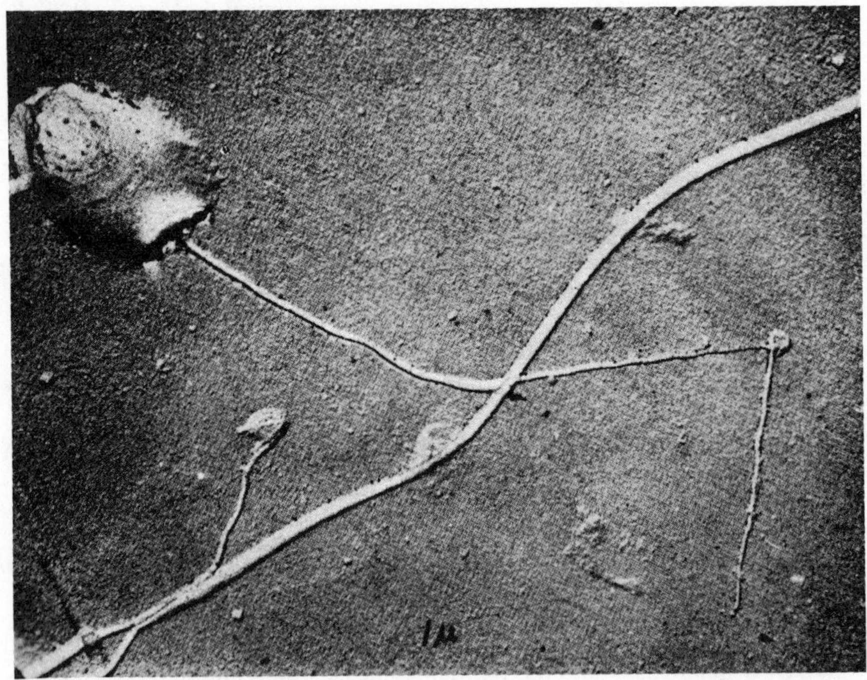


**Fig. 2-10** *Electron micrograph of starfish sperm, treated to release acrosome filament, seen as a long, inverted* L. *(From J. C. Dan, in* Biological Bulletin, *107, by courtesy of Marine Biological Laboratory.)*

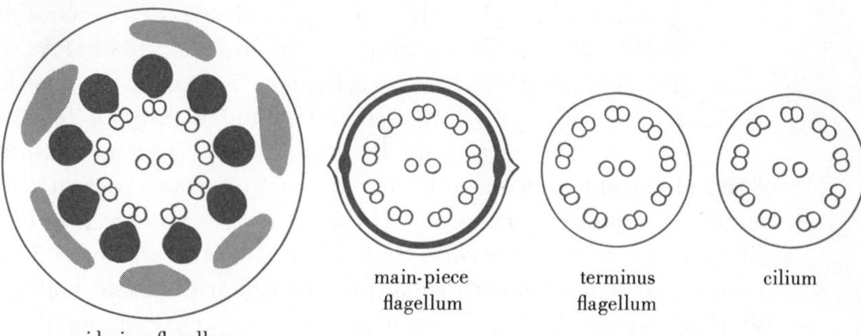


***Fig. 2-11*** *Highly diagrammatic representation of transverse sections of sperm flagellum and cilium; two central and nine double peripheral microtubules are typical of all such motile organelles. Mitochondria (gray) are present in the midpiece. An additional array of nine outermost filaments (solid color) in the midpiece of mammalian sperm extends into the proximal portion of the flagellum. The fibrous sheath of the tail is frequently ribbed as indicated. (From D. W. Bishop, in* Sex and Internal Secretions, *by courtesy of The Williams & Wilkins Company.)*

tip of the acrosome is about 0.002 $\mu^2$, or about one millionth of the surface of the egg.

Proceeding "tailward" from the head, we next encounter the area about which we know least, the short "*neck*," perhaps only 0.5 $\mu$ long, in which the principal elements that make up the mid-piece and tail, or flagellum, arise. The sperm *tail*, like most cilia and flagella except for the bacterial flagellum, is composed of two central microtubules surrounded by a ring of nine double tubules (Fig. 2-11). These microtubules are attached to, and probably originate in, a "basal body" or granule in the neck. In the middle piece, at least one additional major feature must be described: it is in this region that the mitochondria, which we think of as the "power plants" of the cell because of their oxidative and phosphorylative activities, are located; these are arranged spirally around the microtubules.

The sperm tail moves with an undulating snakelike motion. Despite intensive studies of the structure of microtubules and of the chemistry of proteins isolated from sperm tails, nothing has emerged that would clearly define the mechanism of sperm movements. There is no compelling evidence for contraction of the microtubules. There are numerous speculations: microtubules may bend; or they may slide up and down with respect to one another; or there may be a contraction or shortening of other elements in the flagellum running alongside the microtubules. Sperm movements are probably powered by energy generated in the mitochondria, but the manner in which this energy is translated into work is not clear, nor is the mechanism by which these flagellar movements are coordinated.

## ORIGIN OF GAMETES AND FERTILIZATION IN FLOWERING PLANTS

There is no evidence of the continuity of germ plasm in flowering plants. As we shall see in Chapters 5 and 13, the reproductive organs and gametes are not formed until late in development, and any somatic cell of the plant appears to retain the capacity to form tissues that will give rise to reproductive structures. These structures include the *carpel* that will produce ovules in which the female gametes are formed, and the *stamen* that will produce the anthers in which the male pollen arises. We shall bypass the development of the reproductive organs themselves and focus on maturation of the egg and pollen. Figure 2-12 summarizes normal reproduction in flowering plants. Pollen grains are formed in the terminal anther, which is attached to the filament of the stamen. Cells in the center of the anther undergo several mitotic divisions and then

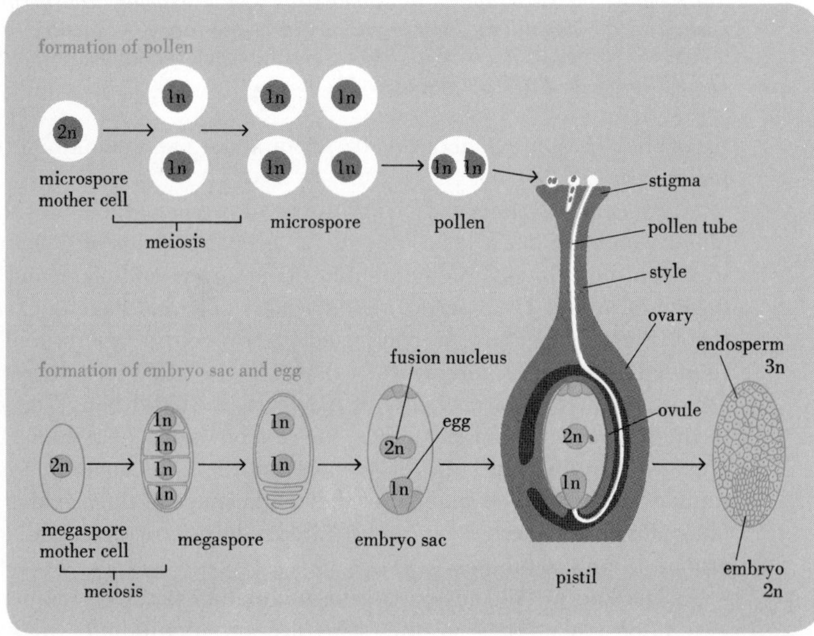


**Fig. 2-12** *Normal reproduction in flowering plants.* (Top left) *Development of pollen after formation in anther, beginning with a* 2n *microscope mother cell and progressing through several divisions until one pollen grain lodges on a stigma and gives rise to a pollen tube.* (Bottom left) *An embryo sac containing an egg similarly develops from a megaspore mother cell. On the pistil, two of the nuclei that have developed in pollen grain start down pollen tube. One nucleus joins the egg to accomplish fertilization; one joins fusion nucleus to create endosperm, a nutritive stimulant to embryo growth. End product is seed containing embryo. (From "The control of growth in plant cells" by F. C. Steward. Copyright © 1963 by Scientific American, Inc. All rights reserved.)*

go into meiosis, each diploid microspore mother cell producing four haploid microspores. The latter, in turn, produce heavy cell walls and become pollen grains. Just as in animal gametogenesis, DNA doubling to the "4c" level occurs in premeiotic interphase and into early first prophase, with no further DNA synthesis occurring during meiosis. Once the microspore is formed, the process is completed by a final nuclear division so that the definitive pollen grain contains two haploid nuclei.

Somewhat similarly, an embryo sac, containing an egg, develops from a megaspore mother cell. The mother cell undergoes meiosis producing four haploid cells, three of which die. The fourth divides, forming the complex embryo sac containing eight haploid cells or nuclei, an egg cell, two synergid cells, three antipodal cells, and two "polar" nuclei (which fuse together to form the "fusion" nucleus).

Once the gametes are formed, they are brought together in two discrete stages: *pollination* and *fertilization.* In the first step, a pollen grain is deposited on the sticky stigma at the tip of the carpel and gives rise to a *pollen tube*, which grows through the style tissues, digesting its way toward the egg. Two male nuclei move down the pollen tube. One fuses with the egg nucleus to form the zygote; the other joins the fusion nucleus to make the nutritive endosperm. The seed contains the diploid embryo and the triploid endosperm.

## FERTILIZATION IN ANIMALS

Nature has set the stage for fertilization in widely different ways, varying with the species and their habits and habitats. Some of the more remarkable patterns of behavior leading up to fertilization are described in *Animal Adaptation*, in this series. But underlying all these variations are the following basic requirements:

1. The life-spans of gametes are limited. Mature eggs quickly become "overripe"; they must be activated promptly if they are to be activated at all. Eggs that are shed into the water, like those of most invertebrates, fish, and amphibians, must be fertilized immediately, or within a few minutes. Eggs that are fertilized within the body of the female generally have a longer life-span; for example, the human egg can be fertilized for at least 24 hours after ovulation.

Sperm also are short-lived; but there are more exceptions to the rule in Nature, and there are notable man-made exceptions. We need not catalog the exceptions; the example most commonly used is the bat, in which viable sperm may persist in the female reproductive tract over the winter, during hibernation. The lifetime of the human sperm in the female tract is of the order of 24 hours. Much of what we have learned about sperm metabolism has come from attempts to preserve sperm. The practice of artificial insemination has great clinical and economic importance. In livestock breeding, stud males ejaculate into a rubber tube or other collecting device; after repeated ejaculations, the semen of a single male can be stored for as long as a week. Such questions as influence of temperature, rate of cooling, preservation, nutritional and buffering content of medium are important in determining just how effective the practice may become. After dilution, the semen is injected into the oviduct or uterus of females in heat.

2. To increase the probability of fertilization, the number of sperm must exceed the number of eggs. It is interesting to contemplate the course of evolution of techniques of bringing eggs and sperm together. A foolproof method that will allow the perfect union of one egg and one sperm without wasting millions of sperm has not been devised. Irrespective of whether the release of gametes is seasonal, or monthly, or irregular, or whether it is influenced by length of day, or phase of the moon, or is under the control of hormones, with or without a copulatory stimulus, or whether it takes place in mid-ocean or within the female tract, sperm must be present in great excess. In mammals, a single ejaculation may range from 0.05 ml in the bat to 2 ml in the ram to 500 ml in the boar, containing 6 million, 2 to 5 million, and 100,000 cells per *micro*liter, respectively. In many species, the male reproductive potential is enormous; the ram, for example, is capable of 20 such ejaculations in a single day owing to the large numbers stockpiled in the epididymis. Apparently, sperm are being generated continuously; there appears to be no "feedback" control of proliferation. In man, the ejaculate averages 3 ml, of which about 20 percent by volume is sperm (100,000 cells per microliter). These large numbers are important; through studies of human fertility, we know that an acceptable sperm count in man is 40 to 50 million cells per milliliter. If the count falls below 20 million cells, fertilizing capacity is poor. In such cases, it is usually necessary to concentrate the sperm and to inseminate artificially to insure pregnancy. Of the enormous numbers of sperm ejaculated, only a tiny fraction reaches the vicinity of the egg. Many parts of the female tract are hostile to sperm; acidity and other environmental factors may be deleterious. Sperm move up the oviduct rapidly, so rapidly in fact that we know that forces other than their own "swimming" capacity are involved; of these forces, the most important are the ciliary and muscular actions of the oviduct. Sperm may reach the upper end of the oviduct in from 15 minutes (mouse) to 3 hours (in man or rabbit). In the rabbit, no more than 50 to 250 sperm may actually reach the site of fertilization.

We have touched on a number of practical problems while considering these important basic ideas. There remains one more such problem: sperm are not homogeneous; in man, a suspension contains two genetic types, those sperm bearing an X and those bearing a Y chromosome. It has often been stated that in suspensions X and Y sperm can be separated by virtue of their behavior in an electric field. Therefore, the possibility of controlling the sex of offspring in artificial insemination is presented; however, such assertions remain unconvincing and, at the moment at least, we have not attained that goal.

***Sequence of Events*** The large numbers of sperm released by the male make it likely that a random collision of eggs and sperm will occur. According to Lillie's *fertilizin–antifertilizin* hypothesis, gametes produce diffusible substances (called by some writers *gamones*) that increase the probability of collision and promote adherence of egg and sperm. Thus, in repeating Lillie's simple, classic experiment, students at the Marine Biological Laboratory at Woods Hole, Massachusetts, allow unfertilized sea urchin eggs to stand in seawater for a few minutes. If, after the eggs are removed, sperm are added to the remaining "egg water," as it was called, they rapidly agglutinate. Lillie, and others after him, argued that a substance in the egg, termed *fertilizin*, interacted in a specific manner with a substance in sperm, *antifertilizin.* That such a substance, a glycoprotein, does indeed exist in the jelly coat surrounding the egg, is clear enough (the evidence for antifertilizin is less clear), but the evidence that fertilizin is *necessary* for fertilization is not conclusive. What can we say about it? First, although there is positive evidence of chemotaxis—that is, of directed movement of sperm toward the egg in response to a chemical gradient from the egg—in some species, especially in plants (brown algae, ferns, and mosses), most egg–sperm collisions appear to be random. Second, there is evidence of activation or stimulation of sperm and their subsequent agglutination in a variety of species, including mammals. And finally, in some species, egg exudates cause sperm to undergo the acrosome reaction—that is, the protrusion of a filamentous projection, the acrosomal filament.

In focusing on the sperm, however, we may foster the erroneous impression that penetration is merely a "boring-in" by the sperm. Rather, it requires a true interaction of egg and sperm. The picture we have drawn is a composite—certain features being emphasized in some species, but absent in others.

Let us examine the association of egg and sperm in one form in more detail, choosing the marine annelid, *Hydroides.* In Figs. 2-13 and 2-14 we see a photograph and diagram of a longitudinal section of the unactivated acrosomal region of the sperm, magnified 84,000 times. The acrosome, which is closely applied to the sperm nucleus, is a vesicle

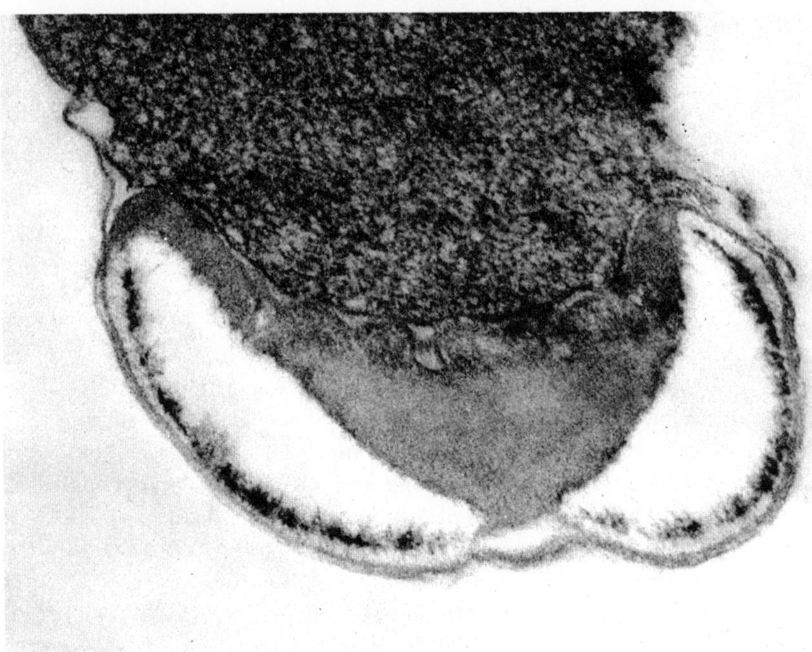

*Fig. 2-13 Electron micrograph of unactivated acrosomal region of* Hydroides *sperm, nearly median longitudinal section. See Fig. 2-14 for details.* × *84,000. (From A. L. Colwin and L. H. Colwin, in* Journal of Biophysical and Biochemical Cytology, *10, by courtesy of the authors and the Rockefeller University Press.)*

bound by a membrane; the vesicle contains a large acrosomal granule.

As we see in Figs. 2-15 through 2-17, when the acrosomal membrane contacts the outer egg envelope the acrosomal tip breaks down; the acrosomal membrane then becomes continuous with the plasma membrane of the rest of the sperm behind it. The sperm, spearheaded by the acrosome, progresses through the egg envelope. As it does, the acrosomal granule breaks down, and at the same time, the egg envelope in front of it disappears. Presumably, breakdown of the granule releases a lytic enzyme, which destroys the envelope; however, the evidence is indirect. From sperm, substances can be extracted that can dissolve egg envelopes, but the precise manner in which they act locally at fertilization is unknown (Fig. 2-18). In mammalian semen, this enzyme usually is a member of the family of hyaluronidases, similar to those that promote bacterial invasion.

Parts of the acrosomal membrane now elongate into tubules, which contact the plasma membrane of the egg. As contact is made, a *fertilization cone* rises in the egg membrane surface. The acrosomal and egg membranes fuse to become one continuous membrane, and soon the interiors of the formerly separate gametes are confluent.

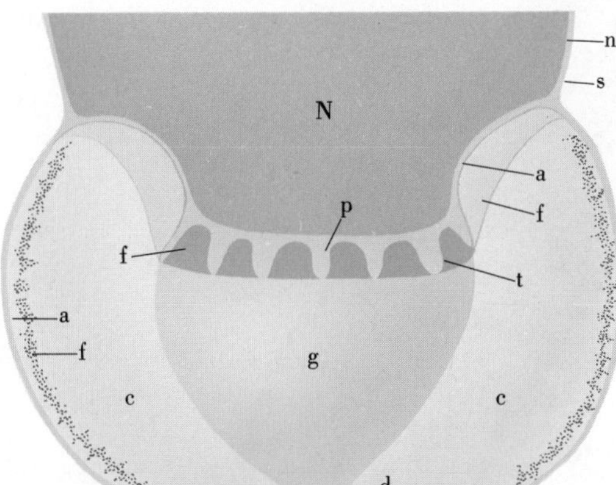


*Fig. 2-14 Diagram of unactivated acrosomal region; median longitudinal section. Acrosomal vesicle embracing apical part of nucleus adjoins nuclear envelope and elsewhere adjoins sperm plasma membrane except at atypical vesicle. Periacrosomal material is in adnuclear region and occupies tubular invaginations of acrosomal membrane.*

*Key for Figs. 2-13 through 2-17:* a: *acrosomal membrane or originally acrosomal membrane;* c: *cavity of acrosomal vesicle;* d: *apical vesicle;* e: *egg plasma membrane or in zygote, part derived therefrom;* F: *fertilization cone;* f: *fine granular material lining acrosomal membrane;* g: *acrosomal granule;* i: *inner layer of egg envelope (vitelline membrane);* k: *acrosomal remnant serving as a marker for identifying sperm plasma membrane;* m: *middle layer of egg envelope (vitelline membrane);* N: *nucleus;* n: *nuclear envelope or membrane components of nuclear envelope;* o: *outer layer of egg envelope (vitelline membrane);* p: *periacrosomal material;* s: *sperm plasma or, in zygote, part derived therefrom;* t: *acrosomal tubule or invagination of adnuclear portion of acrosomal membrane;* v: *egg microvillus. (After A. L. Colwin and L. H. Colwin, in* Journal of Biophysical and Biochemical Cytology, *10, by courtesy of the authors and the Rockefeller University Press.)*

There are a few electron micrographs of the fusion of male and female (both egg and polar) nuclei in ferns and in flowering plants such as petunia and in pine. Such little evidence as is available suggests that there is a fusion between the nuclear envelopes.

Perhaps our discussion emphasizing the stepwise nature of these events has given the impression that fertilization is a slow process. On the contrary, it is usually exceedingly fast. The time from insemination to membrane fusion in *Hydroides* is *less than 9 seconds.* In most species the entrance of a sperm prevents the entrance of others. Although there have been attempts to relate this "block to polyspermy" to morphologic changes at the egg surface, we must say that the mechanism of action is not resolved. In some species, including some birds and reptiles, several sperm penetrate the egg; however, only one sperm nucleus is involved in the culmination of the process: the fusion of egg and sperm nuclei.

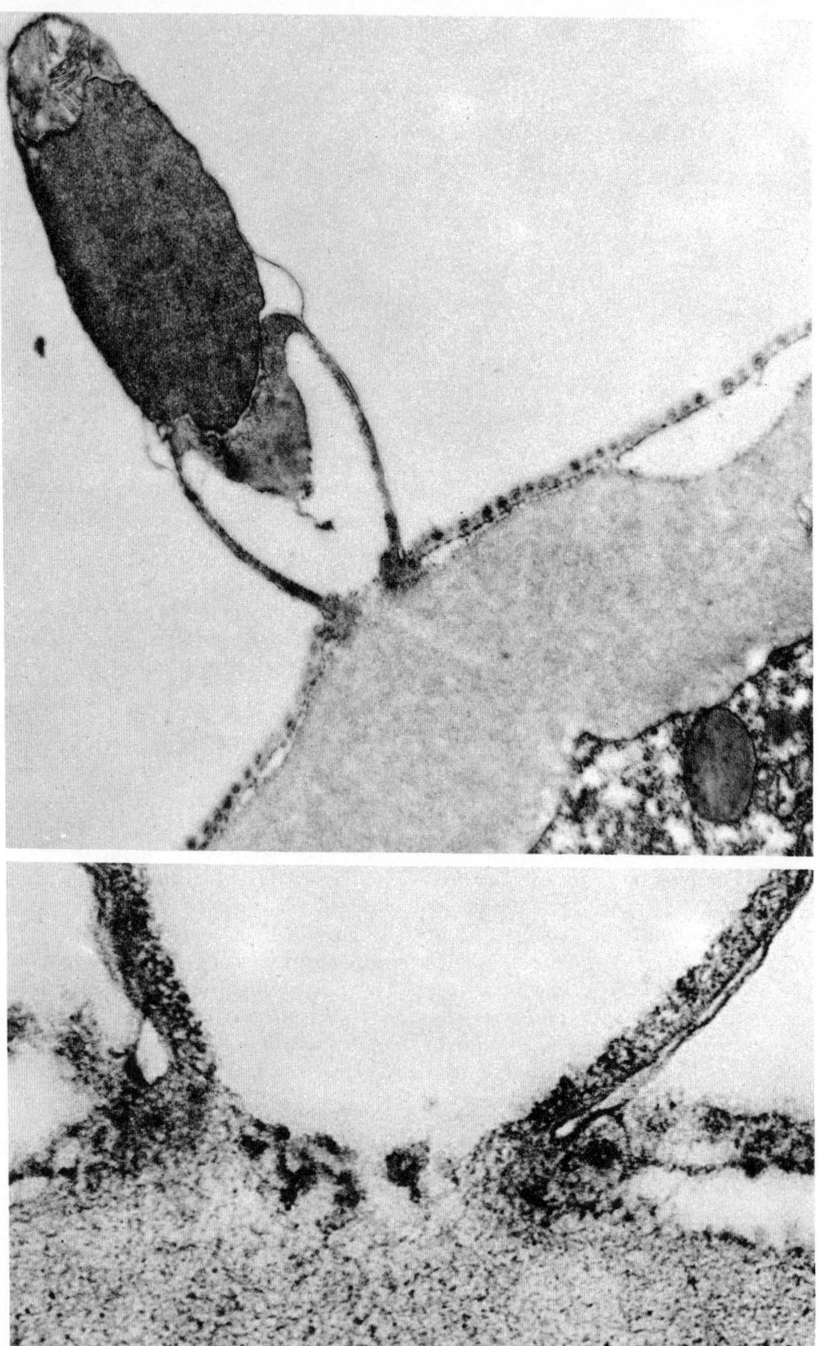

***Fig. 2-15*** *Electron micrographs of initial contact of sperm and egg membrane.* (Top) *Sperm plasma membrane meets egg envelope. Egg plasma membrane is still separated from sperm by egg envelope.* × *37,000.* (Bottom) *Higher magnification of membrane interaction similar to one shown above. Acrosomal membrane is now inserted into sperm plasma membrane.* × *100,000. For details, see Figs. 2-14 and 2-17. (From A. L. Colwin and L. H. Colwin, in* Cellular Membranes in Development, *by courtesy of the authors and Academic Press.)*

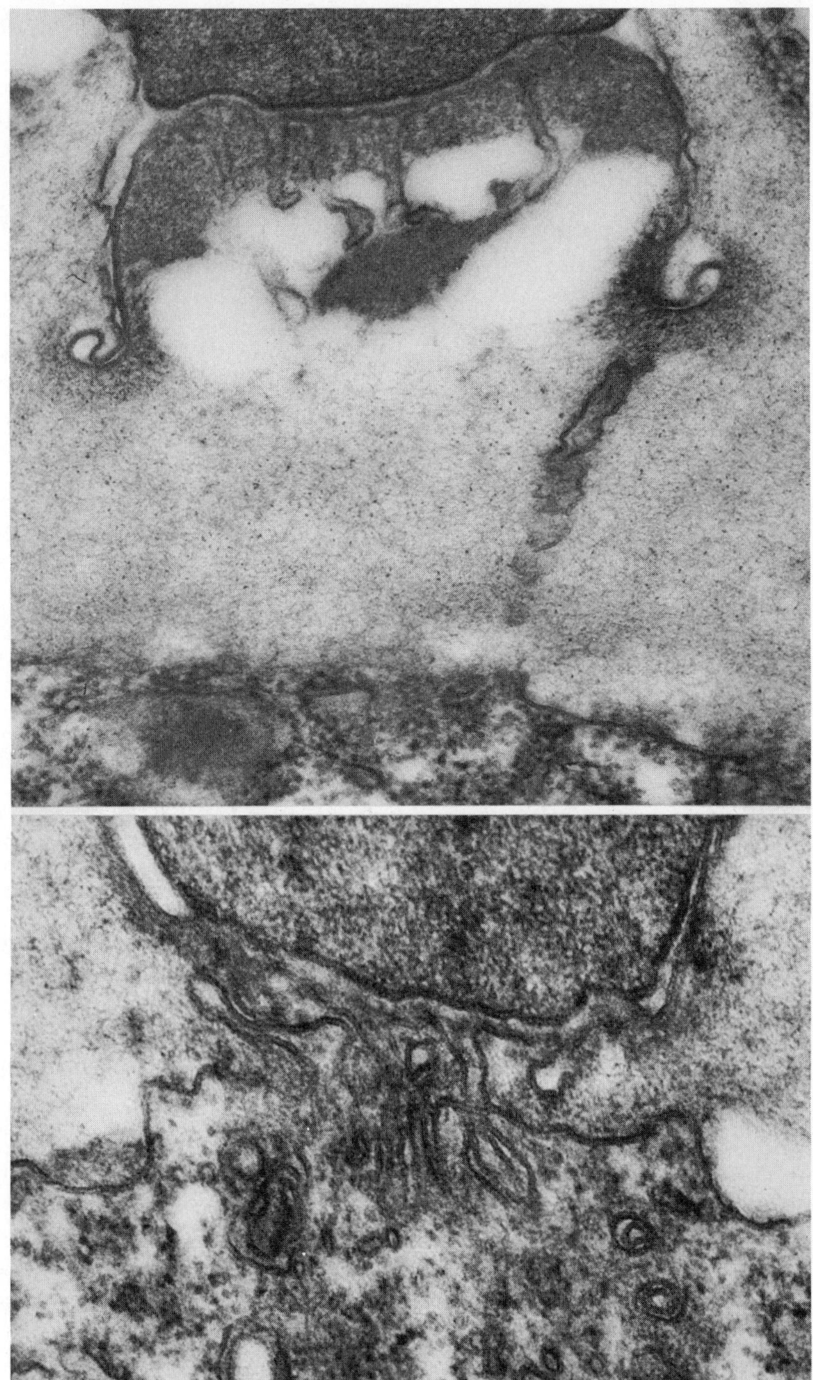

***Fig. 2-16*** *Electron micrographs illustrating stages* (2) *and* (4) *in Fig. 2-17.* (Upper) *Early stages* (2) *of eversion and modification of acrosomal region.* × *82,000.* (Lower) *Stage* (4) *of contact between gamete plasma membranes and formation of fertilization cone.* × *109,000. (From A. L. Colwin and L. H. Colwin, in* Journal of Biophysical and Biochemical Cytology, *10, by courtesy of the authors and the Rockefeller University Press.)*

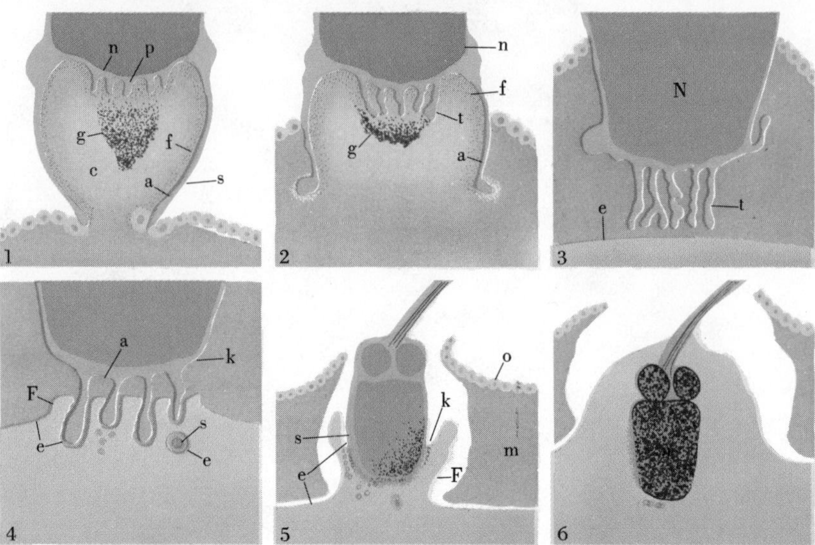


*Fig. 2-17 Principal steps in sperm–egg interaction in* Hydroides. (1) *Soon after the sperm contacts the egg envelope, the sperm tip opens, and the acrosomal and sperm plasma membranes become continuous.* (2) *and* (3) *The acrosomal wall is everted.* (4) *Sperm and egg plasma membranes meet; fertilization cone rises.* (5) *Sperm parts still protrude into egg envelope.* (6) *In zygote, internal sperm parts mingle with egg cytoplasm. (After A. L. Colwin and L. H. Colwin, in* Journal of Biophysical and Biochemical Cytology, *10, by courtesy of the authors and the Rockefeller University Press.)*

Before nuclear fusion occurs, a number of dramatic changes have taken place, most of them being initiated in a matter of seconds. These changes have been catalogued by several generations of investigators led by Lillie, John Runnström, and others in attempts to understand the nature of the activation mechanism. For example, the visible egg contents are rearranged. In some species, discrete granules in the egg cortex break down and contribute to the formation of the fertilization membrane. There are changes in the permeability of the egg membranes. New types of molecules appear. Fertilization is a highly intricate, stepwise process. Yet its *essence* is the activation of DNA synthesis and chromosome replication. During oögenesis the genome has not been completely inactive; major synthetic events have occurred. The unfertilized egg has the innate capacity to proceed, but that capacity cannot be expressed without a stimulus. We will consider the nature of the inhibition of the unfertilized egg more fully in Chapters 7 and 8.

Ultimately, after the membranes have fused and the inner parts of the head and middle piece (only occasionally the tail) of the sperm have entered the egg, there is a movement of the egg and sperm nuclei toward each other. The membrane of the sperm nucleus breaks down, freeing the chromosomes, and a spindle is formed from the material of the mid-piece. On this spindle the egg and sperm chromosomes are aligned. The fertilized egg or *zygote* is now ready for the next major event: *cleavage*.

Before we proceed, one point needs to be clarified—or amplified, for it adds a difficult dimension. As we have considered it, fertilization is a highly intricate, stepwise process. Earlier we related some of the facts of parthenogenesis. Many eggs can be activated artifically: frog eggs by pricking with a needle dipped in frog blood, sea urchin eggs by treatment with acids, rabbit eggs by heat and shock. What have these treatments in common with sperm? All the evidence suggests that the

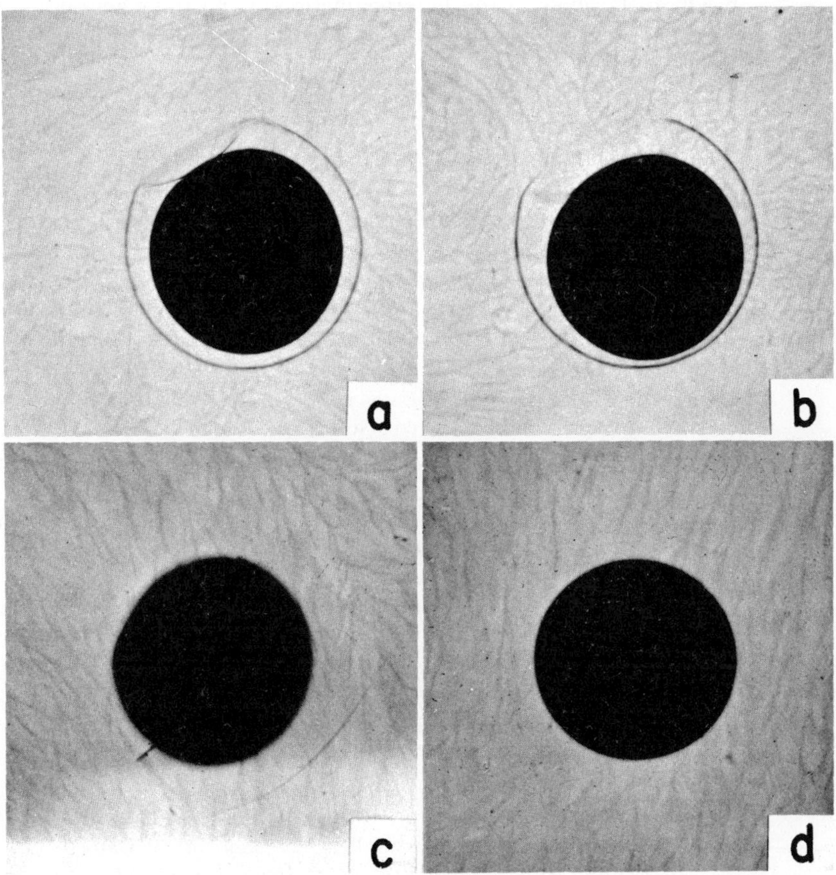


*Fig. 2-18 Photomicrograph of an egg of the keyhole limpet at* (a) *1 minute,* (b) *1¾ minutes,* (c) *2½ minutes,* (d) *3¼ minutes after adding a sperm extract containing an egg membrane lysin.* × *200. (From A. Tyler, in* Proceedings of the National Academy of Sciences, *25, by courtesy of the Academy.)*

sperm triggers some initial step; once set off, the process proceeds apace. The intricacy of the normal process insures a high degree of success, in contrast to the very low frequency of activation with artificial agents. Moreover, one experiment very strongly suggests that the stepwise events must be initiated at the surface, in natural order; in one species, at least, carefully depositing a living, intact sperm *inside* the egg by a micropipette does not initiate development.

After parthenogenesis, eggs do not usually remain haploid. The egg chromosomes are duplicated, often by retention of the second polar body in the final maturation division. Thus, the animal is diploid, but without a chromosomal contribution from the father. Its sex depends on the chromosomal sex-determining mechanism of the species. In rabbits, in which the male is the heterogametic sex, parthenogenetic offspring are females; conversely, parthenogenetic turkeys are all males.

## *FURTHER READING*

Adams, E. C., and A. T. Hertig, "Studies on Guinea Pig Oöcytes. 1. Electron Microscopic Observations on the Development of Cytoplasmic Organelles in Oöcytes of Primordial and Primary Follicles," *Journal of Cell Biology*, vol. 21 (1964), p. 397.

Anderson, E., "Oöcyte Differentiation and Vitellogenesis in the Roach, *Periplaneta americana*," *Journal of Cell Biology*, vol. 20 (1964), p. 131.

Austin, C. R., *Ultrastructure of Fertilization*. New York: Holt, Rinehart and Winston, 1968.

Beams, H. W., "Cellular Membranes in Oögenesis," in *Cellular Membranes in Development*, M. Locke, ed. New York: Academic Press, 1964, p. 175.

Burnett, A. L., and T. Eisner, *Animal Adaptation*. New York: Holt, Rinehart and Winston, 1964.

Colwin, A. L., and L. H. Colwin, "Role of the Gamete Membranes in Fertilization," in *Cellular Membranes in Development*, M. Locke, ed. New York: Academic Press, 1964, p. 233.

Dan, J. C., "The Vitelline Coat of the *Mytilus* Egg. 1. Normal Structure and Effect of Acrosomal Lysin," *Biological Bulletin*, vol. 123 (1962), p. 531.

Fawcett, D. W., "The Structure of the Mammalian Spermatozoon," *International Review of Cytology*, vol. 7 (1958), p. 195.

———, "Changes in the Fine Structure of the Cytoplasmic Organelles during Differentiation," in *Developmental Cytology*, D. Rudnick, ed. New York: Ronald, 1959, p. 161.

Levine, P., *Genetics*, 2d ed. New York: Holt, Rinehart and Winston, 1968.

Machlis, L., and E. Rawitscher-Kunkel, "Mechanisms of Gametic Approach in Plants," *International Review of Cytology*, vol. 12 (1963), p. 97.

Mann, T., *The Biochemistry of Semen and of the Male Reproductive Tract*. London: Methuen, 1964.

Monroy, A., *Chemistry and Physiology of Fertilization*. New York: Holt, Rinehart and Winston, 1965.

Roth, T. F., and K. R. Porter, "Yolk Protein Uptake in the Oöcyte of the Mosquito *Aedes aegypti* L.," *Journal of Cell Biology*, vol. 20 (1964), p. 313.

Szollosi, D., and H. Ris, "Observations on Sperm Penetration in the Rat," *Journal of Biophysical and Biochemical Cytology*, vol. 10 (1961), p. 275.

Telfer, W. H., "Immunological Studies of Insect Metamorphosis. II. The Role of a Sex-linked Blood Protein in Egg Formation by the *Cecropia* Silkworm," *Journal of General Physiology*, vol. 37 (1954), p. 539.

———, "The Route of Entry and Localization of Blood Proteins in the Oöcytes of Saturniid Moths," *Journal of Biophysical and Biochemical Cytology*, vol. 9 (1961), p. 747.

Wallace, R. A. and J. N. Dumont, "The Induced Synthesis and Transport of Yolk Proteins and Their Accumulation by the Oocyte in *Xenopus laevis*," *Journal of Cellular Physiology*, vol. 72, suppl. 1 (1968), p. 73.

# *chapter* 3

# *The Shape of Things to Come: Cleavage and Gastrulation*

*Then at its outsetting that speck grew and, presently tearing its tiny self in two, made an adhering pair. Then they 4, 8, 16, 32 and so on; only to slow down after reaching millions upon millions.*
*Each of the cells from the beginning besides shaping itself takes up for itself a right station in the total assembly according to the state which the assembly has by that time attained. Thus each cell helps to shape, and to construct by design, the total assembly.*

SHERRINGTON, 1951

*CLEAVAGE* In the next step in the panorama of development the fertilized egg begins to divide, or *cleave*. In this phase, which is common to all multicellular animals, but is rarely observed in plants, the single zygote is partitioned into a finite number of smaller cells, or *blastomeres*, which are much nearer the size of normal body cells. Little else visible happens. The embryo does not grow during this period, but there must be a synthesis of nuclear materials at the expense of the cytoplasm and nutrient reserves, as each smaller cell contains a full complement of genes. Nuclear gain is balanced by cytoplasmic decrease. The predominant activity is the synthesis of DNA and the nuclear and cytoplasmic constituents that are required for the process of division.

We have already remarked that cleavage occurs only rarely in the early development of plant embryos; in plants, division is accompanied by growth and cell differentiation. Moreover the second part of Sherrington's description, "... takes up for itself a right station ..." does not hold for the plant embryo in which each cell is surrounded by a rigid wall and cannot move. Thus in this chapter we shall be concerned only with animal development, leaving the analysis of early plant embryogenesis for Chapter 5.

Patterns of cleavage vary widely in the animal kingdom, yet within a given species the divisions may have a striking spatial and temporal pattern. Usually the process is synchronous at the outset; that is, the divisions and the intervening periods proceed in a rhythm affecting all cells at once. Periodically the number of cells is doubled—two, four, eight, sixteen—but the process becomes more irregular with the passage of time. We have already referred to the relation between amount and arrangement of yolk and pattern of cleavage. In eggs having only a small amount of yolk, distributed evenly, the daughter cells are usually of the same size—and cleavages are *equal.* We see this pattern in the cleaving zygote of the sea cucumber (Fig. 3-1).

When there are larger amounts of yolk and when it is distributed inhomogeneously, cleavages are *unequal.* Of the more common examples of the latter type of cleavage, possibly the best known is the frog's egg (Fig. 3-2). The light-colored, vegetal hemisphere contains the bulk of the yolk, which diminishes toward the animal pole. The first two divisions are vertical (longitudinal) and equal, but the third is horizontal, separating the yolk-laden vegetal half from the animal half; the four vegetal cells are larger than the four cells of the animal pole. As cleavage continues, the cells in the animal hemisphere cleave more rapidly than those in the vegetal half, so at the end of cleavage the animal hemisphere contains more, but smaller, cells than does the opposite pole. In the eggs of some species—such as the commonly studied fish, *Fundulus*, and birds such as the chicken, shown in Fig. 3-3—there is so much yolk that only a small cap, or *blastodisk*, cleaves atop the yolk. The yolk is not partitioned. We speak of this pattern of cleavage as incomplete or *meroblastic*, in contrast to the complete or *holoblastic* cleavage seen in the sea cucumber.

But the distribution of yolk is not the only force operating in determining the type of cleavage. The eggs of some species have character-

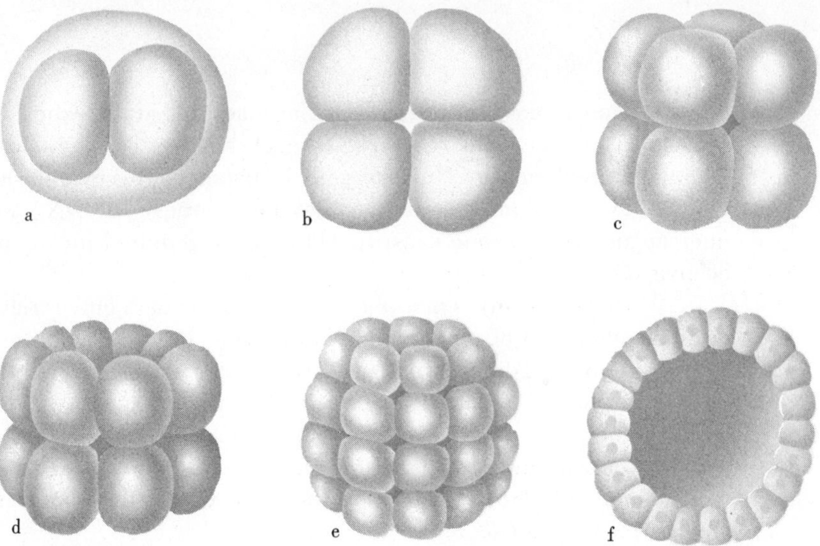


***Fig. 3-1*** *Equal cleavage in the sea cucumber.* (a) *2-cell stage.* (b) *4-cell stage (from animal pole).* (c) *8-cell stage, lateral view.* (d) *16-cell stage.* (e) *32-cell stage.* (f) *Blastula, vertical section. (From B. I. Balinsky,* An Introduction to Embryology, *2d ed., by courtesy of W. B. Saunders Company; after Selenka, from Korschelt, 1936.)*

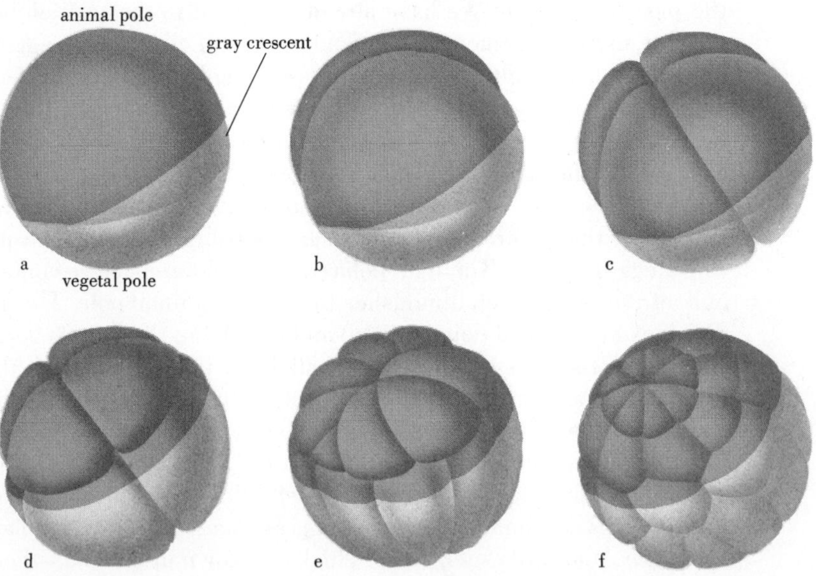


***Fig. 3-2*** *Cleavage in the frog.* (a) *Gray crescent shortly after entrance of the sperm.* (b) *2-cell stage, side view. The plane of division is considered to have bisected the gray crescent and thus corresponds to the future median axis.* (c) *4-cell stage. Blastomeres still equal-sized.* (d) *8-cell stage, showing initial inequality of blastomeres.* (e) *12- to 16-cell stage.* (f) *Late cleavage. (From T. W. Torrey,* Morphogenesis of the Vertebrates, *by courtesy of John Wiley & Sons, Inc.)*

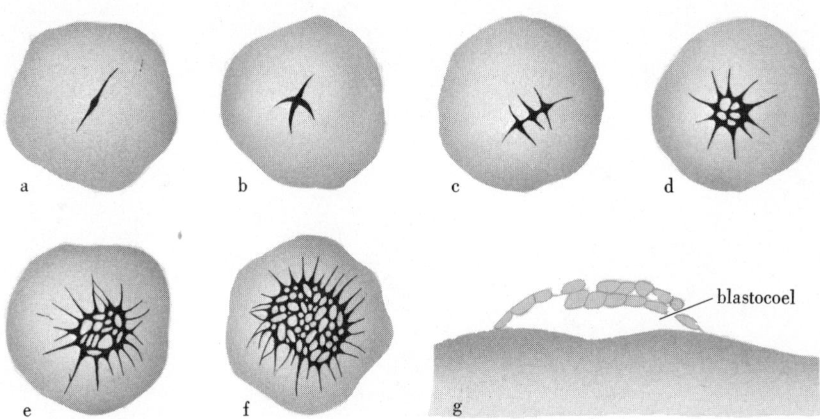


**Fig. 3-3** *Cleavage in the germinal disk of the ovum of the chicken; surface views.* (a) *2-cell stage.* (b) *4-cell stage.* (c) *8-cell stage.* (d) *16-cell stage.* (e) *32±-cell stage.* (f) *154±-cell stage.* (g) *Section through the blastula showing the blastocoel.* (a–f, *after T. W. Torrey,* Morphogenesis of the Vertebrates, *by courtesy of John Wiley & Sons, Inc.;* g, *after C. W. Bodemer,* Modern Embryology, *by courtesy of Holt, Rinehart and Winston, Inc.)*

istic patterns imposed by other variables. It is not unusual to find that a fundamental similarity in pattern characterizes a group of animals, giving evidence of their relationship. A precise pattern of cells, of certain sizes in well-defined positions, may be produced in every individual of the group at a corresponding stage in development. In *radial cleavage* the successive planes pass through the egg, perpendicular to one another and arranged symmetrically around the polar axis of the egg. The sea cucumber (Fig. 3-1) and sea urchin (Fig. 3-4) exhibit this pattern. *Spiral cleavage*, so called because of the oblique direction of many of the divisions, is characteristic of many annelids and mollusks (Fig. 3-5).

Cleavage, then, results in a cluster of cells, called a *blastula*, the shape of the cluster and arrangement of cells in it being characteristic of the species (Fig. 3-6). The cells are not packed tightly together. They may adhere, by a sticky coating on their surface, or they may be kept from falling apart by the outer protective membranes covering the egg. In some species, the new cells temporarily fall apart and only the outer membrane prevents them from being dispersed. Interstices appear between cells; there may be, in fact, an internal, fluid-filled cavity, or *blastocoel*, around which the cells are arranged. In some forms, the surfaces of the cleavage cells are connected by fibers.

Again, however, we are concerned more with principles than with a recital of differences. Fundamentally, the egg has been partitioned and its content of DNA has been increased, since each daughter cell in the

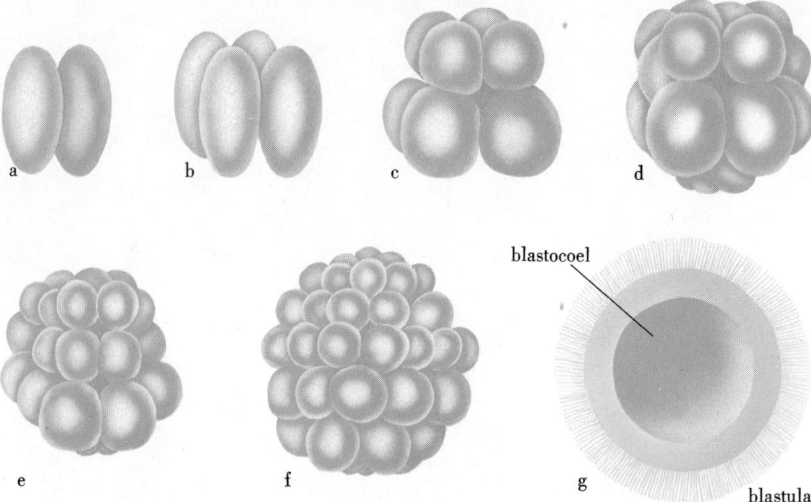


*Fig. 3-4* *Cleavage in the sea urchin egg. The first two divisions* (a *and* b) *produce equal-sized blastomeres. The third division* (c) *produces an upper quartet of blastomeres smaller than the lower quartet.* (d) *16-cell stage, showing small blastomeres at vegetal pole.* (e) *32-cell stage.* (f) *64-cell stage.* (g) *Blastula. (After T. W. Torrey,* Morphogenesis of the Vertebrates, *by courtesy of John Wiley & Sons, Inc.)*

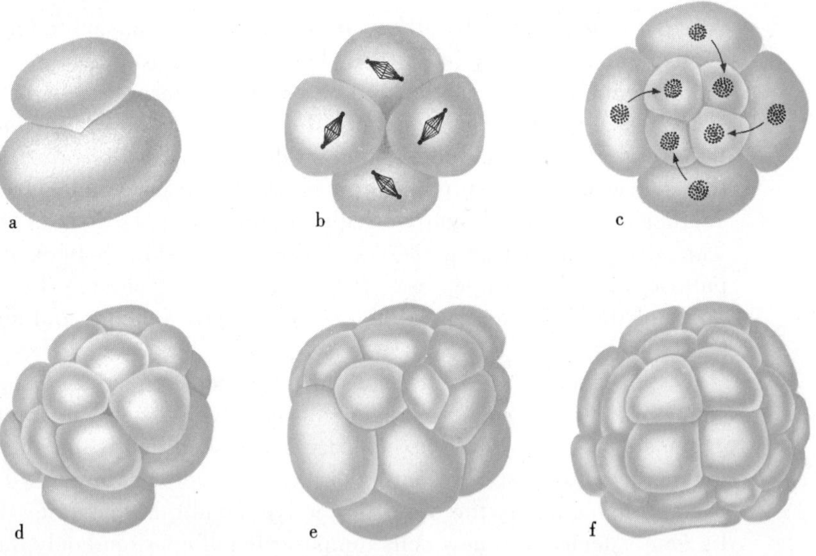


*Fig. 3-5* *Spiral cleavage in the egg of an annelid,* Nereis. (a) *2-cell stage, polar view.* (b) *4-cell stage, polar view.* (c) *8-cell stage, polar view.* (d) *16-cell stage, polar view.* (e) *16-cell stage, viewed from right side.* (f) *32-cell stage, polar view. Arrows in* (c) *indicate the directional shift in position of the upper cells in relation to the larger, lower cells. (After T. W. Torrey,* Morphogenesis of the Vertebrates, *by courtesy of John Wiley & Sons, Inc.)*

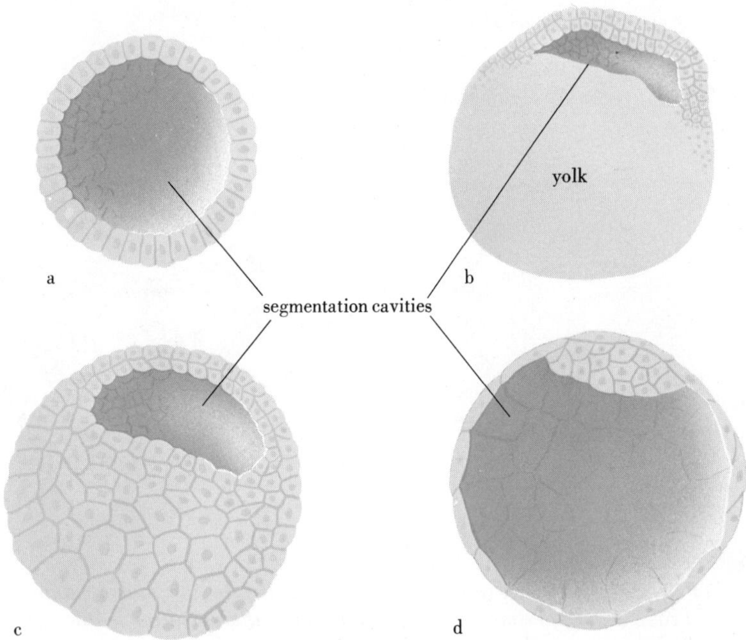


**Fig. 3-6** *A diagrammatic comparison of chordate blastulae, demonstrating the basic anatomical pattern of the blastula stage. These drawings are not made to scale. Each blastula is composed of many cells surrounding (or partially surrounding) a central or eccentric segmentation cavity, sometimes termed a blastocoel.* (a) Amphioxus, *representing primitive chordates;* (b) *teleost;* (c) *amphibian;* (d) *mammal. (After C. W. Bodemer,* Modern Embryology, *by courtesy of Holt, Rinehart and Winston, Inc.)*

blastula is diploid. Along with the increase in DNA with increasing cell numbers there has been a rapid increase in *cell surfaces*; regional cytoplasmic differences in the egg have been "packaged," and clearer discontinuities have emerged.

Is cleavage *necessary*? Development does not proceed normally in animal embryos without cleavage. As we shall see, the relations between cell proliferation and differentiation are complex, but at this point it may be asserted that when cleavage is suppressed in animal embryos, relatively little development is possible. In some well-known older experiments, when cleavage was suppressed in eggs of the marine segmented worm *Chaetopterus*, some of the superficial structures such as cilia appeared. These were somewhat reminiscent of the normal embryo (Fig. 3-7). Even in this abortive "differentiation without cleavage," however, it was shown that some DNA synthesis occurred.

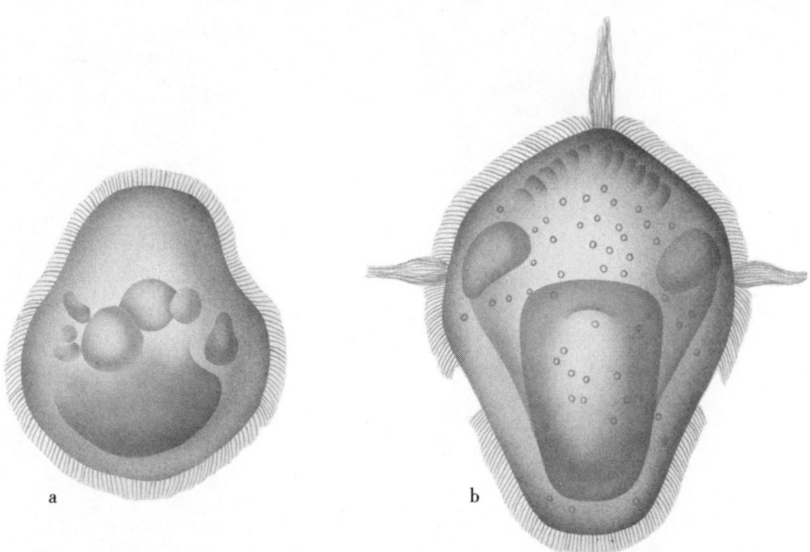


***Fig. 3-7*** *Differentiation without cleavage in the marine annelid* Chaetopterus. (a) *Experimental larva; optimal differentiation without cleavage.* (b) *Control larva. (After F. R. Lillie.)*

***GASTRULATION*** Following cleavage, cells are regrouped, forming new associations as a result of a profound, and as yet poorly understood, process called *gastrulation.* It is but one of a series of orderly formative movements. Our approach will be to describe the movements in gastrulation and then to begin asking questions about the forces directing them. Only after we have studied other patterns of morphogenesis and have a better awareness of the importance of these fundamental processes will we probe more deeply into mechanisms of cell association and interaction in Chapter 10.

At the end of cleavage, cells in certain regions of the embryo begin to show at their surfaces intense activity, preceding and accompanying their movement. The cell movements are undertaken both individually and by groups of cells, and must be highly integrated. In most of the invertebrates and all the vertebrates, these movements result in the transformation of the blastula into the three-layered gastrula—the outer cell layer being the *ectoderm*, the inner layer the *endoderm*, and the intermediate layer the *mesoderm.*

Before we examine patterns of gastrulation in several representative species, let us pose several questions to which we shall return repeatedly. *First*, what is the significance of gastrulation? As we shall see, cells of the blastula are differentiated, to a degree, but their further differentiation requires their intimate association with other cells in new ways. It is clear that cell interaction is a cardinal principle of development. These interactions are made possible by the new associations formed first during gastrulation. Somehow, as a consequence of these interactions, the synthetic machinery of the cells is set in train on a larger scale; then at varying intervals after gastrulation, depending on the species, ribosomes are again elaborated, on which enzymes and other proteins will be synthesized. Gastrulation is a "critical" or sensitive period in the sense that the processes are disrupted easily, further development being blocked. *Second*, what initiates gastrulation? We don't really know. The first sign we have been able to detect is pulsatory activity at the surfaces of a few cells. *Third*, what integrates the cell movements of gastrulation? Suggestions will emerge as we examine the process and its consequences in three forms: sea urchins, amphibians, and birds. For all of these forms, the techniques are essentially the same: (1) thorough study of the external and internal form of the embryo, as revealed in the microscopic examination of whole embryos and sections at frequent intervals; (2) making observations by means of time-lapse cinematography; and (3) tracing cells "tagged" with carbon particles, dyes, or radioactive molecules; occasionally, combinations of these approaches are used.

In the sea urchin embryo, which has been studied intensively by time-lapse cinematography, cells at the lower or *vegetal pole* undergo great surface pulsatory activity, leading to and accompanying their migration into the cavity of the blastula (Figs. 3-6, 3-8). Before gastrulation begins, the epithelial cells of the blastula are arranged in a column, those at the vegetal pole being somewhat fuller. As we have just remarked, the inner surfaces of these cells begin to "bubble," sending out irregular pseudopods. As the cytoplasm is so expanded and protruded, the shapes of the cells change, and the relations of any given cell with its immediate neighbors must be altered. The cells are no longer held together, and some of them, the *primary mesenchyme cells*, are forced into the blastocoel. As one consequence, the vegetal pole is indented; the indentation continues to deepen, forming the cavity of the gastrula or *archenteron* (primitive gut); the opening of this cavity to the exterior is the *blastopore*.

Next, a few cells send out long, filamentous processes across the blastocoel to attach to its inner wall. These *filopodia* migrate to and "feel about" the inner surface of the ectoderm, gradually collecting at or near the upper or *animal pole*. Then they contract, drawing the archenteron tip up the animal pole. The highly active cells at the tip of the archenteron are detached to form the *secondary mesenchyme*, or definitive mesoderm.

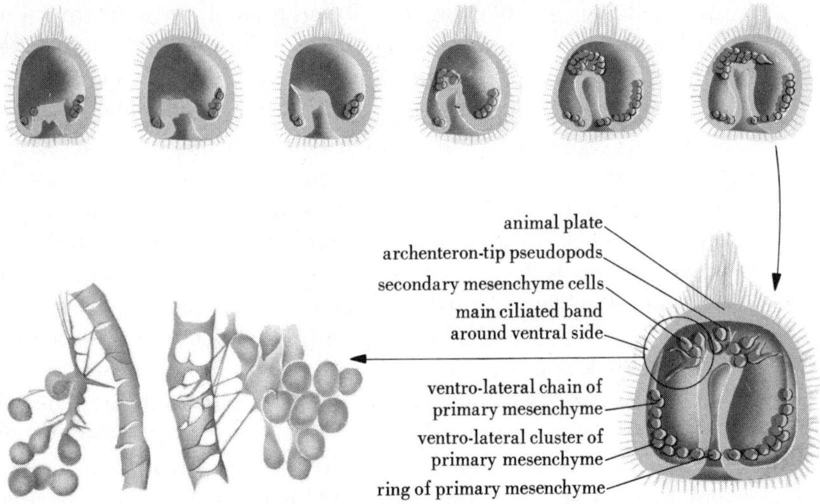

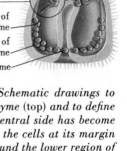

*Fig. 3-8 Morphogenesis in the sea urchin gastrula. Schematic drawings to illustrate interactions of ectoderm and primary mesenchyme* (top) *and to define some of the regions of an advanced gastrula where the ventral side has become flattened as the result of an increase in contact between the cells at its margin* (bottom right). *The increase in cell contact in a zone around the lower region of the ectoderm is also indicated. The primary mesenchyme cells line up along zones with high contact between the ectoderm cells with the exception of the animal plate.* (Bottom left) *The attachment of primary mesenchyme pseudopods to the ectoderm. The left portion illustrates ventrolateral mesenchyme cable in a slightly compressed larvae; the right portion illustrates ventrolateral mesenchyme cluster (note bridges between ectoderm cells between the two main levels of contact). (After T. Gustafson, in* Experimental Cell Research, *32, by courtesy of Academic Press.)*

Thus, the movements of gastrulation in the sea urchin appear to involve at least three properties of cells: random motility (the capacity of individual embryonic cells to move), differential adhesiveness (displayed by the selective attachment of migrating cells to the inner surfaces of ectodermal cells of the gastrula wall), and contractility (displayed by filopodia).

In *amphibian* embryos, cells are redistributed by a series of movements that have been known since the classic work of Vogt who, by marking small groups of cells with vital dyes, followed their movements through gastrulation. Let us take the frog as an example. As we have seen (Fig. 3-2), a series of cleavages results in a blastula. Initially the divisions are holoblastic and equal; gradually the pattern becomes unequal, resulting in a blastula in which the cells of the animal pole are very small, while the yolk-laden cells at the vegetal pole are much larger. A blastocoel has formed, largely in the animal hemisphere.

Now a dark cleft appears on the surface of the egg, just below the equator, at the junction of the animal and vegetal hemispheres (Figs. 3-9 and 3-10), This cleft marks the *dorsal lip of the blastopore*, through which cells move to the interior. The fact that the egg is pigmented and opaque, in contrast to the clear egg of the sea urchin, makes photographic studies of the actual movements ineffective. However, marking techniques reveal that cells originally in the surface move toward and through the blastopore. The ectoderm must stretch to cover the surface of the sphere. At first the archenteron is small, but as cells continue to migrate, it increases in size (Fig. 3-10). Gradually the blastopore connecting the archenteron with the exterior is complete; however, it is "plugged" by yolk (Figs. 3-9 and 3-10). As the archenteron increases in size, the blastocoel is gradually obliterated.

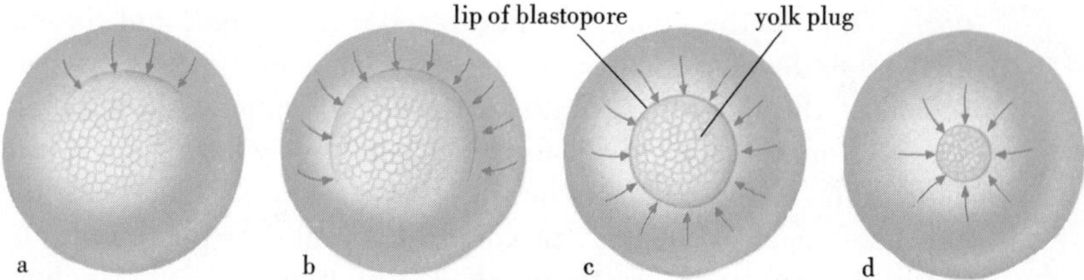


*Fig. 3-9 Germ layer formation in the amphibian egg.* (a–d) *are surface views looking toward the blastopore. Arrows indicate direction of morphogenetic movements. (From T. W. Torrey,* Morphogenesis of the Vertebrates, *by courtesy of John Wiley & Sons, Inc.)*

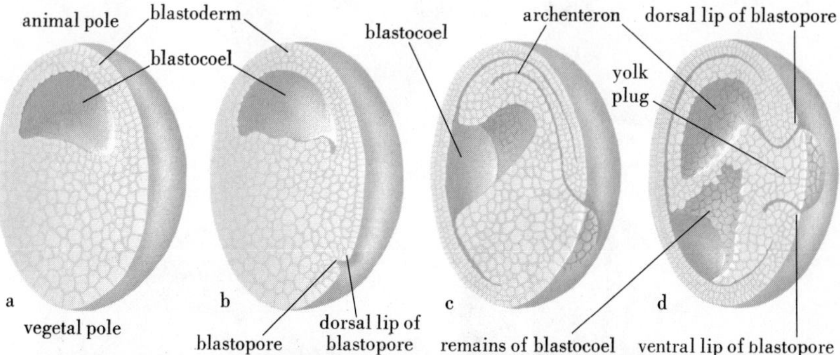


*Fig. 3-10 Germ layer formation in the frog embryo.* (a) *At the late blastula stage;* (b) *at the very beginning of gastrulation;* (c) *in the middle of gastrulation;* (d) *at the late gastrula stage (semidiagrammatic). Embryos viewed at an angle from the dorsal side* (a, b, c) *or from posterior end* (d). *(From B. I. Balinsky,* An Introduction to Embryology, *2d ed., by courtesy of W. B. Saunders Company)*

It is difficult to reconstruct these formative or morphogenetic movements by the study of whole and sectioned embryos alone. It is necessary to label groups of cells and determine their positions at successive intervals during development, following them to their ultimate positions. In amphibian embryos vital stains have been used with profit. By a vital stain we mean a relatively nontoxic dye, one that when used in low concentrations will mark living cells. Neutral red and Nile blue sulfate are commonly used. Figure 3-11 not only illustrates the method but adds another key observation to our analysis of the amphibian egg and the onset of gastrulation. About 30 minutes after fertilization in the amphibian egg, the pigmented surface layer or cortex rotates relative to the underlying yolk-rich cytoplasm. In doing so, it uncovers a crescent-shaped area on the side opposite the point of sperm entry. The resultant *gray crescent* is one of the first evidences of differentiation prior to cleavage. It marks the site of the future dorsal lip of the blastopore. Thus, if the center of the crescent is stained, and the embryo is allowed to develop further, the mark persists in the early gastrula in the region of the dorsal lip. With further development, these labeled cells are invaginated, or carried into the embryo, where they are found to contribute to the head endoderm of the primitive gut. This technique has been used extensively in amphibian embryos for the production of "fate maps," such as those shown in Figure 3-12, which give the ultimate relations between areas of the early gastrula. The boundaries

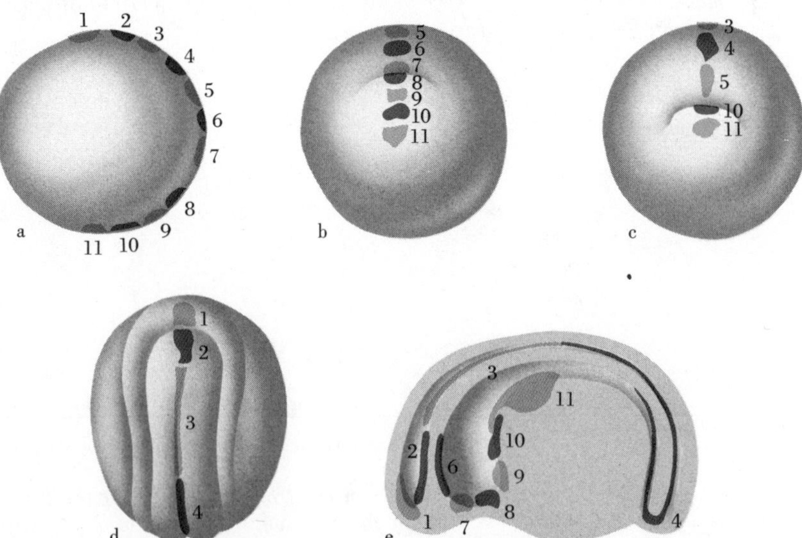


***Fig. 3-11*** *Vital staining of the newt gastrula. Marks were placed along the dorsal median surface.* (a–d) *Surface views;* (e) *embryo dissected in the medial plane to show stained regions in the interior. (From Vogt, 1929.)*

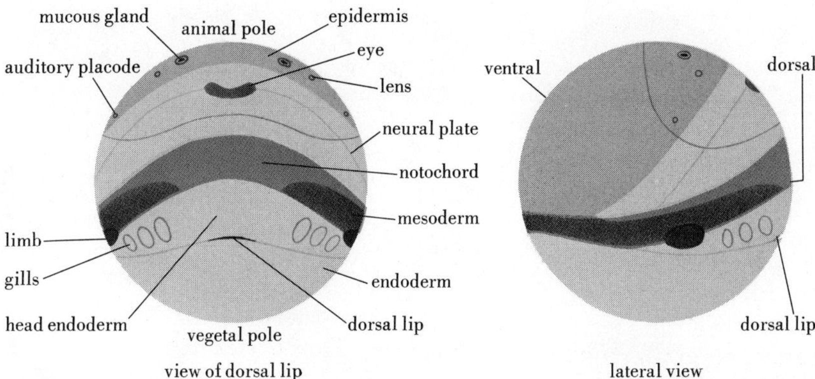


*Fig. 3-12* *A map of the prospective values of the early gastrula of the European toad,* Bombinator, *as shown by vital staining. (From C. W. Bodemer,* Modern Embryology, *by courtesy of Holt, Rinehart and Winston, Inc.)*

of given areas may differ from species to species, but these maps are more than adequate to illustrate the technique and generalizations derived from it.

What can we say of the mechanisms of these movements? As in the sea urchin (and, in fact, as in any species), very little. There are clearly two fundamental series of events, closely intertwined: (1) at the dorsal lip, cells become active, undergo striking changes in shape, and move into the interior; (2) at the same time, the ectoderm undergoes an active expansion. Although both sets of events are related, they can be performed independently if parts of the embryo are isolated. For example, if a small piece of the dorsal lip is transplanted to a distant part of another embryo, well removed from the site of the normal dorsal lip, it will invaginate and form a new archenteron, independent of the host's archenteric cavity. Moreover, a fragment of dorsal lip placed on a piece of endoderm *in vitro* invaginates into it. In contrast, a piece of ectoderm from an early gastrula placed atop a piece of endoderm spreads to enclose it.

Just as in the sea urchin, the movements of cells during amphibian invagination are related to changes in shape of individual cells. Epithelial cells expand at their internal ends and contract at their external ends. The cells become "bottle" or "flask" shaped (Figs. 3-13 and 3-14). The attenuated neck of the bottle keeps the cells in touch with the surface as the ends of the cells move inward. As a consequence the entire layer is bent inward (Fig. 3-15).

Although there must be specific forces signaling a few cells in the region of the blastopore to start their inward trek, we know only that the movement is a property of individual cells. They are not "propelled" inward by forces at the surface.

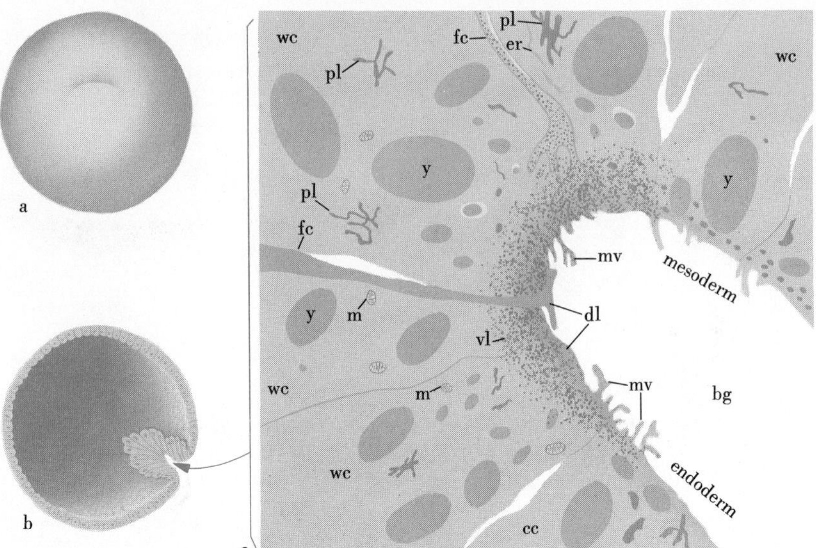


*Fig. 3-13 Fine structure and morphogenetic movements in the amphibian gastrula.* (a *and* b) *Surface view and sagittal section of an early gastrula. Flask and wedge cells border the blastoporal groove.* (c) *Drawing based on a montage of electron micrographs of a section of the groove* (bg). *Flask* (fc), *wedge* (wc) *and cuboidal* (cc) *cells line the groove. (From P. C. Baker,* Journal of Cell Biology, *24, 1965, by courtesy of the author and the Rockefeller University Press.)*

As we have said, there is a great variety in the course of these movements in different species due at least in part to variations in the distribution of yolk. Let us now consider the *chick* embryo. The egg, after fertilization, completes its journey down the oviduct, where it is invested with its protective mantle, the shell, shortly before it is laid. During this period, cleavage, into a disk containing many cells, is also completed. Then, at laying, development stops until the egg is incubated. In Fig. 3-16 (stage 1) we see such an embryo, removed from its normal position on the surface of the yolk, where it appears as a tiny white disk, about 2 mm in diameter, containing some 60,000 cells. In section (Fig. 3-17) we see no striking landmarks, only an apparently undifferentiated mass of cells, formed into two layers separated by a cavity.

We do not know precisely how these two discrete layers were formed; presumably the lower layer or *hypoblast* split off from the upper layer or *epiblast*. The hypoblast gives rise to the yolk sac endoderm. At this stage the epiblast is composed of cells that will contribute to ectoderm, mesoderm, and endoderm.

After a few hours of incubation the first visible signs of polarity appear (Fig. 3-16, stage 2), a thickening in what will become the pos-

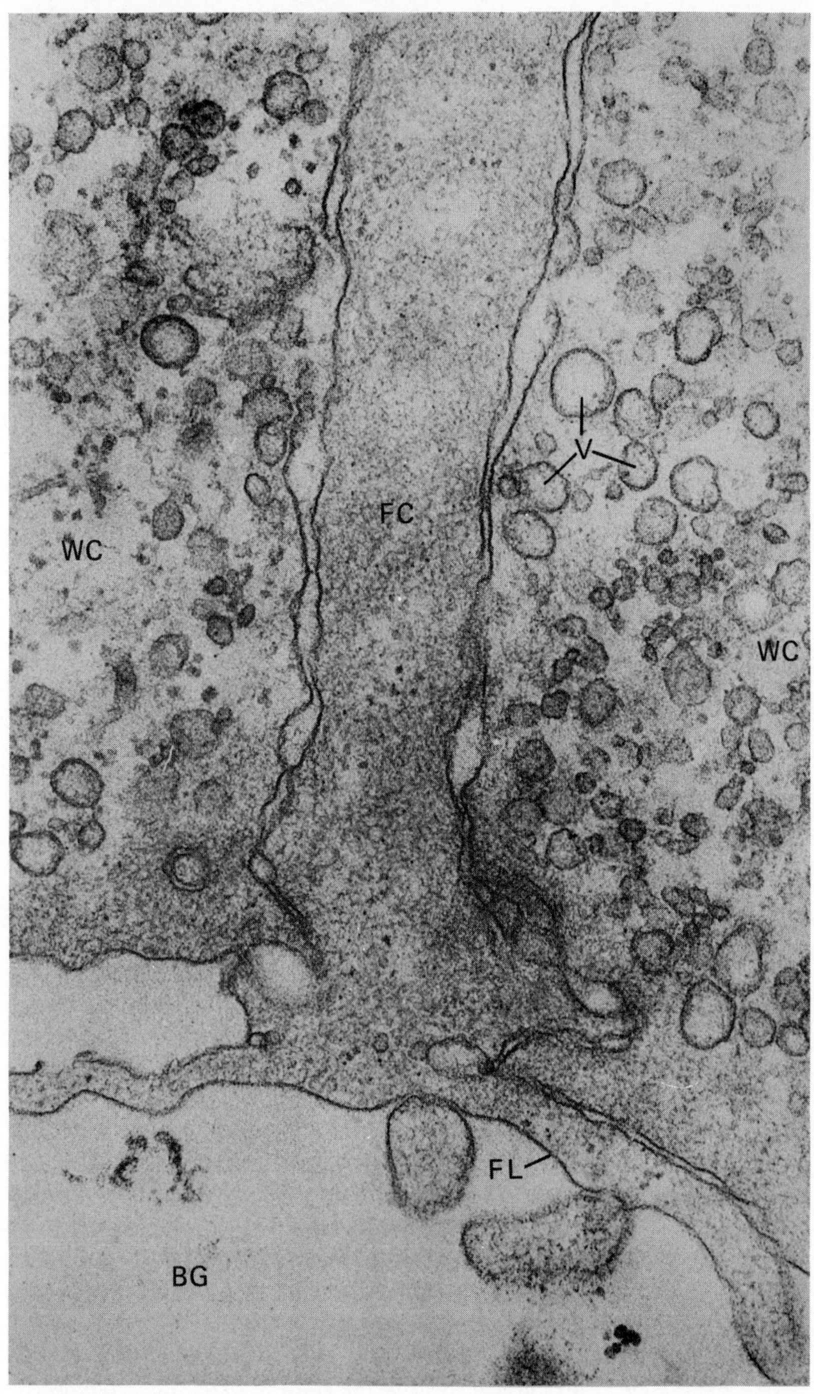


***Fig. 3-14*** *Distal end of neck of flask cell, showing lip* (FL) *of cell and interlocking of flask* (FC) *and wedge* (W) *cells.* × *65,000. (From P. C. Baker,* Journal of Cell Biology, *24, 1965, by courtesy of the author and the Rockefeller University Press.)*

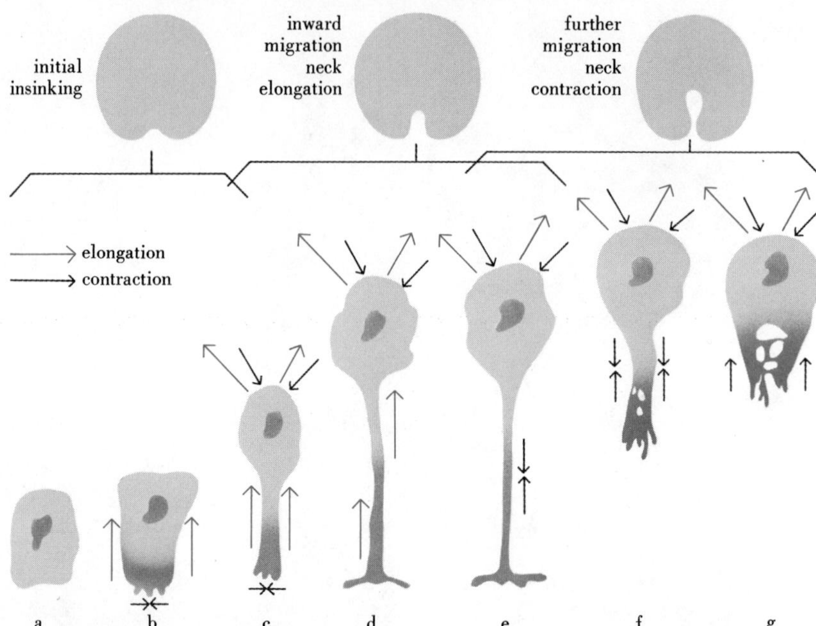


**Fig. 3-15** *Theoretical scheme for the transformations of a blastoporal cell during gastrulation, based on the hypothesis of contraction and expansion of the dense layer. The end of the cell neck remains securely adhered to adjacent cells.* (a) *The cell as it appears on the surface prior to gastrulation.* (b) *The distal surface contracts causing initial insinking of the blastoporal pit.* (c and d) *The proximal cell surface begins to migrate inward, deepening the groove; the distal surface continues to contract and the neck elongates.* (e–g) *The neck contracts after it reaches maximum elongation. This shortening, together with the pull exerted by the migrating inner ends, furthers invagination and pulls adjacent cells into the groove. Some flask cells bypass stages* (d) *and* (c). *(From P. C. Baker,* Journal of Cell Biology, *24, 1965, by courtesy of the author and the Rockefeller University Press.)*

terior end of the embryonic axis resulting from an increase in the number of cells. It marks the posterior end of the *primitive streak* or elongated, modified blastopore. At this time the hypoblast is separating from the epiblast in the posterior region. In the epiblast cells are moving, posteriorly and toward the midline, and the primitive streak takes shape (Fig. 3-16, stages 3 and 4). The movements of these cells have been studied in embryos labeled not with vital stains, which

**Fig. 3-16** *Early development of the chick embryo. Stage 1: prestreak; stage 2: initial streak (6–7 hours of incubation); stage 3: intermediate streak (12–13 hours); stage 4: definitive streak (18–19 hours). The primitive streak has reached its maximal length. Stage 5: head process (19–22 hours). The notochord or head process extends forward as a rod of condensed mesoderm. (From H. L. Hamilton,* Lillie's Development of the Chick, *Holt, Rinehart and Winston, Inc.)*

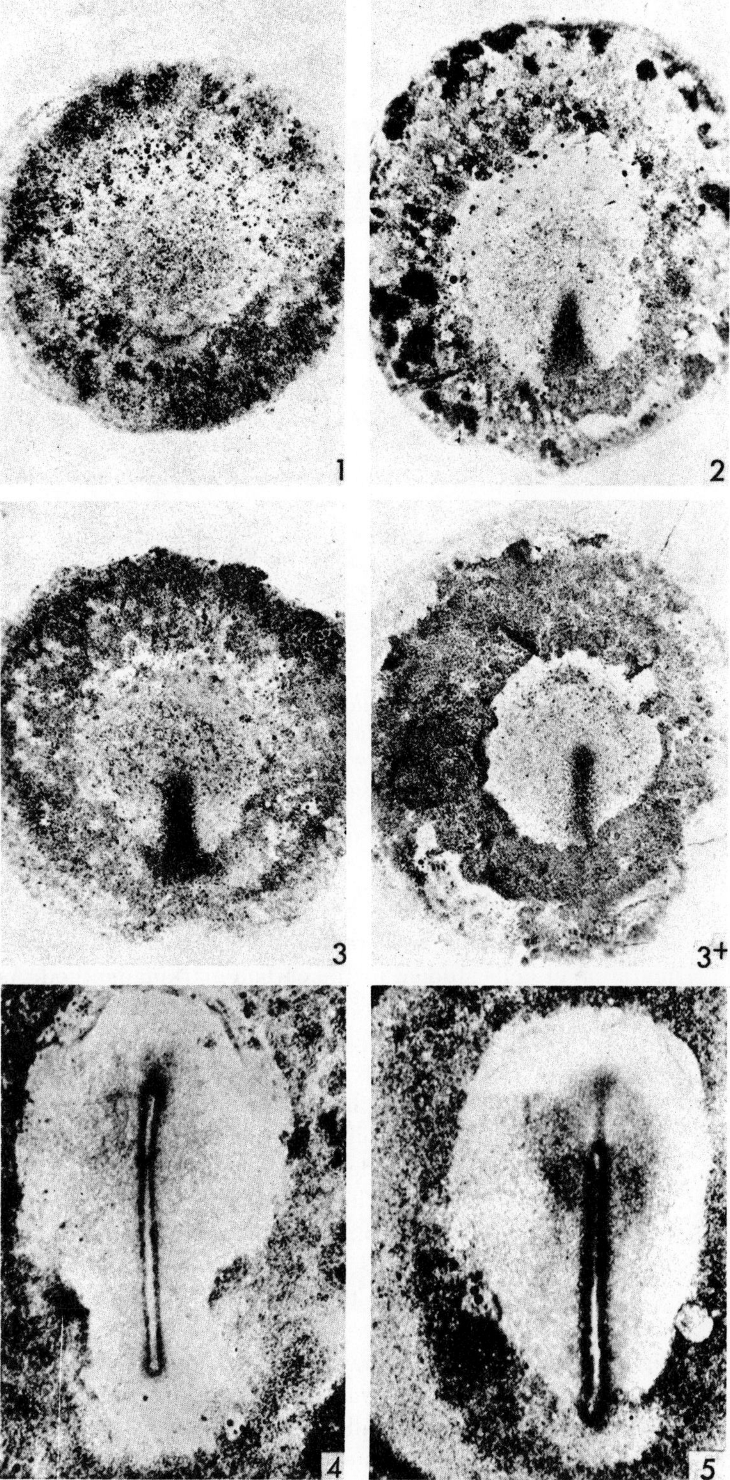
1
2
3
3+
4
5

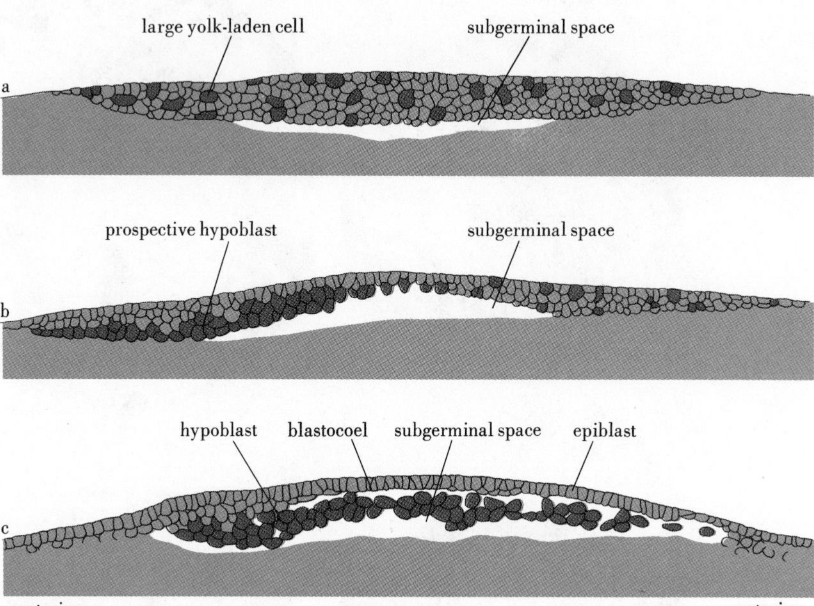


*Fig. 3-17 Longitudinal sections through chick blastoderms of successive ages showing origin of hypoblast.* (a) *Larger yolk-laden cells intermixed with smaller cells.* (b) *Accumulation of larger cells in subsurface position, notably at the rear.* (c) *Organization of larger cells to form hypoblastic layer, concomitantly with rise of blastocoelic spaces and thinning of epiblast. (From T. W. Torrey,* Morphogenesis of the Vertebrates, *by courtesy of John Wiley & Sons, Inc.)*

diffuse too readily in the chick embryo, but with finely powdered carbon particles and with radioactive tracers. Figure 3-18 shows the migration of cells studied by the carbon-marking method. As the streak grows, a median *primitive groove* appears, clearly delimited by lines of cells moving into it. The groove ends anteriorly in a deep pit forming the center of *Hensen's node.* Cells move up to and through the patent streak. The entire posterior half of the early epiblast moves toward the streak, while the surface of the anterior half spreads to cover the entire surface of the blastoderm.

Cells destined to contribute to both endoderm and mesoderm pass through the primitive streak. Cells moving through the anterior end of the streak in its early stages become embryonic endoderm, as marking experiments using tritiated thymidine most clearly confirm. Thus the endoderm has a dual origin, some of the cells being derived from the original hypoblast, others moving through the streak from the epiblast.

Larger numbers of invaginating cells become mesoderm, moving away from the streak laterally after they have moved through it. Cells moving into the streak from one side may move out of the streak in the deeper layers on the same side, or on the opposite side. Although the

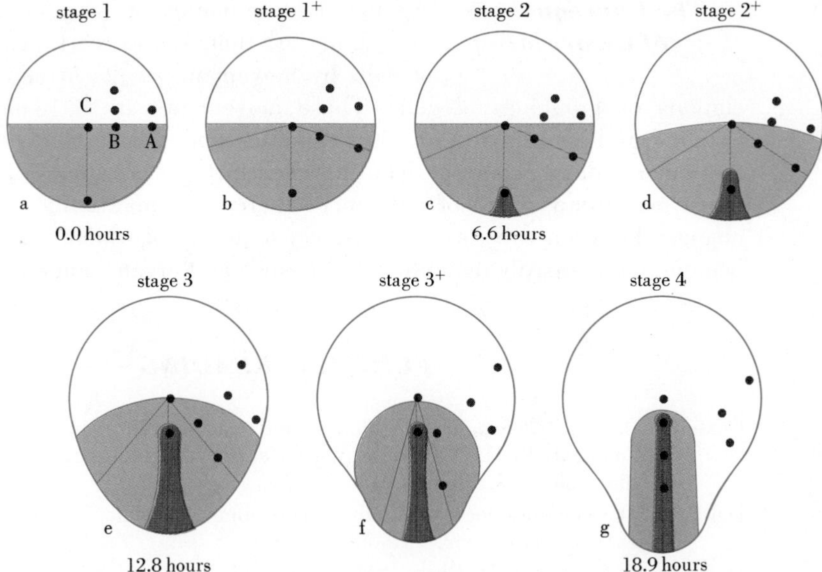


*Fig. 3-18 Diagram of cellular movements in the epiblast associated with the development of the primitive streak. Note especially the movements of points* A *and* B *in successive stages. See Fig. 3-16. (From N. T. Spratt, Jr., in* Journal of Experimental Zoology, *103, by courtesy of The Wistar Institute.)*

amount of mixing or crossover of cells is not entirely clear, it is known that some mixing of cells does occur.

Cells destined to give rise to the notochord, the elastic turgid rod supporting the neural tube, accumulate in the deeper parts of Hensen's node and move forward in front of the node underneath the surface of the ectoderm (Fig. 3-16, stage 5; Fig. 3-19). This mass of notochordal cells is known as the *head process* (or notochordal process).

In summary, the primitive streak may be regarded as a blastopore through which cells immigrate from the surface to form endoderm and mesoderm. Hensen's node may be considered homologous to the dorsal lip. The principal difference between the amphibian and chick blastopores is that the formation of the latter does not lead to the formation of an archenteron.

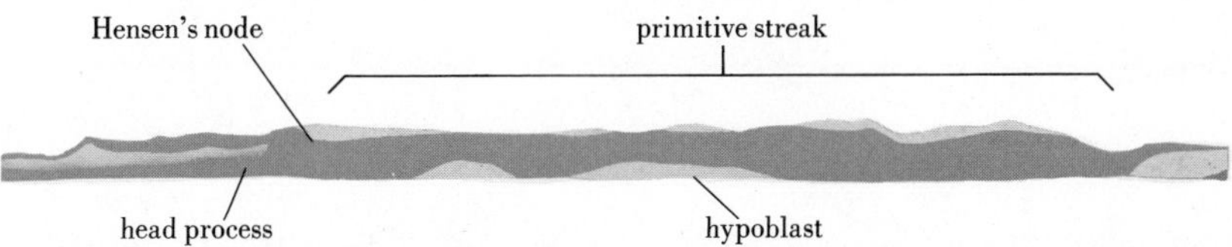


*Fig. 3-19 Longitudinal medial section of a chick embryo in the head-process stage.*

**The Consequences of Gastrulation** We may restate one of our key generalizations thus far; that is, the rearrangements of cells by movement results in new associations or groupings of cells. These movements are in some way initiated and coordinated by discontinuities and cell surface changes that emerge during cleavage. Thus new cell surfaces are apposed, and—if we may jump ahead of our story—there are opportunities for exchanges between cells, as a further consequence of which cell groups become more sharply delimited, and tissues and organs emerge.

## FURTHER READING


Costello, D. P., "Cleavage, Blastulation and Gastrulation," in *Analysis of Development*, B. H. Willier, P. Weiss, and V. Hamburger, eds., Philadelphia: Saunders, 1955, p. 213.

Dan, K., "Cyto-embryology of Echinoderms and Amphibia," *International Review of Cytology*, vol. 9 (1960), p. 164.

DeHaan, R. L., and J. D. Ebert, "Morphogenesis," *Annual Review of Physiology*, vol. 26 (1964), p. 15.

Gustafson, T., and L. Wolpert, "The Cellular Basis of Morphogenesis and Sea Urchin Development," *International Review of Cytology*, vol. 15 (1963), p. 139.

Hamburger, V., *A Manual of Experimental Embryology*. Chicago: University of Chicago Press, 1960.

Mercer, E. H., "Intracellular Adhesion and Histogenesis," in *Organogenesis*, R. L. DeHaan and H. Ursprung, eds., New York: Holt, Rinehart and Winston, 1965, p. 29.

Trinkaus, J. P., "Mechanisms of Morphogenetic Movements," *ibid.*, p. 55.

———, "Morphogenetic Cell Movements," in *Major Problems in Developmental Biology*, New York: Academic Press, 1966, p. 125.

Weiss, P., *Principles of Development*. New York: Holt, Rinehart and Winston, 1939.

Wilson, E. B., "Cell Lineage and Ancestral Reminiscence" (1898). Reprinted in *Foundations of Experimental Embryology*, B. H. Willier and J. M. Oppenheimer, eds. Englewood Cliffs, N.J.: Prentice-Hall, 1964, p. 53.

Wolpert, L., "The Mechanics and Mechanism of Cleavage," *International Review of Cytology*, vol. 10 (1960), p. 164.

*chapter 4*

# Tissue Interactions in Animal Organogenesis

It is tempting to turn at once to the repertoire of new ideas on the nature of cell interactions, including cell adhesions and surface interactions and those in which an exchange of materials between cells is involved. However, thus far we have touched only upon the problems of morphogenesis in early embryos; before attempting to analyze these mechanisms, we need to take a closer look at morphogenesis on a larger scale—in the formation of multicellular tissues and the organs of vertebrates.

Organs may be highly complex; they may be composed of many cells of diverse origins and functions. But despite the complexity of these processes, and the variety of patterns expressed from organ to organ, we can recognize several basic principles. The most effective way of revealing the recurring principles in organ formation is by looking at the processes themselves.

## *PROGRESSIVE DETERMINATION IN THE NERVOUS SYSTEM*

Today the well-appointed embryology laboratory must contain a spectrophotometer, an ultracentrifuge, a fraction collector, and a scintillation counter; the techniques of molecular biology require them. However, we are so accustomed to thinking that research can be accomplished only through use of such techniques that we may be inclined to overlook simple experiments. We must bear in mind that many of the significant advances in understanding mechanisms of development were made with techniques of utter simplicity, the only requisites being glass needles and hair loops, physiological saline, and the ability to pose the right questions.

We shall continue to stress the importance of cellular interactions, for the key to an understanding of development lies in an appreciation of the changes with time in the relations between the embryo and its parts, and of the influence of the parts upon one another.

These relations may be probed by *defect*, *isolation*, and *recombination* experiments. How will an embryo at any stage—cleavage, blastula, or gastrula—develop if some of its cells are removed? How will the cells removed behave if they are cultivated in isolation? Such defect and isolation experiments are complementary. Suppose that in isolation a group of cells fails to develop as it does normally in the embryo. The failure may indicate that in the absence of specific interactions, development cannot proceed. However, negative evidence of this nature may be misleading. It may mean only that the conditions of cultivation in isolation are unsatisfactory. We must then turn to a recombination experiment.

We begin our study of organogenesis by considering the emergence of the nervous system. This is a logical starting point, for several reasons: first, the formation of the primary neural structures is a direct consequence of changes that occur during gastrulation; second, neurogenesis embraces, hence serves to illustrate all, or nearly all, of the basic mechanisms with which we are concerned; and finally, the study of neurogenesis has attracted some of the field's most gifted experimenters. In retracing their classical experiments we become acquainted with what Twitty has called some of "embryology's finest hours."

In the preceding chapter we saw how it is possible to construct a fate map of an early amphibian gastrula. In Fig. 3-12, we saw that a given region of the embryo is destined to become ectoderm, another chorda mesoderm, another endoderm, and so on. However, fate maps tell us nothing about either the time at which a given population of cells is irrevocably fixed to a given pathway of development or the steps involved.

We speak of the restriction of a cell or population of cells to a given pathway of differentiation as *determination*. It is illustrated in the following experiment, depicted in Fig. 4-1. It will be recalled that prospective neural cells are mapped in the region of the animal hemisphere adjacent to the dorsal lip, whereas prospective epidermal cells are on the opposite side of the embryo. If the upper part of an early amphibian gastrula is cut off, rotated 180° and replaced, the cells of each region will come into new relations with those of the lower hemisphere. Once the graft has healed, normal development continues. Neural tissues develop in their usual position; epidermal tissues in theirs. This means, however, that neural tissues of the operated embryo are formed from what had been prospective epidermis, and epidermal tissues from prospective neural cells. In a closely related experiment, tissues were exchanged between early gastrulae of two different species of newts. Such *heteroplastic* combinations, as grafts between different species are called, are especially useful because one can assess the contributions of donor and host, especially if they are differently pigmented, or if their cells contain other markers. A small piece of pros-

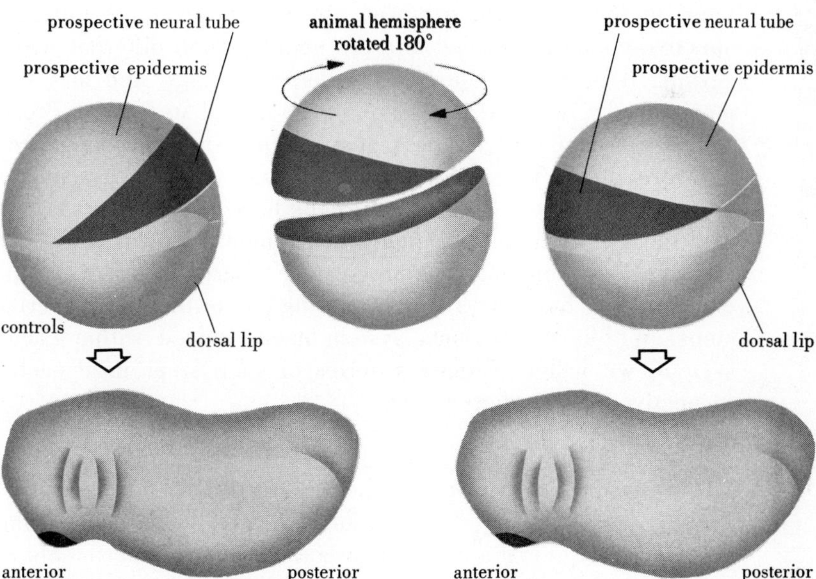


***Fig. 4-1** Rotation of the animal hemisphere of an early amphibian gastrula results in a normal embryo.*

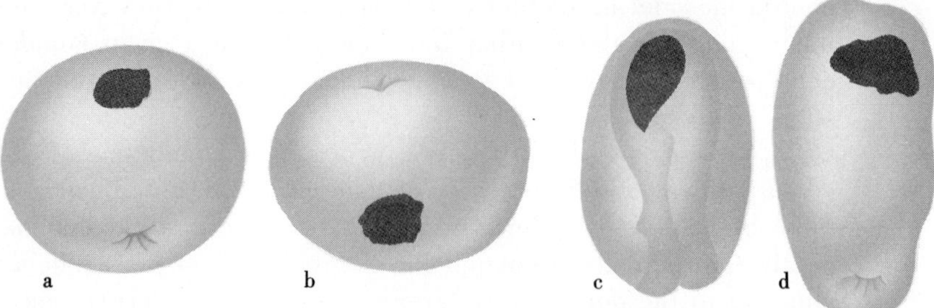


*Fig. 4-2 Exchange of prospective epidermis and prospective neural plate in the early gastrula stage of newt embryos differing slightly in density of pigmentation.* (a *and* b) *Embryos immediately after the operation.* (c *and* d) *The same embryos in the neural plate stage.* (c) *The graft forms part of the neural plate.* (d) *The graft forms part of the anteroventral epidermis. (After H. Spemann,* Embryonic Development and Induction, *by courtesy of Yale University Press.)*

pective epidermis of one species was transplanted into the prospective neural region of the other, with the reciprocal graft also being performed (Fig. 4-2). In each combination the result was the same: the grafted material developed in conformity with its new position.

In short, in the early gastrula the cells in these two areas, prospective epidermis, and prospective neural tissues have not yet been determined. They are not yet restricted to a single pathway.

However, if the same experiments are performed just two days later, at the *end of gastrulation,* the results are quite different. Prospective epidermis placed into the neural region differentiates as epidermis. It does not conform to its surroundings, and may interfere, mechanically, with the formation of the neural structures. Prospective neural tissue placed in the region of epidermis develops into brainlike structures. Thus within this time span, during gastrulation, the capacities of these two groups of cells have been restricted. They are said to be determined. In fact, development involves a series of progressive determinations, of progressive restrictions of developmental capacity. We have just been discussing one of the early determinative steps. Within the nervous system itself, indeed within each organ system, we will encounter a series of such steps until each of the respective parts is determined.

***Embryonic Induction*** We have observed that progressive determination of the nervous system accompanies gastrulation. But is it a result of gastrulation? The experiments we have described show that in some way a cell's surroundings influence its differentiation. How have the surroundings of prospective neural and

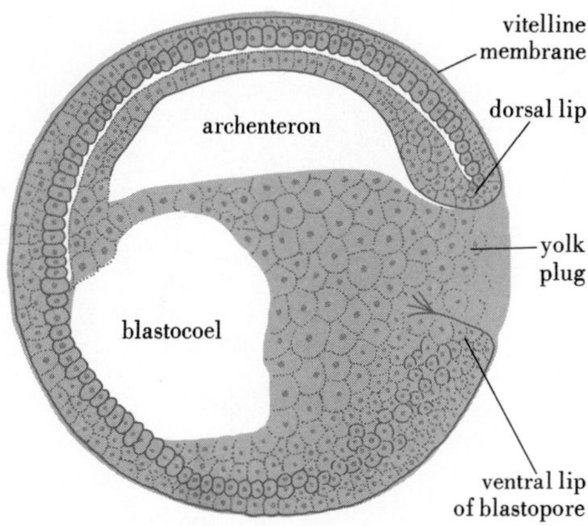


*Fig. 4-3 Longitudinal section of a late amphibian gastrula, having a circular blastopore composed of dorsal, lateral, and ventral lips enclosing a yolk plug. (After L. G. Barth,* Embryology, *Holt, Rinehart and Winston, Inc.)*

epidermal cells changed as a result of gastrulation? Looking at a section of a late gastrula (Fig. 4-3) we see that the cells of the prospective neural plate at the animal pole are now underlain by the chorda mesoderm, whereas those of the epidermis are not. We may ask, then, whether some interaction between the ectoderm and chorda mesoderm results in the development of the ectoderm into neural structures. How might this hypothesis be tested?

First, if the interactions were prevented, would neural tissues develop? This question has been posed for many species, and the answer is clear. The normal development of the nervous system is dependent upon normal contact between the chorda mesoderm and the ectoderm. If one removes the membranes from an amphibian egg and grows it in a hypertonic solution, or in a solution containing lithium chloride, normal gastrulation fails to occur. Instead the embryos *exogastrulate.* The prospective chorda mesoderm and other mesodermal and endodermal cells do not invaginate, but grow outward instead (Fig. 4-4). Under these conditions, endodermal and mesodermal structures develop to a great extent, but the ectoderm fails to form neural structures.

This conclusion is confirmed by a culminating and decisive experiment, one of the enduring, great experiments of modern biology. Over a period of about 20 years, the German embryologist, Hans Spemann, explored the mechanisms of progressive determination in the amphibian egg. Early experiments convinced him of the importance of the interactions of tissue layers, and in seeking the factors underlying the formation of the neural structures, he increasingly narrowed his search to the invaginating chorda mesoderm in the region of the dorsal lip. If the dorsal lip does interact with ectoderm, *inducing* it to become neural tissue, why not graft it to a different position in the embryo and see whether it influences the surrounding tissues to differ-

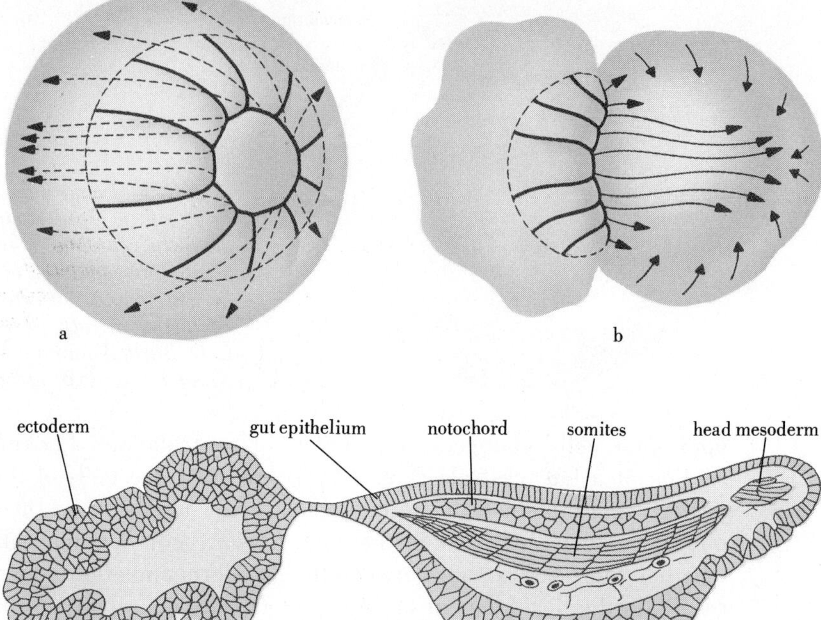


***Fig. 4-4*** *Diagram illustrating the morphogenetic movements* (a) *of normal gastrulation and (b) of exogastrulation.* (c) *Differentiation in an exogastrulated embryo. (After J. Holtfreter and V. Hamburger, in* Analysis of Development, *by courtesy of W. B. Saunders Company.)*

entiate in directions other than their normal fates? The young blastopore and the tissue adjacent to it were cut out and grafted to the prospective belly region of another young embryo (Fig. 4-5). Here in this new position, in Twitty's words, "a new embryo developed face to face with its host."

In the definitive work, published in 1924, Spemann and his student, Hilde Mangold, used donor and host embryos of differently pigmented species of newts to prove conclusively that both graft and host cells participate in the formation of the secondary embryo. The graft "self-differentiates" (that is, it forms the very structures it would have given rise to in its normal setting), but it induces the adjoining host tissue to form spinal cord and other structures, including somites and kidney tubules.

When a large number of such combinations are examined, a broad spectrum of results is observed. In some, the transplanted dorsal lip has stimulated the formation only of a second neural tube. In others, almost an entire embryo is found in the region in which the transplant

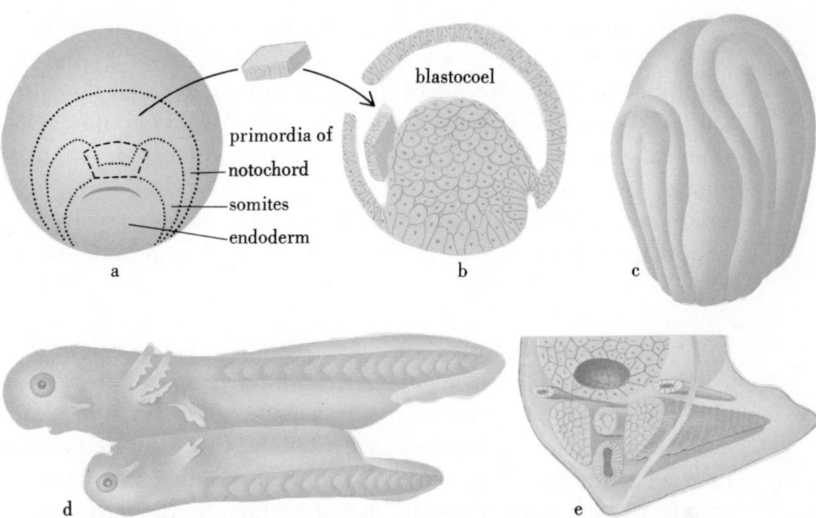


*Fig. 4-5 Diagram of the transplantation of a piece of the upper blastoporal lip into another gastrula* (a *and* b) *and the self-differentiation plus inductions of the graft* (c–e). *(From J. Holtfreter and V. Hamburger, in* Analysis of Development, *by courtesy of W. B. Saunders Company.)*

was made. The secondary embryos produced in the best of these experiments have dual origins, many parts being contributed by developing donor cells, others being induced in host cells. Commonly the neural tube is an induced structure. Many parts may be produced synergistically, with interactions of donor and host cells being clearly involved. Spemann and Mangold spoke of the dorsal lip as the *organizer*, or *primary organizer* because it caused the development of a complete secondary embryo. As we have seen, however, not all of the parts of the secondary embryo are induced, in the strict sense. Before concluding this section, therefore, we should make two generalizations:

1. We now know that the interaction of the chorda mesoderm and ectoderm is but one of many such interactions during development. To be sure, it is "primary" in the sense of being the first in a sequence of inductions. However, it is not an "organizer" in the sense attributed to it by its discoverer. The organization of the secondary embryo results from a series of both inductive interactions and self-differentiative changes in host and donor tissues. Today, therefore, the more general terms *embryonic induction* or *inductive interaction* are preferred. We shall be considering examples of such interactions in several tissues in invertebrates and vertebrates, and later we shall consider their cellular and molecular bases.

2. Although we have discussed the induction of the neural axis by the dorsal lip and its derivative, the roof of the archenteron, only in amphibians, these structures have the same role in all the vertebrates, and in some of the lower chordates as well. Ascending the phylogenetic scale, the dorsal lip is active in neural induction in *Amphioxus*, cyclostomes (especially the lampreys), bony fishes, and frogs, newts and salamanders. Its homologue, the primitive streak, especially the anterior end of the streak, is similarly effective in birds and mammals.

***Neurulation*** As gastrulation ends, *neurulation*, the process of the primary differentiation of the nervous system, begins; in the frog or salamander, the neural plate induced as a consequence of gastrulation is transformed into a neural tube within a day. The neural plate itself—an ovoid, flattened plate of ectoderm—has formed on the dorsal surface by the movement of cells from more lateral regions. Although the transformation of the plate into a tube can be described, the forces underlying the process are not understood. A *neural groove* results from the depression of the center of the plate and the folding of its edges; ultimately the folds come together and fuse in the midline, forming the *neural tube*. When the raised folds meet from the two sides, epidermis fuses with epidermis and neural layer with neural layer (see Figs. 4-6 and 4-7).

What underlies the differences between center and edges of the neural plate? These folding processes can proceed in isolated plates, and thus we must look for causes within the plate itself. But none of the hypotheses advanced has yet proved correct. Differential water uptake and differential cell division appear to have been ruled out. The inti-

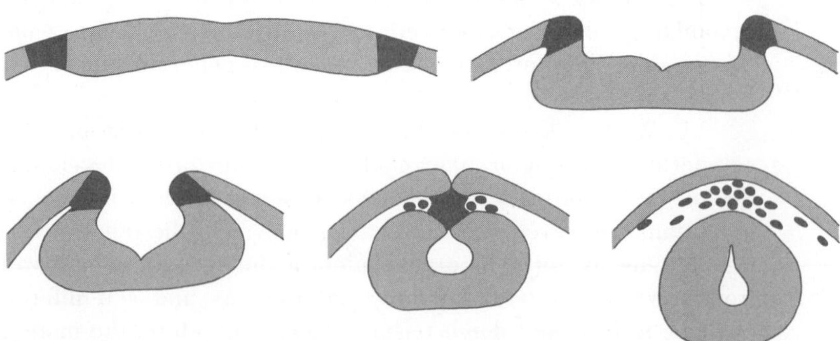

**Fig. 4-6** *Stages in the formation of the neural plate and neural tube in amphibians. Transverse sections (diagrammatic). Neural crest cells are shown in dark color. (From B. I. Balinsky,* An Introduction to Embryology, *2d ed., by courtesy of W. B. Saunders Company.)*

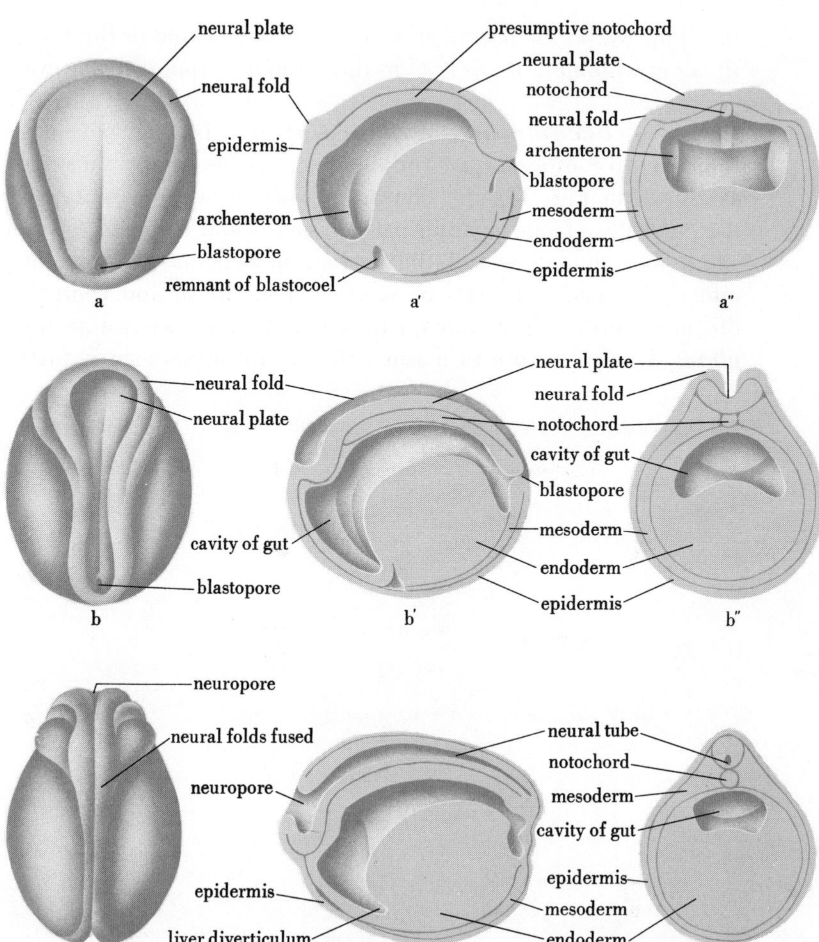


***Fig. 4-7*** *Three stages of neurulation in a frog embryo. The drawings on the left show whole embryos in dorsal view. The drawings in the middle show the right halves of embryos cut in the median plane. The drawings on the right show the anterior halves of embryos cut transversely.* (a, a′, a″) *Very early neurula;* (b, b′, b″) *middle neurula;* (c, c′, c″) *late neurula with neural tube almost completely closed.* (c) *shows the blastopore closed. (From B. I. Balinsky,* An Introduction to Embryology, *2d ed., by courtesy of W. B. Saunders Company.)*

mate molecular distinctions within the cells of the plate have not been resolved.

In some vertebrates the neural tube is laid down as a hollow tube, as described; in others (the bony fishes, for example) it develops as a solid rod of cells. However, one of the initial processes in both hollow tube and solid rod is the secretion of fluid by some of the cells. In the tube this fluid maintains the distention of the brain cavity, supporting the limp walls; in the rod it appears to play a role in the formation of the central lumen, as well as in the ultimate support of the walls.

***Changes accompanying neurulation*** During neurulation the separation of other organ rudiments proceeds apace. In amphibians, as shown in Fig. 4-7, the chorda mesoderm becomes subdivided into the centrally placed, cylindrical notochord and the mesoderm on both sides. The dorsal part of the mesoderm becomes further subdivided into a series of segments or *somites*. The lateral and ventral parts of the mesoderm, which remain unsegmented, are known as the *lateral plates*. During somite formation, the lateral mesoderm is further split

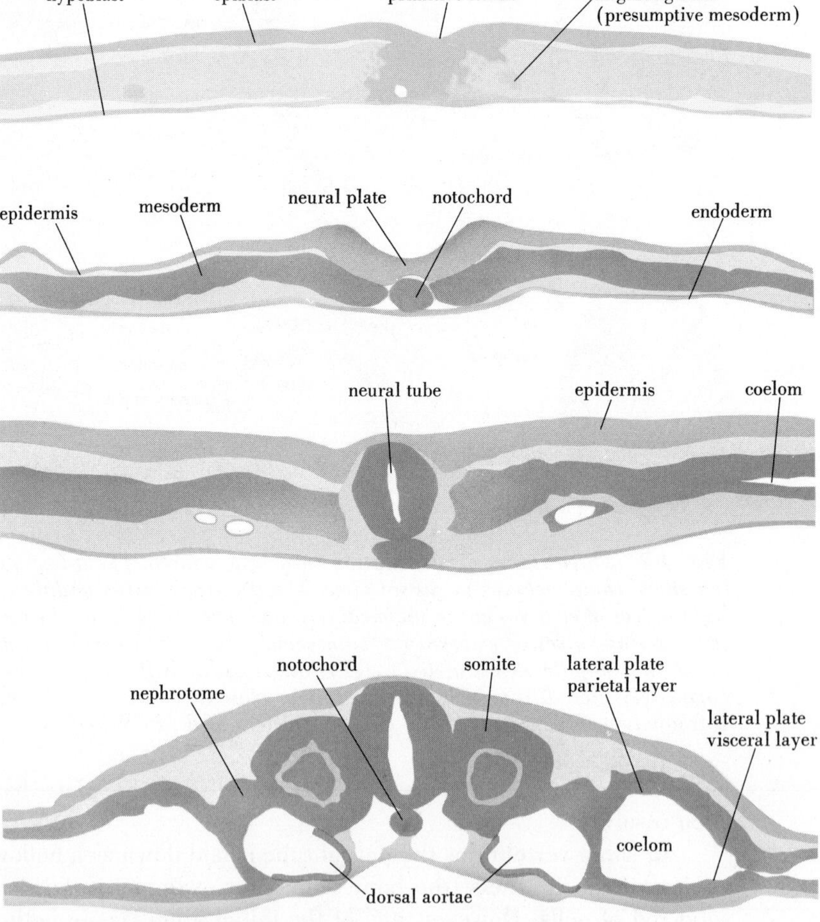


***Fig. 4-8*** *Stages of early development of the chick (transverse sections).* (Top) *Primitive streak;* (upper middle) *neural plate stage;* (lower middle) *neural tube stage;* (bottom) *primary organ rudiments.*

into an external or *parietal* layer, associated with the ectoderm, and an inner, *visceral* layer, applied to the endoderm. The cavity between the two layers is the *coelom*, which ultimately becomes the body cavity of the adult.

The subdivision of the mesoderm is shown very clearly in the chick embryo (Fig. 4-8), in which neurulation and the sequence of formation of the mesodermal rudiments are very similar to that observed in amphibians.

At the same time, the primary endodermal structures are also taking shape. In amphibians, the *foregut*, or anterior portion of the alimentary canal, is completely enclosed by endoderm. Further back, the *midgut* is becoming enclosed, closure proceeding progressively from anterior to posterior. The strictly anteroposterior pattern observed in the formation of the gut in amphibians differs from that seen in birds and mammals. The chick embryo will serve as another example. A sheet of endoderm lying flat on the yolk folds *downward* (Fig. 4-8), the edges of the folds ultimately fusing to form the gut. The inner surfaces of the folds contribute to the floor of the gut, whereas the outer surfaces are continuous with the lining of the yolk sac. This process does not proceed exclusively from the anterior to the posterior end of the embryo. Fusion, and separation from the yolk sac cavity occurs first anteriorly, forming a foregut. Soon thereafter the canal is enclosed at its posterior end (the *hindgut*). Between the foregut and hindgut a gap remains, progressively diminishing as the foregut and hindgut move together. The progressively migrating openings of the foregut and hindgut are known as the *anterior* and *posterior intestinal portals*, respectively. Eventually the gap is reduced to the opening of the *yolk stalk* connecting the cavity of the gut to the cavity of the yolk sac.

## *INTERACTIONS OF NERVE FIBERS AND PERIPHERAL TISSUES*

By a series of intricate processes, involving interactions of the originally simple neural tube with the developing skull capsule and spinal column on the outside and the turgor of the fluid on the inside, the configuration of the tube changes; three general regions, the *forebrain* or prosencephalon, *midbrain* or mesencephalon, and *hindbrain* or rhombencephalon emerge. As the walls of the brain begin to show local differentiations, the forebrain and hindbrain are further subdivided, the former into prosencephalon and diencephalon, the latter into metencephalon and myelencephalon, the last being continuous with the spinal cord (Fig. 4-9). We do not know the mechanisms responsible for the emergence of these outpocketings, fissures, and folds. Now, over a period of weeks, depending on the species, the brain and spinal cord become a mosaic

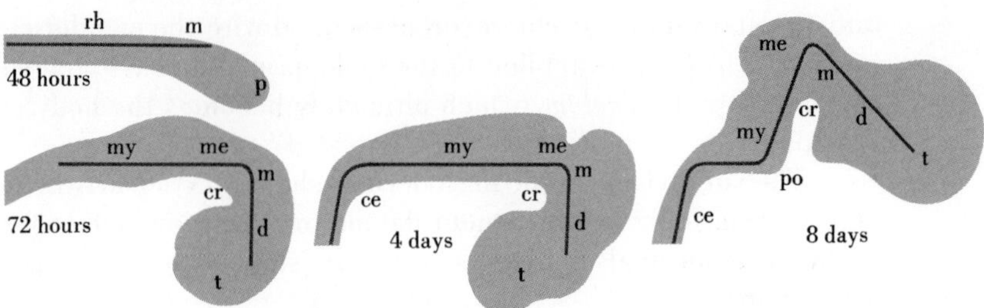


*Fig. 4-9 Drawings illustrating the development of the flexures of the brain. At 72 hours the cranial flexure* (cr) *develops; the cervical flexure* (ce) *develops at four days; and in the eight-day embryo the pontine flexure* (po) *has developed. Noted, diencephalon;* p, *prosencephalon;* m, *mesencephalon;* me, *metencephalon;* my, *myelencephalon;* rh, *rhombencephalon;* t, *telencephalon. (From C. W. Bodemer,* Modern Embryology, *by courtesy of Holt, Rinehart and Winston.*

of specialized areas, and regional differences emerge at different levels along the longitudinal axis.

At first, the wall of the neural tube is made up of elongated cells extending from the lumen toward the surface. Adjacent to the lumen, mitotic figures are abundant. Originally it was thought that these cells represented an active "germinal" layer. It is now clear, however, that there is no special germinal layer. When a cell in the tube divides, its nucleus moves toward the lumen. At division, the daughter cell nuclei again move toward the surface. Measurements of DNA content and studies using radioactively tagged nucleotides show that DNA synthesis may occur at different levels (depths) of the neural epithelium, depending on the species. The daughter cells are now ready to undergo structural differentiation. We will take as our principal example the differentiation of motor neurons.

Motor cells (Fig. 4-10) originate in the ventral part of the tube, as small spindle-shaped cells. At this stage in its life history each motor

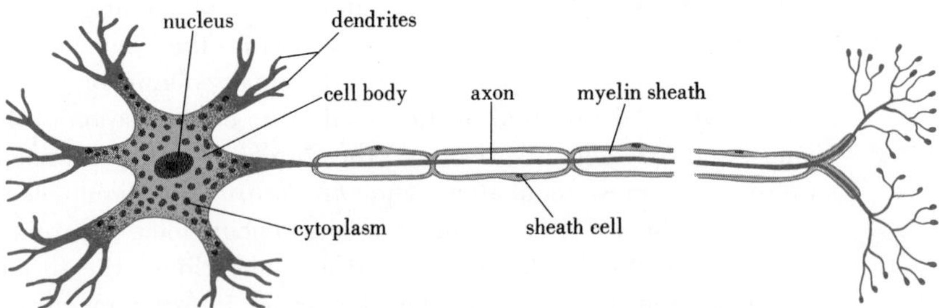


*Fig. 4-10 Motor neuron showing the myelin sheath formed from sheath cell membranes rolled several layers thick around the axon. See also Fig. 4-31. Many neurons are much longer, relative to the diameter of axon and cell body, and many have much thinner myelin sheaths or only a single cell membrane surrounding the axon. (From D. R. Griffin,* Animal Structure and Function, *Holt, Rinehart and Winston, Inc.)*

cell already has acquired the capacity to migrate and the capacity to spin out a long axon. Thus far, we have been unable to define the subtle events that have changed this cell into a nerve cell. The young motor neurons gather in compact columns (Fig. 4-11), and now spin out their axons.

It was Ross G. Harrison who first proved that nerve fibers are outgrowths of single neuroblasts. The brilliance of his experiments is not fully appreciated unless one knows the alternatives. Three theories had been held: the *cell-chain theory*, which postulated that the fiber was laid down by the chain of cells that sheath the nerve fiber; the *plasmodesm theory*, which suggested that the fiber was produced along preformed protoplasmic bridges; and the *outgrowth theory*, which held that the fiber is an outgrowth of a single neuroblast, which becomes the neuron.

In his crucial experiments Harrison developed what has become one of our more powerful techniques, the cultivation of living tissues outside the body. He followed the formation of axons in isolated fragments of neural tube. His own words tell the outcome: "These observations show beyond question that the nerve fiber develops by the outflowing of protoplasm from the central cells. This protoplasm retains its amoeboid activity at its distal end, the result being that it is drawn out into a long thread which becomes the axis cylinder."

Each branching axon terminates in a "growth cone," from which pseudopodial processes emerge, penetrating peripheral tissues. The first emerging fibers are called *pioneering fibers*, which reach out into the embryo. Other fibers follow, and soon a bundle of fibers reaches from the sedentary cell bodies in the cord to the periphery. Now a subtle interaction occurs; for the continued development of the motor neuron, in fact its very survival, depends on the conditions the axon

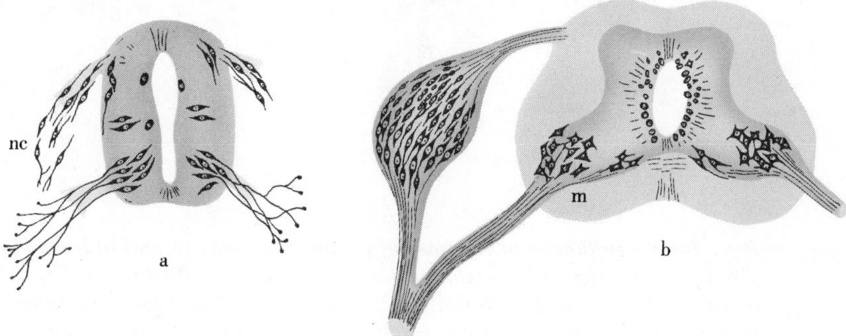


*Fig. 4-11 Embryonic neural tube and spinal cord.* (a) *Neural tube and neural crest* (nc) *of two-day chick embryo.* (b) *Spinal cord and spinal ganglion of eight-day embryo.* m: *lateral motor column. (From V. Hamburger, in* Journal of Cellular and Comparative Physiology, *60, by courtesy of The Wistar Institute.)*

encounters at the periphery. Normally, there are massive accumulations of motor neurons in the spinal cord, at the level of the limbs. Suppose, as shown in Fig. 4-12, that the developing leg bud is removed from a chick embryo at 2½ days of incubation. When the pioneering fibers reach the scar, their further growth is blocked and a tumorlike fibrous mat is formed. Almost immediately, there is a "feedback" to the motor neurons. Within three days after the operation, the motor cells at that level undergo a rapid breakdown. The reverse effect, a stimulation that leads to an increase in the number of neurons, is shown in Fig. 4-13. When the peripheral area is enlarged by implanting an extra wing bud within reach of the outgrowing fibers, the *ganglia* (groups of nerve cell bodies outside the spinal cord) that supply the wings are increased in size.

How can these reactions be explained? We are just beginning to understand some of the mechanical and metabolic bases. Not only are axons "spun out" from the center, but once they are established, there is a constant flow of materials from the nerve cell body to the periphery.

The cell body of the neuron is the center of the synthesis for the entire cell. No matter over what distance an axon may extend, the bulk of its synthetic activities takes place in the cell body. The flow of materials from the cell body down the axon has been demonstrated in several ways; for example, the movement of radioactively labeled

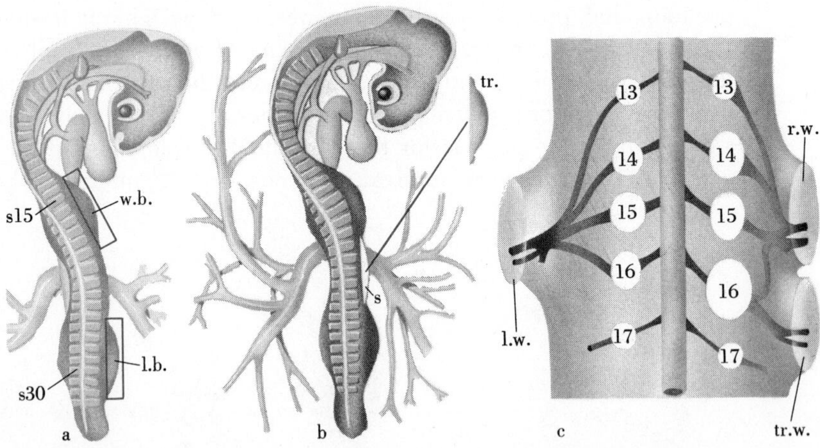


**Fig. 4-12** *Influence of increase of peripheral load.* (a *and* b) *Limb-bud transplantation in the chick embryo.* (a) *donor embryo;* (b) *host embryo (both 34 somites, 72 hours of incubation);* l.b. = *leg bud;* s = *slit for the reception of the transplant;* s15, s30 = *somites 15, 30;* tr. = *transplant;* w.b. = *wing bud. (From V. Hamburger,* A Manual of Experimental Embryology, *by courtesy of The University of Chicago Press.)* (c) *Nerve distribution in normal* (r.w.) *and transplanted* (tr. w.) *of wing of chick embryo. Note increased size of ganglia 14–16, which supply wings. (After V. Hamburger, in* American Scientist, *45, by courtesy of Society of Sigma Xi.)*

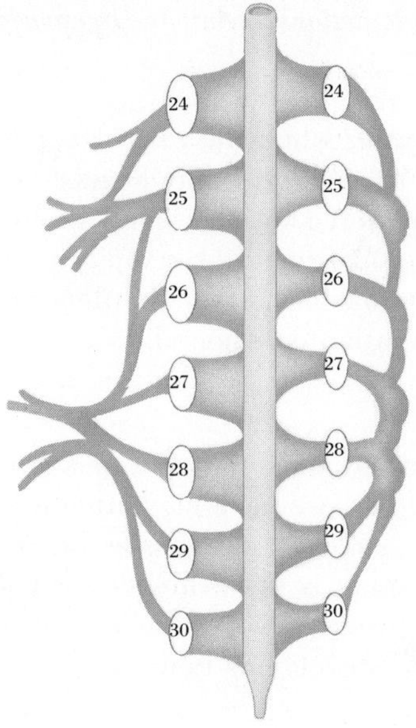


*Fig. 4-13 Influence of reduction of peripheral load. Nerve distribution in leg level of a six-day chick embryo whose right leg has been extirpated at 2½ days. All segmental nerves on right side terminate in a neuroma. Figures indicate segmental numbers of ganglia, which are reduced in size on right side. (After V. Hamburger, in* American Scientist, *45, by courtesy of Society of Sigma Xi.)*

molecules has been followed along the axon. However, this more recent approach has only served to confirm and extend an earlier ingenious experiment by Weiss and Hiscoe, diagrammed in Fig. 4-14. Peripheral nerve fibers are capable of regeneration. After the axon shown in (a)

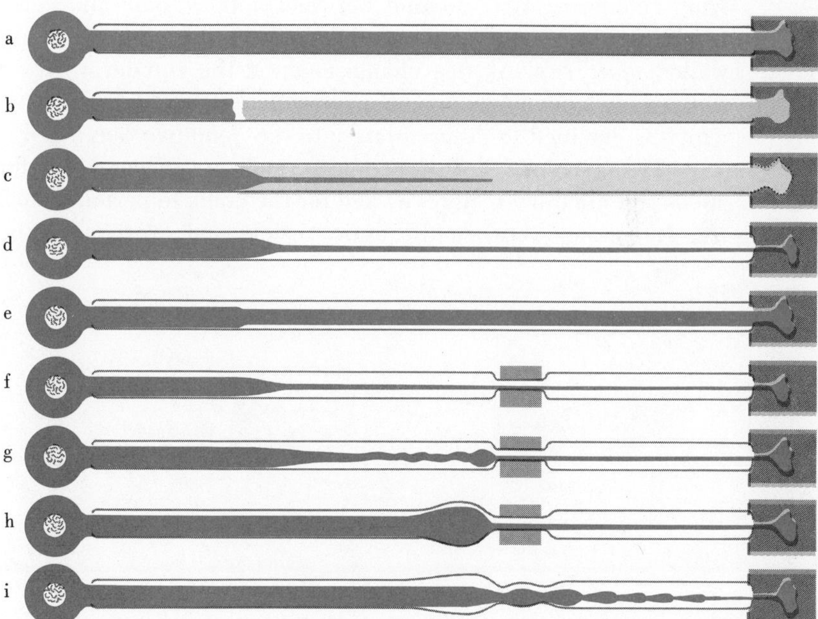


*Fig. 4-14 Damming of axoplasm in constricted nerve fibers.* (a–e) *Consecutive stages of unimpeded regeneration;* (f–h) *consecutive stages of regeneration with "bottleneck;"* (i) *after release of constriction. (After P. Weiss and H. Hiscoe, in* Journal of Experimental Zoology, *107, by courtesy of The Wistar Institute.)*

is cut, the *distal* portions degenerate, while the *proximal* portion of the axon (nearest the cell body) survives (b) and begins to regrow (c, d) until the complete axon is restored (e). However, when the regenerating axon is obstructed or constricted (f), axoplasmic flow is impeded, and materials pile up proximal to the bottleneck (g, h). When the dam is broken (i), regeneration is resumed.

## "RECIPROCITY" IN TISSUE INTERACTIONS

Thus far, although we have spoken of tissue *interactions*, we have paid attention only to the influence of one tissue on a second tissue. What is the evidence for reciprocal action? Is interaction a two-way street? Two examples will be considered: the developing lens and the developing limb.

### Induction in Lens Development

The lens is formed in the ectoderm of the head as a result of contact with the underlying forebrain. As the forebrain takes shape, it bulges laterally on each side, forming the *optic vesicles*, which expand and make contact with the inner surface of the overlying ectoderm. After contact between optic vesicle and ectoderm is established, the wall of the vesicle pushes inward to form a double-walled *optic cup*. As this change occurs, the ectoderm over the cup thickens to form a *lens placode*, forerunner of the discrete *lens vesicle*, which is destined to differentiate into the definitive *lens* (see Fig. 4-15).

The evidence for induction is sufficiently complete for many species of amphibian embryos and for the chick to permit us to generalize. If contact between the optic vesicle and overlying ectoderm is

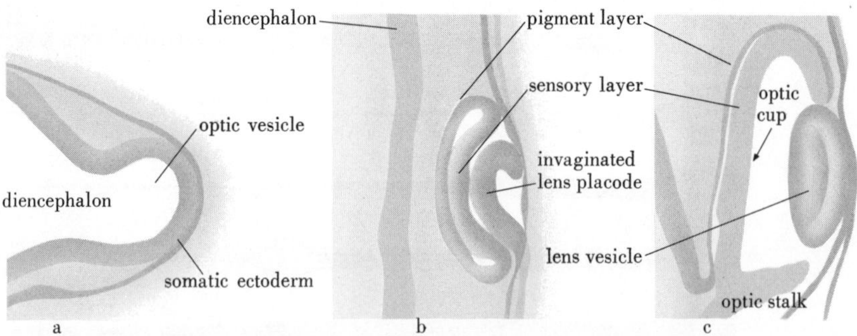


***Fig. 4-15*** *Diagrams of sections through the developing eye of the chick embryo.* (a) *Optic vesicle of 33-hour embryo.* (b) *Optic cup and early lens of 48-hour embryo.* (c) *Optic cup and lens of 72-hour embryo. (After T. W. Torrey,* Morphogenesis of the Vertebrates, *by courtesy of John Wiley & Sons, Inc.)*

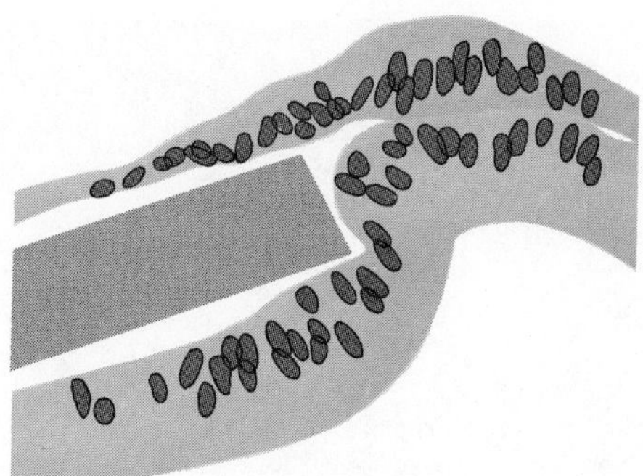

*Fig. 4-16 Camera lucida drawings of a section through the eye of a 20-somite chick embryo in which contact has been prevented between a part of the retina and overlying lens ectoderm by interposing a piece of cellophane at the 5-somite stage. Note orientation of nuclei. (From M. McKeehan, in* Journal of Experimental Zoology, *117, by courtesy of The Wistar Institute.)*

prevented, the lens fails to form. When a barrier is interposed between the optic vesicle and ectoderm, the cells of the ectoderm fail to accomplish even the first step toward the formation of the lens. In the chick embryo between the 10- and 20-somite stages the optic vesicle and head ectoderm are normally in contact. As seen in Fig. 4-16, if a piece of cellophane is interposed between the two layers at the 5-somite stage, the lens does not begin to form in the obstructed part of the ectoderm. One of the first visible signs of lens induction is a change in the orientation of the ectodermal cells, which elongate and become oriented in the same plane as the underlying cells. In the obstructed areas, this realignment fails to occur; when the barrier does not exist, the normal orientation is observed.

This kind of evidence for an inductive interaction is compelling; moreover, it is further supported by transplantation experiments. When optic vesicles are removed from their normal sites, freed of their own epidermis, and placed under the epidermis in other parts of the embryo, lenses are developed in the epidermis at the new position. Similarly, if the epidermis that normally forms the lens is excised and replaced with epidermis from the head or belly, the new epidermis responds to the underlying optic vesicle and forms a lens.

But the evidence pertinent to the question of *reciprocity* is derived from the following type of experiment. If the optic cup of a large, rapidly growing species of salamander is covered by ectoderm of a smaller slow-growing species, the optic cup and lens exert a *mutual influence* on one another, with the result that the chimeric eye is intermediate in size and harmoniously proportioned.

The formation of the eye involves a series of tissue interactions; in a sense it begins with the primary induction of the neural plate by underlying chorda mesoderm; it continues with the induction of the lens by the optic vesicle; and finally, in this sequence, the corneal epithelium is produced by the further interaction of ectoderm and the lens vesicle (Fig. 4-17).

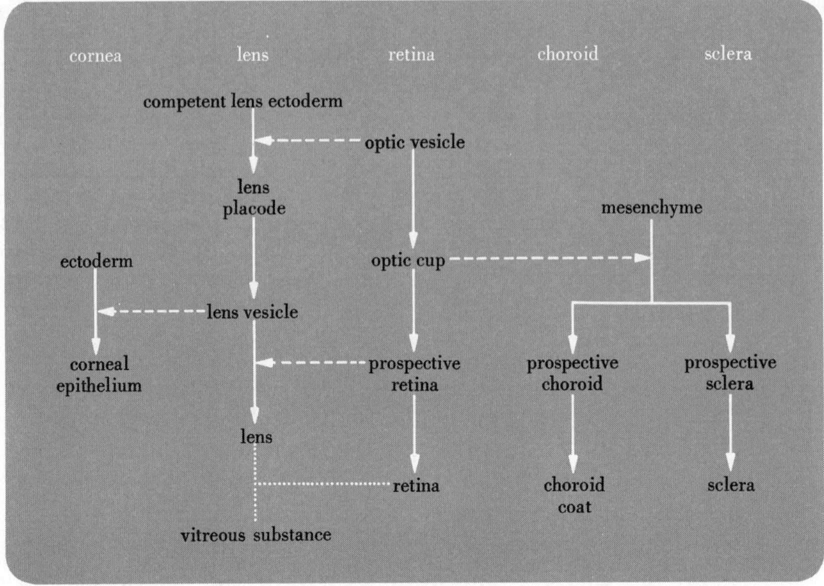


*Fig. 4-17 Tissue interactions in the development of the eye. (By courtesy of A. J. Coulombre.)*

**Reciprocal Interactions in the Developing Limb**

When they first appear on the third day of embryonic development, chick limb buds consist of two components: an inner core of condensed mesenchyme derived from the lateral mesoderm and an outer covering of ectoderm. Along the free edge of the bud, the ectoderm is specialized in the form of a thickened, crestlike *apical ectodermal ridge* (Fig. 4-18). If the ridge is removed surgically, limb outgrowth ceases; the implication is that the ectodermal tip is necessary for the establishment of proximodistal outgrowth. Is its role "inductive"? If so, we would expect to find that mesoderm ordinarily destined to become part of the thigh would form digits if it were placed in the distal region of the limb and combined with the ectodermal ridge. It does.

Before we investigate the evidence for a reciprocal relationship, let us pause to introduce two new experimental approaches. First, the tissue layers of embryos may be separated by chemical agents. We shall take up the mechanism of action of these reagents and examine the implication of their action for the study of the organization of tissues in Chapter 10. Here it will suffice to make the following points. When a limb bud is treated with a dilute solution of the enzyme trypsin, the ectoderm and mesoderm are separated. The mesodermal core may be damaged but the ectoderm is viable. Conversely, the chelating agent, ethylenediamine tetraacetic acid (EDTA) separates the layers, damaging

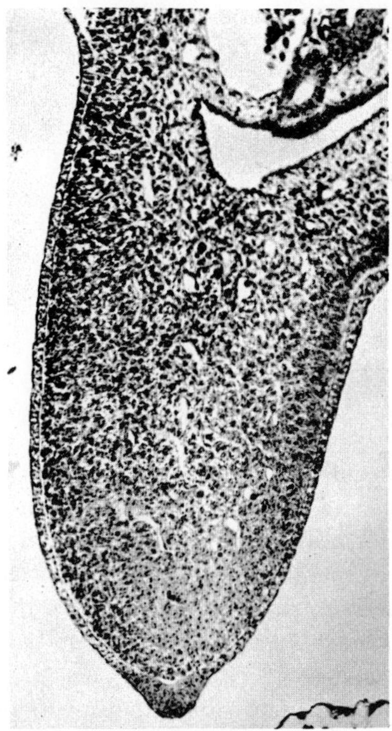

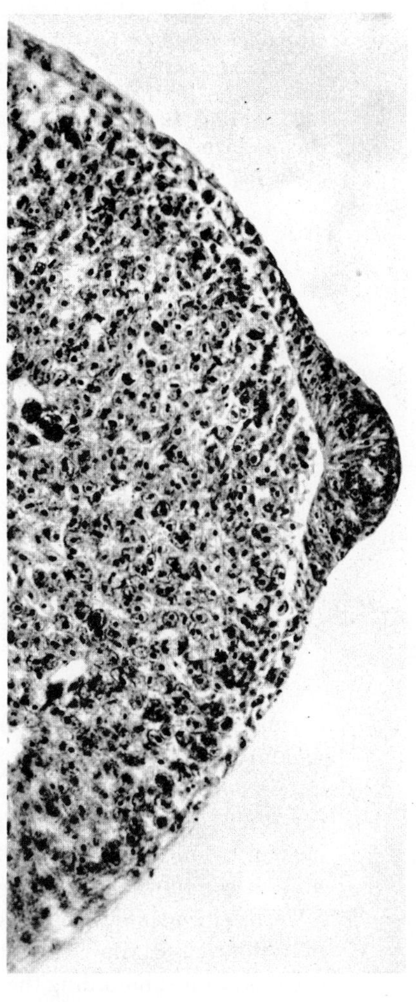

***Fig. 4-18*** (Left) *Photomicrograph of cross section of the right wing bud of the chick embryo having 40 to 43 somites.* (Right) *Higher magnification of apex of the left wing bud of an embryo at the same stage. The apical cap forms a definite ridge or crest of tightly packed cells. (From J. W. Saunders, Jr., in* Journal of Experimental Zoology, *108, by courtesy of the author and The Wistar Institute.)*

the ectoderm but leaving the mesoderm unscathed. Thus it becomes possible to carry out recombination experiments using ectoderm of one limb and mesoderm of another.

The second approach involves the use of embryos in which, because they are *genetically* different from the normal, the normal pattern of limb development does not occur. One such *mutant* in the chick is *wingless*. Wing buds are formed, and develop for a time; then the ectodermal ridge degenerates and development of the wing stops. Thus these mutants are almost exactly like the embryos in which apical ridges were removed surgically.

If the tissue layers of normal and *wingless* limb buds are dissociated by the techniques just described, and the normal ectoderm

*Fig. 4-19 Interchange of mesoderm and ectoderm between genetically normal and polydactylous limb buds. The polydactylous condition develops from the combination of mutant mesoderm and normal ectoderm. (After E. Zwilling, in* Cold Spring Harbor Symposia on Quantitative Biology, *21, by courtesy of the publisher.)*

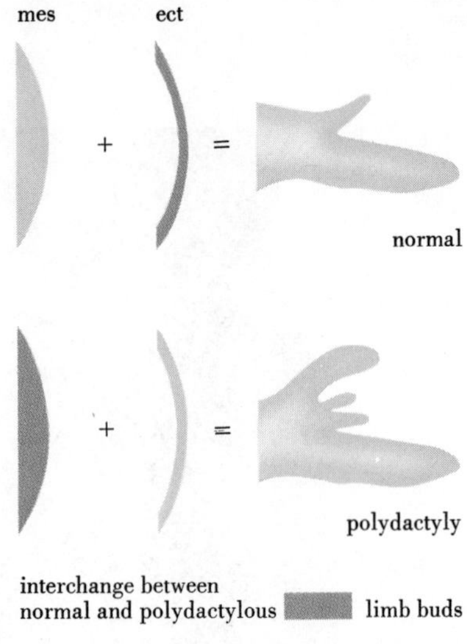


combined with *wingless* mesoderm, the mesoderm begins to respond to the influence of the ridge. Shortly, however, the normal ridge begins to degenerate and outgrowth stops. Thus, the normal ridge appears to depend on the mesoderm for its maintenance. In fact, the existence of a *maintenance factor* has been postulated. Although such a factor has not been identified, the evidence does indicate that a search for it is warranted.

Other evidence for reciprocity comes from experiments in which normal ectoderm is combined with mesoderm from a bud from a *polydactylous* mutant—that is, a mutant in which more than the normal number of digits is formed. As shown in Fig. 4-19, the *polydactylous* condition develops from the combination of mutant mesoderm and normal ectoderm. The reciprocal combination—normal mesoderm and mutant ectoderm—produces a normal limb.

Reciprocal dependence is not restricted to eye or limb. Details may differ, but similar relations have been described in a number of organ systems, including the development of tail fins in salamanders and feathers in chickens. In fact, it is the interaction of the skin ectoderm and the underlying mesoderm that leads to the formation of all the characteristic regional specializations of the epidermis in birds: not only the feathers, but also the scales, spurs, and beak. In mammals, similar interactions are involved in the formation of hair.

## FOLDING AND DEFORMATION OF CELL SHEETS IN THE DEVELOPING HEART

In embryos of all vertebrates, from fishes to man, the heart, a mesodermal organ, arises from the ventral edges of the lateral plate. The primary rudiment of the heart is paired: that is, either half is capable of forming a whole heart. In the normal course of events, however, the fused halves form a single organ.

In the chick embryo, which we shall take as our principal example, the two halves of the heart begin differentiating independently, as a pair of delicate tubular primordia arranged on either side of the *anterior intestinal portal*, the opening from the yolk into the developing foregut. Folding movements of the ectoderm and endoderm in the formation of the head fold and foregut bring the heart rudiments together in the ventral midline, where they fuse (see Figs. 4-20 and 4-21). Each of the

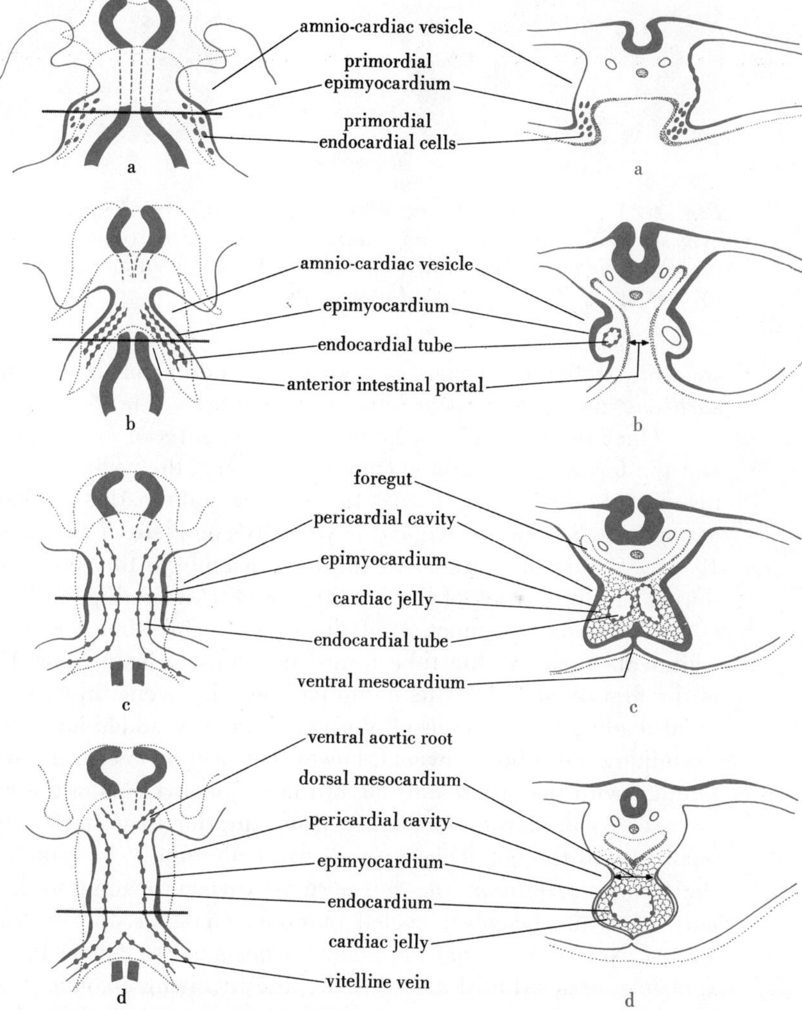


***Fig. 4-20*** *Diagrams of chick embryos, showing the origin and subsequent fusion of the paired primordia of the heart, the establishment of the pericardial cavity, and the elongation of the foregut. Figures at left, ventral views; at right, corresponding transverse sections at the levels indicated by heavy lines across ventral-view figures. Time interval: 25th to 30th hour of incubation. (From T. W. Torrey,* Morphogenesis of the Vertebrates, *by courtesy of John Wiley & Sons, Inc.)*

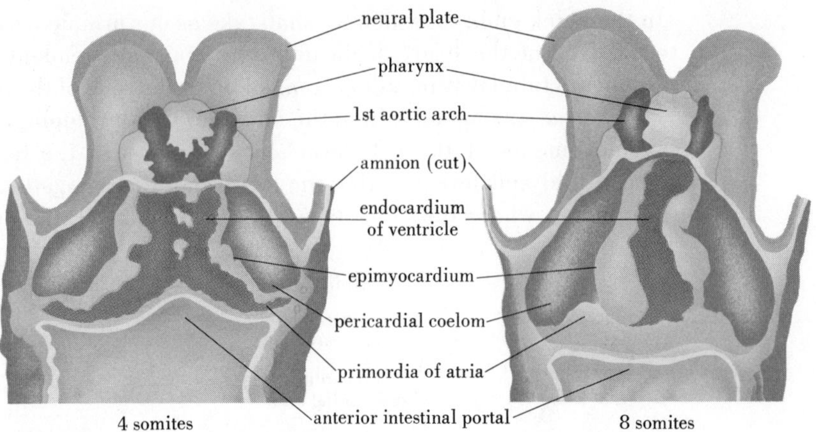


*Fig. 4-21 The human heart, at the stage of fusion of the paired primordia, exposed as a transparent organ in ventral view (Patten).* (Left) *At four somites;* (right) *at eight somites. (After L. B. Arey,* Developmental Anatomy, *6th ed., by courtesy of W. B. Saunders Company.)*

separate rudiments consists of an inner endothelial lining, the *endocardium*, and other *myocardium* or muscle layer (Fig. 4-20B and C).

Once the body folds have undercut the anterior end of the embryo and the foregut is separated from the yolk sac, the rudiments meet and fuse in the midline (Fig. 4-20D). In the chick embryo, this process begins at the 7- to 8-somite stage and is virtually complete in embryos having 19 or 20 pairs of somites. In the human embryo this process begins during the third week of life, by the 8-somite stage (Fig. 4-21).

The definitive function of the heart begins almost as the paired rudiments fuse. As the tube forms, it begins to beat. Since the heart is the first organ to play its definitive role, the events in early development leading up to the onset of function take on added interest. Before examining the primary events, however, it will be necessary to become familiar with the origin and role of the major divisions of the heart.

The paired rudiments fuse from anterior to posterior, from the head toward the tail. The first regions of the tube to be established are the *truncus arteriosus*, the anteriormost division leading to the ventral aorta, and the heavily muscled pumping chamber, the *ventricle* (Fig. 4-22A). Next the *atrium*, the chamber designed to deliver blood to the ventricle, arises (Fig. 4-22B). Finally, the receiving chamber for venous blood, the *sinus venosus*, is established (Fig. 4-22C and D). A number of investigators, beginning with Florence R. Sabin, have analyzed the origin of the heartbeats. The first twitching can be seen early in the second day of development. The slow but rhythmical beat begins along the right side of the ventricle and gradually involves the whole ventricu-

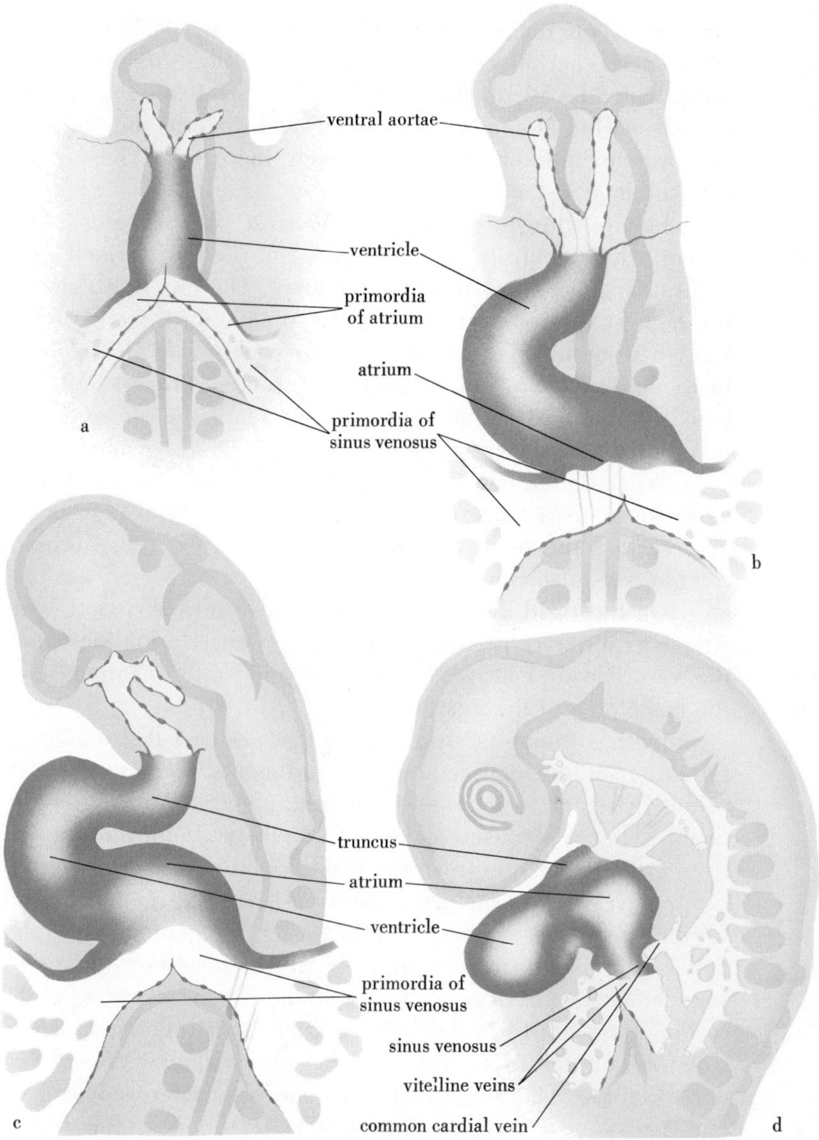

ventral aortae
ventricle
primordia of atrium
atrium
primordia of sinus venosus
a
b
truncus
atrium
ventricle
primordia of sinus venosus
sinus venosus
vitelline veins
common cardial vein
c
d

*Fig. 4-22 Progressive fusion of the paired primordia of the heart of the chick embryo.* (a) *At 9-somite stage (approximately 30 hours). Truncoventricular region established; primordia of atrium and sinus venosus still paired.* (b) *At 16-somite stage (approximately 40 hours). Atrium established.* (c) *At 19-somite stage (approximately 46 hours). Primordia of sinus venosus beginning to fuse.* (d) *At 26-somite stage (approximately 56 hours). Sinus venosus established. (From B. M. Patten,* Foundations of Embryology, *by courtesy of McGraw-Hill Book Company.)*

lar wall. Soon the entire muscle of the ventricle is contracting synchronously—periods of pulsation alternating with periods of rest. Meanwhile, the atrium has been forming. As it takes shape, it too begins to contract but at a more rapid rate, which governs the rate of the heart as a whole, the ventricular rate being increased. These contractions set the blood in motion.

Finally, the *pacemaker* or *sinoatrial node* develops. When this region, which controls the contractions of the fully formed heart, starts contracting, the whole heart accelerates. The fact that the region with the highest rate of contraction sets the pace for the entire organ gives rise to a number of interesting experiments. If the regions of the heart are cut apart and isolated, each tends to revert to its characteristic rhythm. If they are recombined, again the slower is increased to keep pace with the faster.

What synchronizes these early contractions? Nerve fibers, which, as we have seen, grow out from the central nervous system, have not yet reached the heart. By what mechanism are these primitive contractile cells also conductile?

On the third day of incubation, the ventricle, now U-shaped, assumes its ultimate position behind the atrium. During the following day the ventricle and atrium each are divided into chambers completing the basic plan of the heart.

Only so much is visible to the eye. But what events presage the formation of the paired primitive rudiments? In prestreak and early primitive streak stages, cells with the potential to form heart are not confined to any localized region of the epiblast. If one cultures fragments from any part of the prestreak blastoderm, poorly organized, pulsatile tissue develops. As the primitive streak is established, cells with heart-forming potential are limited to the posterior half of the blastoderm. The distribution of "heart-forming cells" in the head-process stage embryo was determined by Rawles. In her studies, embryos were cut, by means of fine glass or steel needles, into tiny fragments, each of which was isolated and grown on the vascular chorioallantoic membrane of an older embryo. The masses were sectioned for microscopic examination and the tissues formed from each were recorded. In this way a map of organ-forming areas was constructed. In Fig. 4-23 we see such a map for the whole embryo at the head-process stage; Fig. 4-24 shows only the heart-forming areas. These studies suggest that heart-forming cells in the epiblast move from the periphery of the embryo to and through the primitive streak, reassembling as mesodermal regions on either side of the head process.

A more accurate method for mapping the prospective fate of embryonic cells used extensively by DeHaan and his colleagues, Rosenquist and Stalsberg, is that of transplanting fragments from a radio-

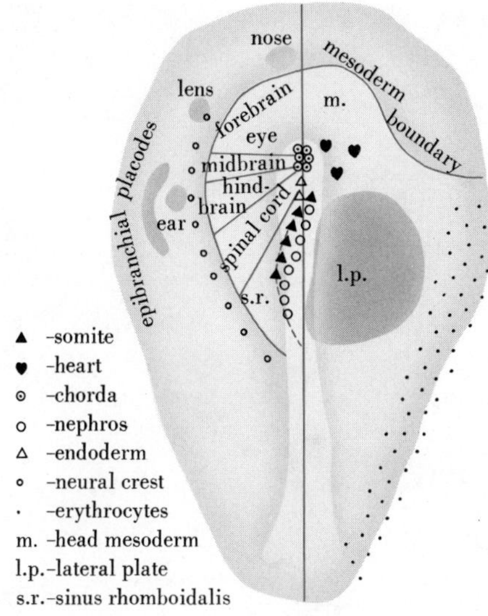


**Fig. 4-23** *Map of prospective areas in the chick embryo at the primitive streak stage. (After D. Rudnick, in* Quarterly Review of Biology, *19, by courtesy of the Editor.)*

actively labeled donor embryo to an unlabeled recipient. We see the results of such an experiment in Fig. 4-25. Each labeled fragment is grafted to the matching site in an unlabeled host, where it heals rapidly. Later the host embryo is fixed, sectioned serially, and the distribution of labeled cells determined by autoradiography. As seen in Fig. 4-25, the implanted cells appear to participate normally in the development of the host. This technique, along with those of carbon-marking and time-lapse cinematography, permits us to trace the movements of precardiac cells with greater precision than heretofore possible.

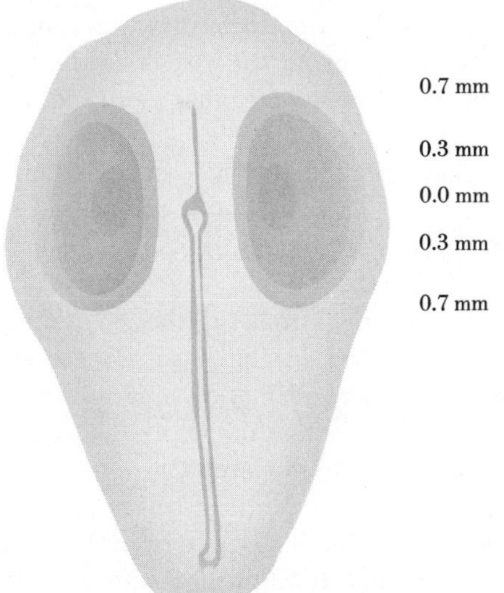


**Fig. 4-24** *Map of heart-forming areas in head-process-stage chick blastoderm. Numerals at right show distances in millimeters from the level of the primitive pit. (After M. E. Rawles, in* Physiological Zoology, *16, by courtesy of University of Chicago Press.)*

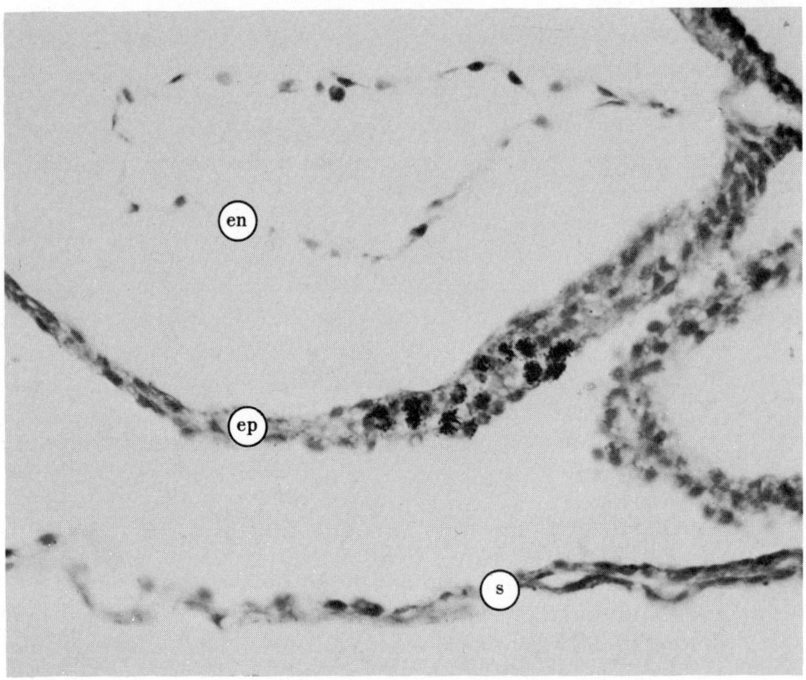


*Fig. 4-25 Autoradiographed section through the ventricle of a stage 12 host chick embryo, illustrating the labeled implant in the epimyocardium with well-defined boundaries.* en, *endocardium;* ep, *epimyocardium;* s, *splanchnopleure.* × *410. (From H. Stalsberg and R. L. DeHaan,* Developmental Biology, *19, by courtesy of the authors and Academic Press.)*

In the midstreak stage, these cells lie in paired regions about midway down the length of the streak, extending from the midline about halfway to the edge of the embryo. From the streak to the periphery, they are arranged as follows: prospective conus, ventricle, and atrial and sinus cells.

Labeled cells enter the streak and move into the mesoderm. Mixing occurs, for implants made on one side or the other contribute cells to both sides of the heart. Preendocardial cells appear to move independently of premyocardial cells.

The last sinus cells enter the streak at the beginning of the head-process stage. Thus by the end of the head-process stage, the preepimyocardial mesoderm is organized in two separate regions with a gap of noncardiogenic mesoderm of about 0.8 mm between them. Preendocardial cells are organized into similar areas.

It is also possible, by further refinements of this approach, to map the heart-forming areas in greater detail, by determining the localization within them of the cells destined to form the major subdivisions of the heart. It is not necessary to present the evidence in detail, for it

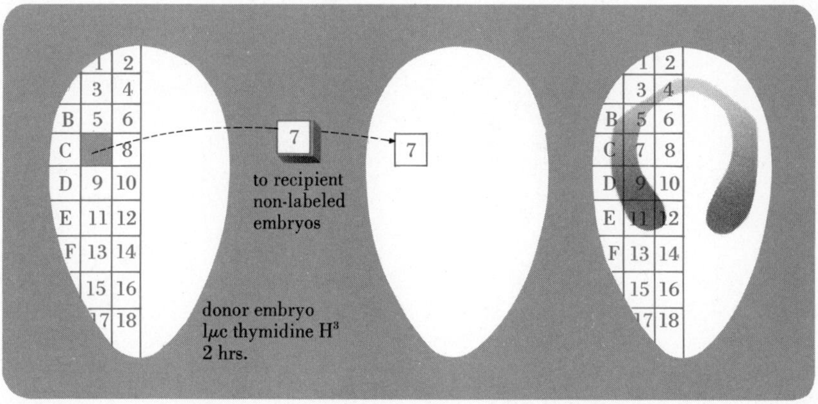

**Fig. 4-26** *Cardiogenic areas as revealed by the autoradiographic transplantation mapping technique.* (Left) *Fragments of endoderm-mesoderm are transplanted from a thymidine-$^3$H-labeled donor embryo into nonlabeled recipient embryos. After development of the host to stage 12, the position of labeled cells in the heart is determined by autoradiographic analysis.* (Right) *Regions shaded on the embryo are those that contributed cells to the stage-12 heart. Dark shading: myocardium and endocardium; light shading: endocardium only. (From R. L. DeHaan, in* The Emergence of Order in Developing Systems, *by courtesy of the author and Academic Press.)*

largely confirms what we have already said about the order of movement of cells through the primitive streak. One point, however, warrants special mention. Labeled endocardial cells are often scattered as single cells or clustered in small groups, suggesting a "freedom of movement" of endocardial cells and substantial intermingling of graft and host cells. However, myocardial cells tend to be found in well-defined areas with no indication of intermingling of graft and host cells, suggesting that the integrity of the boundaries of this sheet of mesoderm are retained during the further steps in cardiogenesis.

How is the flat sheet of precardiac mesoderm of the embryo at the head-process stage transformed into the heart? The autoradiographic and other marking analyses just described, coupled with electron microscopic studies of the mesoderm, show that the mesoderm behaves as a coherent sheet. It condenses, deforms, and folds, but it does not lose its integrity. In contrast to views held only recently, it does not break up into cells or cell clusters which migrate independently.

In the head-process stage embryo the mesoderm is a thin, flat sheet with the character of an epithelium, that is, the cells are held together tightly (rather than a mesenchyme or loose network of cells). As seen in Fig. 4-27B, as the ectoderm and endoderm begin to fold, forming the head fold and foregut, the anterior portion of the mesoderm folds between them to enclose the roof and walls of the foregut.

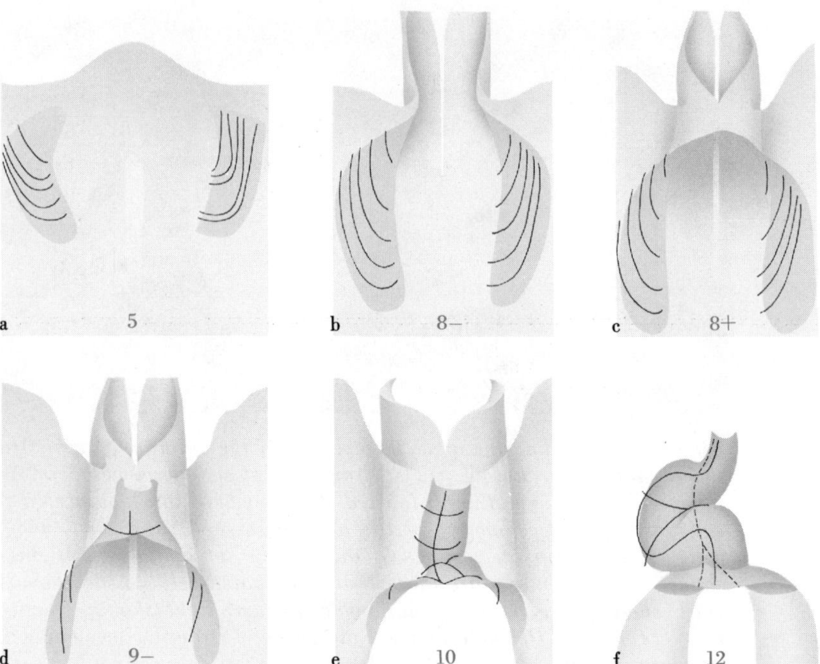


*Fig. 4-27 Heart formation from stage 5 to stage 12. Only mesoderm is shown.* (a) *The premyocardial subdivision map is superimposed on the mesoderm layer, and* (b, c) *is appropriately distorted in shape to fit the forming myocardial troughs and elongating myocardial tube* (d, e). (f) *Subdivision of the stage-12 heart from which division lines were obtained. (From H. Stalsberg and R. L. DeHaan,* Developmental Biology, *19, courtesy of authors and Academic Press.)*

In these regions the mesoderm begins to thicken and condense. Cells become more closely apposed and change from flat to columnar in shape. As the folding proceeds (C, D), the foregut is enclosed by mesoderm. Finally (E, F), the myocardial folds swing together establishing the ventral midline of the heart tube.

Thus far we have said nothing about inductive tissue interactions in the formation of the heart. Bear in mind, however, that the transplanted fragments with which the mapping experiments were done included cells of all three layers.

There is evidence that in amphibians the heart-forming regions are determined by interactions with neighboring tissues. Prospective heart mesoderm comes into contact with endoderm during gastrulation. In newt embryos in which the endoderm is removed surgically, the heart is completely absent. Thus the endoderm appears to exert a positive inductive influence. Conversely, anterior neural plate material in some way represses heart formation.

Evidence for inductive influences on the precardiac mesoderm of birds and mammals is unsatisfactory. There are no relevant experiments on mammalian embryos. Although experiments have been performed on chick embryos, critical evidence is lacking.

However, there is additional evidence that the association between endoderm and mesoderm is intimate, for disturbances in their normal relations lead to modifications in cardiogenesis. For example, in chick embryos treated with low concentrations of sodium citrate or ethylenediaminetetraacetic acid (EDTA), agents that act by disturbing the intercellular contact relations of the preheart mesoderm and the endoderm, normal fusion of heart primordia is prevented. Treatment of the primitive streak stage results in double-hearted embryos. The production of embryos with two complete hearts is not new. Any means of obstructing the fusion of the two rudiments, by removing the wedge of tissue lying between them or by inserting a barrier, results in the formation of two hearts (Fig. 4-28).

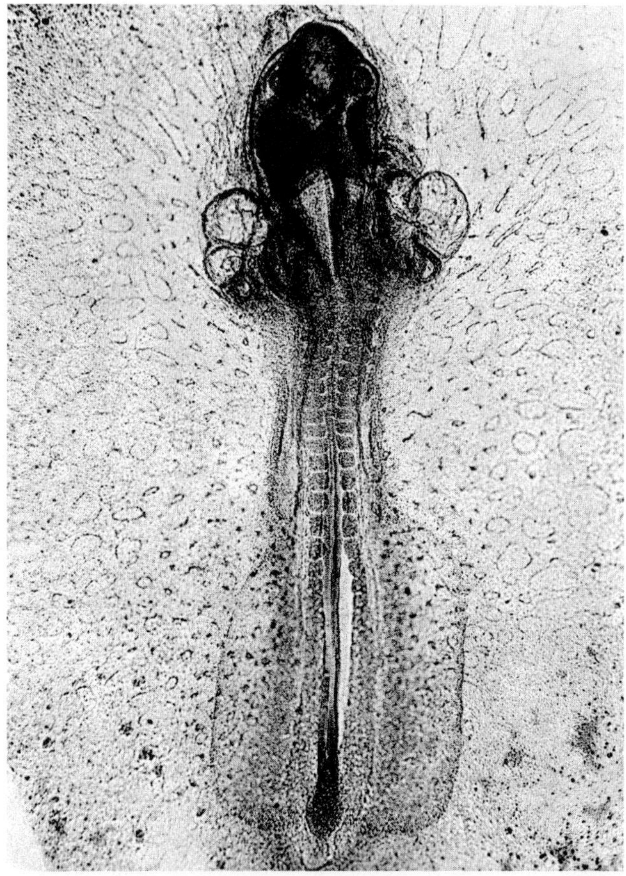

**Fig. 4-28** *A 21-somite double-hearted embryo, produced by cutting the floor of the foregut at stage 7+. (From R. L. DeHaan, in* Developmental Biology, *1, by courtesy of the author and Academic Press.)*

But the advantages of this simple chemical technique are twofold: first, as we shall see later, it helps us understand the nature of interactions between cells; and second, it can be applied at any stage and to any degree. By altering the concentration of sodium citrate applied to the endodermal surface of a series of embryos at the head-process stage, the endodermal layer may be almost completely removed, leaving the mesoderm and ectoderm relatively intact and able to continue development. Several pairs of somites and a distinct medullary plate are formed, and tubulation movements and fusion of the neural folds are prevented. Many small spontaneously twitching vesicles of heart tissue are formed, without further migration, at each location at which a mass of precardiac mesoderm is left by the disaggregated endoderm (Fig. 4-29).

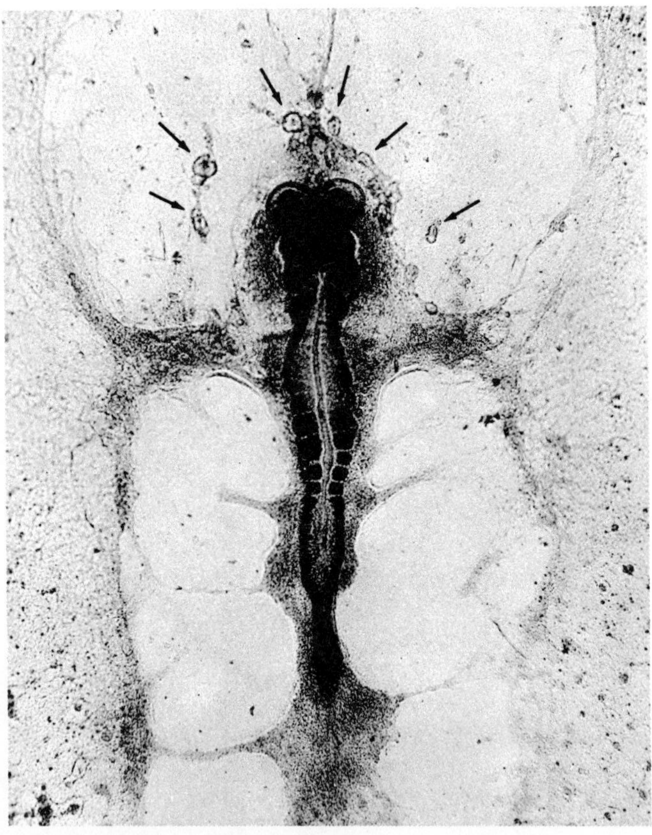

**Fig. 4-29** *Chick embryo, whole-mount preparation, treated with sodium citrate at stage 6+; shows endoderm denuded from the area pellucida and a crescentic array of small cardiac vesicles (see arrows) at the anterior end of the embryo. (By courtesy of R. L. DeHaan and Carnegie Institution of Washington.)*

## *THE MOVEMENTS OF CELLS OVER LONG DISTANCES*

Thus far we have considered the interactions of adjacent cells and tissues and, as we saw in the outwandering axon, the interaction of the extension of a cell with its environment. However, several key steps in development require that cells migrate over long distances to reach their ultimate destinations.

### *Neural Crest*

During the closure of the neural tube (page 68), a mass of cells "escapes" into the space between the tube and the overlying ectoderm. These cells, which have their origin along the edges of the neural plate, make up the *neural crest* (Fig. 4-6). They are among the more intriguing cells of the embryo, for they are destined to differentiate, to realize their potential so to speak, only after they "leave home" and take up new locations. Thus, depending on their original location along the longitudinal axis of the neural tube, from head to tail, and the new environment in which they come to lodge, they may become specialized as cartilage, spinal ganglia, and sensory nerves; or as sheath cells (which cloak the axon); or as cells of the adrenal medulla, or sympathetic nervous system; or as pigment cells.

We know the fate of these cells through several kinds of experiments. For example, in a typical defect experiment, if the dorsal portion of the neural tube, including the adjacent neural crest, is excised in a frog or salamander embryo, spinal ganglia and sensory nerves fail to develop (Fig. 4-30). Motor nerves develop, as you would expect, recalling that they grow out from the ventral part of the spinal cord; but the motor fibers lack sheath cells. Moreover, not only are the associated sheath (*Schwann* or *satellite*) cells lacking, but the motor fibers lack their typical multilayered myelin membranes as well. Clarification of the relations between Schwann cell, "myelin," and axon required the application of electron microscopy. We now know that the membranes

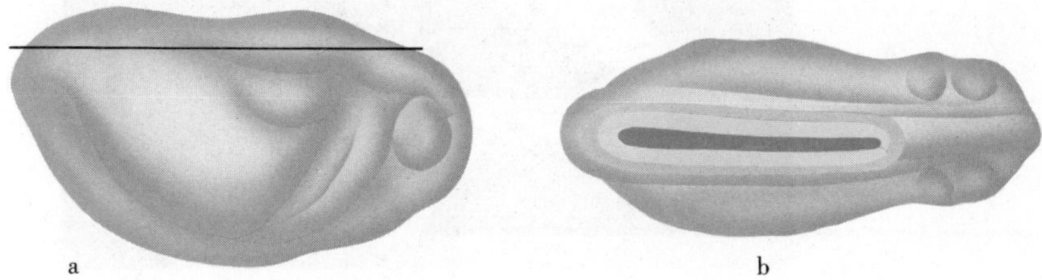


**Fig. 4-30** *Removal of the neural crest in the frog embryo.* (a) *Profile view; line indicates incision.* (b) *Dorsal view showing the open neural tube. (From R. G. Harrison, in* Journal of Comparative Neurology, *37, by courtesy of The Wistar Institute.)*

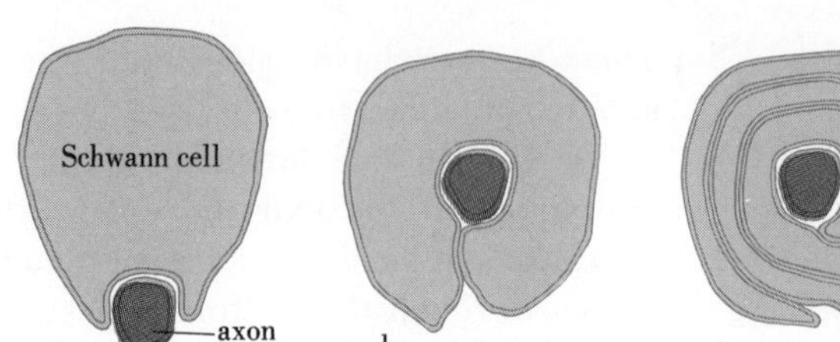


***Fig. 4-31*** *Schematic representation of the progressive development of an axon by the membranes of the Schwann cell, as described by Betty Geren. Such an axon is said to be myelinated. (From C. P. Swanson,* The Cell, *2d ed.,* © *1964 by permission of Prentice-Hall, Inc., Englewood Cliffs, N.J.)*

of Schwann cells progressively envelop the axon. Thus, an axon is said to be myelinated when it is wrapped in a spiral consisting of several layers of plasma membrane produced by associated Schwann cells (Fig. 4-31).

Pigment cells are derived from the same source. Melanoblasts (prospective pigment cells) from a chick embryo of a pigmented breed, say a Barred Plymouth Rock, may be transplanted into the base of the wing bud of a White Leghorn embryo. When the chick hatches it will have a wing with a pigment pattern that is characteristic of the breed from which the implanted neural crest cells were taken (Fig. 4-32).

***Fig. 4-32*** *White Leghorn pullet which received in its right wing bud at 75 hours' incubation a graft of melanoblasts from an early embryo of a pigmented breed. Note the barred pattern of the juvenile wing plumage characteristic of the donor. (By courtesy of M. E. Rawles.)*

The neural crest is a *transitory* embryonic structure whose cells begin to disperse almost as soon as the crest is formed, in a head to tail "wave" along the embryonic axis. We know with some degree of certainty the origin of the crest and the inventory of its final products. However, a difficult question remains. Do cells move out and migrate randomly, their ultimate fate depending on the locations in which they come to lie? Or, conversely, is the direction of movements of a given cell related somehow to its state of determination? A few clues are beginning to emerge from the use of radioactively tagged cells. Neural crest grafts are made from donor chick embryos labeled with tritiated thymidine into unlabeled hosts. By following these marked cells, we may detect two patterns of movement.

One stream, probably the normal migratory pathway of pigment cells of the integument, moves within the ectoderm. Another major stream moves into the mesenchyme between the neural tube and the mesoderm that will form muscle. Certainly, these two populations of cells are not grossly different at this time, but might we find that cells migrating in the ectoderm contain precursors of melanin pigments? The methods for enabling us to answer such questions are now at hand.

### *Primordial Germ Cells*

When we discussed oögenesis and spermatogenesis (Chapter 2), we took as our starting point the germ cells (oöcytes and spermatocytes) in the ovary and testis, respectively. Yet it is one of the remarkable features of the development of these organs in vertebrates that the primordial germ cells originate outside the gonad and migrate into it only secondarily. In human development, primordial germ cells are first distinguishable about 24 days after fertilization, when they are found in the wall of the yolk sac (Fig. 4-33). In increasingly older embryos they are found in the wall of the hindgut, in its mesentery, and finally, after about a week, in the primordium of the gonad, which only then is beginning to take shape as a *genital ridge* or longitudinal mesodermal thickening on each side of the body, closely associated with the developing kidney. In experimental animals, living germ cells may be identified by behavioral, in addition to cytological, characteristics. Time-lapse cinematography reveals that in mammals they migrate from the yolk sac to the germinal ridge by ameboid motion. The observations on living mammalian germ cells confirm earlier histochemical studies in which their distribution was mapped at successive stages, the identification being aided by a special stain for the enzyme, alkaline phosphatase, in which germ cells are particularly rich (Fig. 4-33). Other special staining techniques applied to amphibian embryos suggest that the germ cell lineage can be traced all the way back to cells found

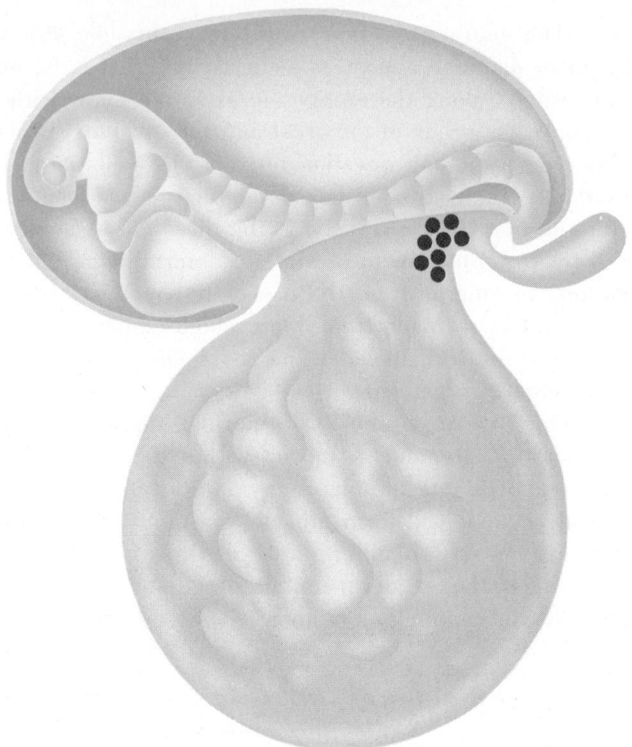

*Fig. 4-33 In the human embryo, primordial germ cells (black dots) are first seen during stage XI (13 to 20 somites, 24 to 26 days) when they are in the yolk sac. They spread into the hindgut by stage XII (21 to 29 somites) and up the mesentery to its root by stage XIII (4–5 mm). (Original drawing by B. G. Böving after E. Witschi,* Contributions to Embryology, *32, by courtesy of Carnegie Institution of Washington.)*

first in the prospective endoderm of the blastula. In birds, the primordial germ cells, which are first observed in the anterior border of the extraembryonic region, in the *germinal crescent* (Fig. 4-34), are distributed initially via the bloodstream. Descriptive studies, again employing selective staining (Fig. 4-35), show that these cells are first transported to all parts of the embryo at the time the blood begins to circulate. The germ cells are observed in the heart and great vessels, and also in the notochord, neural ectoderm, and endoderm. By the end of the second day of incubation, however, they are concentrated in the future gonadal region. The pattern appears to be, first, widespread distribution by the vascular route, followed by active ameboid movements of individual cells. But why do germ cells persist only in the gonadal region? Do they move out of other tissues? Do they die if they fail to reach the gonad? Do they differentiate in other directions?

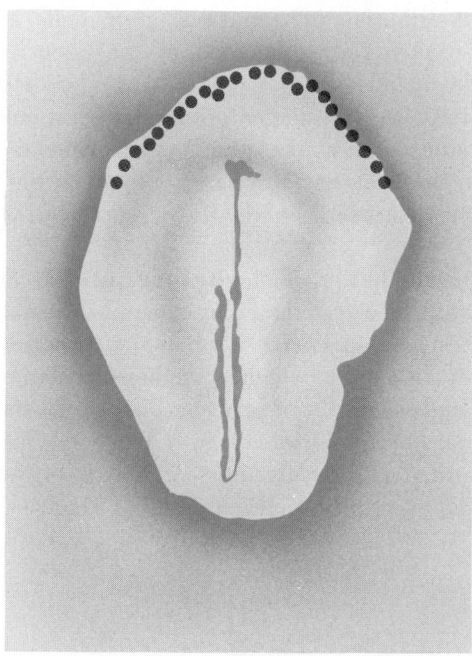

***Fig. 4-34*** *Surface view of a chick embryo in the head-process stage, showing the original position of the primordial germ cells in the margin between the area pellucida and the area opaca. (From Swift, 1914.)*

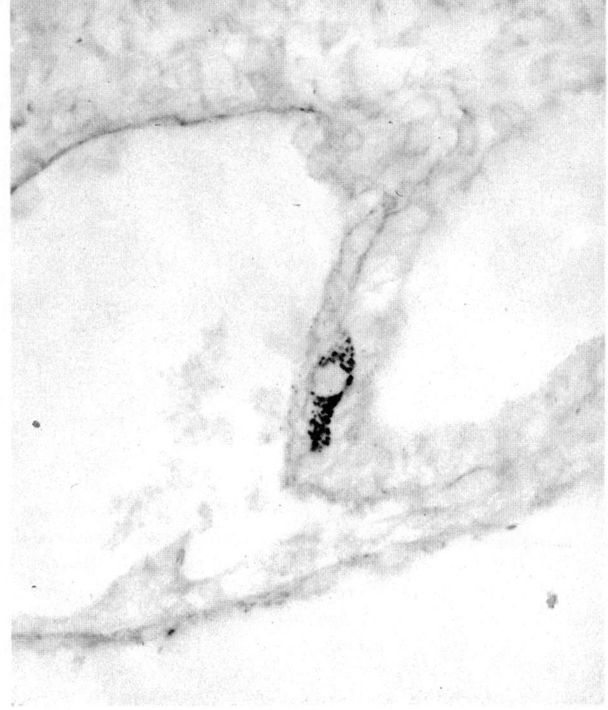

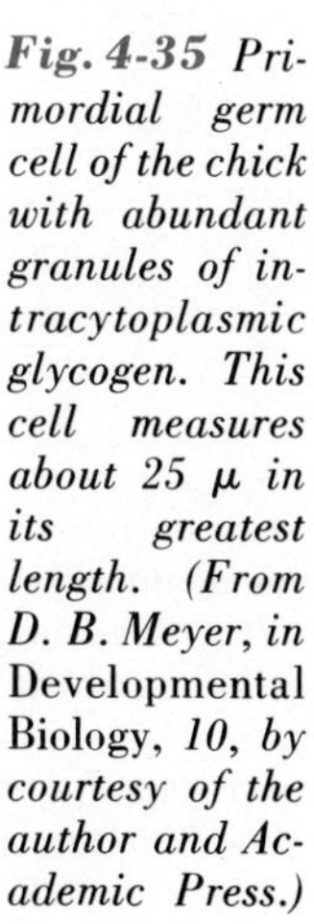

***Fig. 4-35*** *Primordial germ cell of the chick with abundant granules of intracytoplasmic glycogen. This cell measures about 25 μ in its greatest length. (From D. B. Meyer, in* Developmental Biology, *10, by courtesy of the author and Academic Press.)*

The widespread distribution of primordial germ cells at these early stages provides a possible explanation of the occurrence of *teratomas*, tumors made up of heterogeneous mixtures of cell types, which are probably derived from misplaced primordial germ cells.

But descriptive studies such as these call for experimental proof; by now, the next step in the analysis should be apparent. If the source of primordial germ cells is destroyed, whether by surgery or irradiation, a sterile gonad results.

This proof leads to another. Evidence for the vascular dissemination of primordial germ cells was obtained by Simon, who established common circulatory pathways between normal and experimentally produced sterile embryos. In pairs in which a common circulation could be demonstrated, the gonads of sterile embryos become "reprovisioned" with germ cells from their normal partners (Fig. 4-36).

But are the primordial germ cells the *sole* source of definitive germ cells? Do the primordial germ cells themselves give rise to the

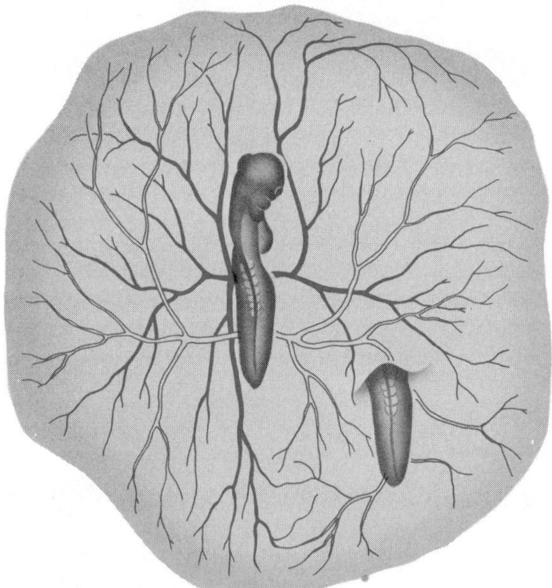

**Fig. 4-36** *Vascular dissemination of primordial germ cells in the chick embryo. Diagram of an experiment by Simon in which the posterior region of a chick embryo was grafted into the extraembryonic region of another embryo. Both donor and host had less than ten somites; thus neither had primordial germ cells in the region of the gonads at the time of operation. Later the supplementary gonads became populated with primordial germ cells as a result of their migration from the single germinal crescent through the blood stream into both host and donor gonads. (From D. Simon in* Archives d'Anatomie microscopique et de Morphologie experimentale, *49, 1960, by courtesy of the author and Masson et Cie.)*

definitive gametes, or do they stimulate other cells in the gonad to develop into gametes, or are the gametes derived both directly and indirectly? All of these possibilities have been championed in the past, but experimental evidence has become available only recently.

As we observed on page 16, there is abundant evidence from studies using tritiated thymidine that in the mouse the proliferative phase of oögenesis is confined to the period of intrauterine life. All of the eggs to be ovulated by a mature mouse during her lifetime are derived from oöcytes laid down during embryogenesis. We can also say for *Xenopus laevis*, the South African clawed toad, that all definitive gametes may originate from primordial germ cells. In A. W. Blackler's experiment, transplantations of endoderm containing primordial germ cells were carried out between two subspecies of *Xenopus*, the two subspecies being distinguishable by several markers. The grafted animals were allowed to grow to sexual maturity and their gametes were examined. *Some experimental toads produced only gametes of graft origin.* Moreover, after successful grafting, females of one subspecies laid eggs like those of other subspecies. Thus, the genotype of the primordial germ cell specifies the characteristics of the egg, which are independent of the ovarian environment.

## *INTERACTIONS OF MAMMALIAN EMBRYOS AND THEIR ENVIRONMENT*

The mammalian embryo does not develop in pond or seawater, or even in a closed environment as do the embryos of birds and turtles. It develops in the interior of another living organism and depends on interactions with it for its very life. George W. Corner, one of the fathers of modern reproductive physiology, put the problem of the human embryo this way: "Accepting the shelter of the uterus, it also takes the risks of maternal disease or malnutrition, and of biochemical, immunologic and hormonal maladjustment. Even before it strikes its roots in the living tissues of the endometrium (lining of the uterus) it has a week's journey to make, as long as a submarine takes to pass beneath the polar ice cap. Like the U-boat it has to carry most of its supplies with it in its trip down the oviduct and the uterine lumen, and to add to its difficulties it is surrounded by a far more variable and chemically active medium than ice-cold sea water." Let us trace this journey, centering the discussion on events as they occur in the rabbit, with occasional comparisons with man and mouse.

After fertilization in the upper reaches of the oviduct, there is a delay, as long as 36 hours in the rabbit, before transport to the uterus begins. Cleavage begins during this time and continues as the egg makes its way to the uterus. Transport itself requires another 24 hours, so the

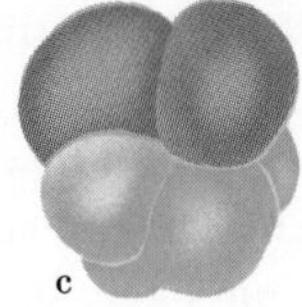
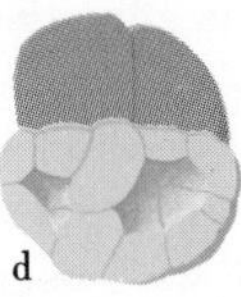
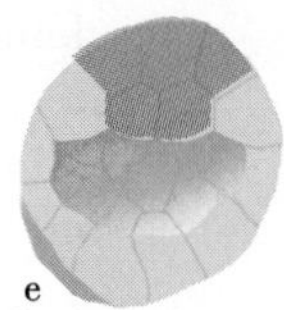


**Fig. 4-37** *Cleavage stages and blastocyst of the pig (Heuser and Streeter).* × *240. The enveloping zona pellucida has been omitted.* (a) *Two blastomeres;* (b) *three blastomeres;* (c) *six blastomeres;* (d) *hemisected morula (20 cells), with early cavities appearing;* (e) *hemisected blastocyst (30 cells).*

rabbit egg enters the uterus about 60 hours after it is fertilized. The human egg also passes to the uterus about 3 or 3½ days after ovulation, which is about average for many mammals. The cleaving egg is propelled by muscular action of the oviduct and uterus, and the rate of propulsion is influenced by maternal hormones.

Cleavage of the egg is complete but may be highly irregular and asynchronous. One finds cleaving eggs with three, five, six, or seven blastomeres (Fig. 4-37). In most mammals, cleavage results in a solid mass of cells or *morula*. In the morula a distinction emerges between the internal and enveloping (outer) cells. The cells lying in the interior are known as the *inner cell mass* (Fig. 4-38). They will constitute the main body of the embryo; the cells of the enveloping layer will contribute largely to the extraembryonic membranes. This layer is lifted off the inner cell mass, remaining attached on only one side. The enveloping layer establishes connections with the uterine lining, interposing between that lining and the embryo proper a layer known as the *trophoblast*. At this stage the embryo is known as a *blastocyst*.

We may now consider *implantation*, the process by which the blastocyst is attached to the uterine wall.

Within the human uterus, the egg usually attaches in a restricted region—the implication being that there are mechanisms that influence its location. Within the rabbit uterus (which consists of two long, tubular "horns") a solitary egg usually lodges near the midpoint of the horn containing it, whereas two or more eggs become evenly spaced along the horn (Fig. 4-39). It has been argued that from each end of the horn and wherever the horn becomes distended by a growing egg within it, propulsive contractions arise and are propagated in each available direction. The contractions are presumed to "milk" the eggs along the uterine horn, with the ultimate result that each egg has caused its neighbors to be repelled to a maximum, and consequently equal, distance.

But what "turns off" the transport mechanism? Experiments suggest that the expansion of the egg, which stimulated the spacing

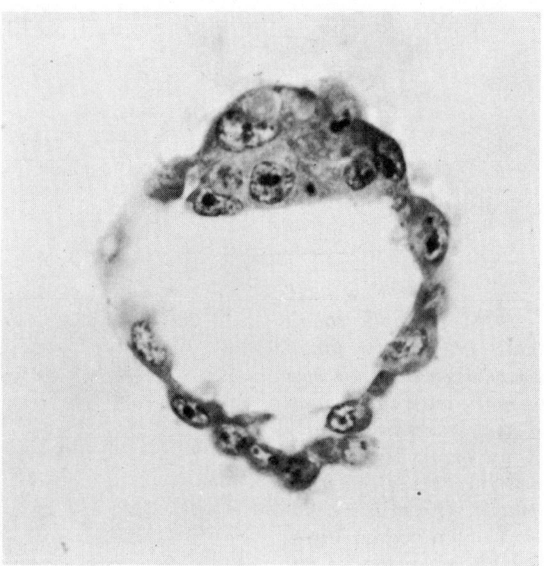

**Fig. 4-38** *Photomicrograph of section of 107-cell (about five-day) human embryo, showing inner cell mass* (top) *and trophoblast.* × *600. (From A. T. Hertig, J. Rock, E. C. Adams, and W. J. Mulligan, in* Contributions to Embryology, *35, by courtesy of Carnegie Institution of Washington.)*

mechanism, simply continues and becomes the inhibitor of propulsion. Whereas small glass beads placed in uteri of pregnant rabbits are moved along randomly and occasionally are even expelled, large beads are not moved at all, and the critical size is about that of eggs at the time when they come to rest.

The expansion of the rabbit egg continues even after it comes to rest, and one wall of the uterus balloons out to accommodate it. Since it is always the same wall (the most thin and free wall, farthest from the supporting structures of the uterus) one may suspect that the ballooning out has something to do with the egg's always attaching to that same wall of the uterus. Using the technique of looking into the uterus through a window and manipulating the egg inside, one gains the distinct impression that the egg is grasped within the ballooned-out part of the uterus and is thereby denied contact with, hence chance of attachment to, the other walls of the uterus.

Now the egg attaches, and each mammalian species exhibits a characteristic orientation. Not only is the privileged geographic region or wall of the uterus consistent, but it is always the same pole of the egg that attaches there. The polarity of the essentially spherical egg

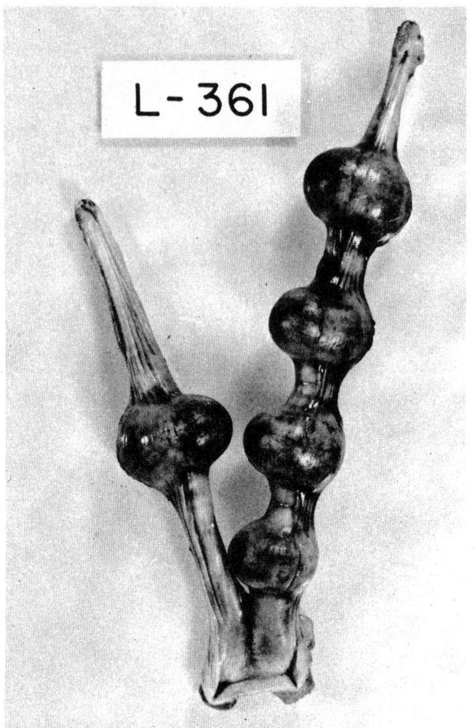


*Fig. 4-39 A single rabbit conceptus is usually found midway between the ends of a uterine horn* (left); *large numbers of conceptuses are usually spaced by nearly equal distances* (right). *Seventeen days after mating. (From B. G. Böving, in* Conference on Physiological Mechanisms Concerned with Conception, *1963, by courtesy of the author and Pergamon Press.)*

is defined by the region where the *inner cell mass* remains, after fluid accumulates within the previously solid mass of cells into which the egg has divided (Fig. 4-38). The pole is called the *embryonic pole*, because the inner cell mass is destined to form the embryo. The trophoblast is peripheral and is destined to invade the uterus. The human ovum begins attaching at its embryonic pole; the rabbit ovum at the opposite pole. A possible explanation for the latter phenomenon is suggested by the fact that a rabbit ovum removed from the uterus shortly before attaching exhibits a remarkably high alkalinity in the region that attaches first. Moreover, if such an ovum is placed in a very alkaline solution, it becomes sticky—but sticky all over.

On microscopic examination, one finds that the earliest attachments are indeed adhesions, but, although they are restricted to the region capable of giving the alkaline reaction, they do not involve the whole region. They occur selectively where there is a maternal blood vessel beneath the uterine epithelium; thus, elicitation of the alkaline reaction and consequent adhesion are restricted by the maternal system to just that microenvironment most favorable to chemical transfer. The localized transfer involves discharge of bicarbonate from the ovum, which has a concentration three times that of the mother's blood; the attendant rise in alkalinity probably results from passage of carbon dioxide into the mother's blood while alkaline carbonates are left behind in the uterine epithelium. The occurrence of adhesion selectively over maternal blood vessels has an interesting consequence. Since it is the first stage of attachment, it chooses the sites for the immediately following invasive stage and, in effect, "aims" each invasion at a maternal vessel.

There are several theories on how the trophoblast then penetrates the uterine epithelium and reaches the vessels: by fusing with it, by ameboid motion, by ingesting it, by digesting it, or by differential dissociation. The last interpretation is simply that the localized alkalinity, which promotes stickiness between the egg and the surface of the uterine epithelium where there is mucus, promotes a loss of cohesion between the cells of the uterine epithelium where there is no mucus. Thus, the cellular epithelium simply comes apart over blood vessels, whereas the adjacent trophoblast does not, thanks to the fact that the specialized invading portions have developed a syncitial structure, a multinucleate mass lacking cell boundaries. These nondissociating parts of the trophoblast then move along the path of loosened epithelium by a mechanism whose investigation has only recently begun.

What makes the trophoblast invasions stop instead of going on indefinitely? One probable contributing factor is that the uterine epithelium that remains uninvaded soon loses the membranes between its cells and thus becomes just as resistant to dissociation as the trophoblast, with which it fuses. Another factor is that the concentration of bicarbonate in the ovum decreases gradually to that of the maternal blood as a result of the improving chemical exchange that occurs during implantation.

Having considered implantation, let us return only momentarily to the development of the embryo proper. Although there are numerous differences in detail between embryogenesis in birds and reptiles on the one hand, and mammals on the other, the overall patterns and processes are fundamentally alike: an epiblast and hypoblast are clearly delimited, and a transient primitive streak is formed, with a node at its anterior end. Neurulation and somite formation closely resemble the processes in the chick embryo (Fig. 4-40). Since we are concerned more with principles than with detailed comparison, we turn next to the formation of extraembryonic structures. Here it is difficult to generalize, for Nature has experimented widely with ways of forming the extraembryonic membranes; therefore we shall confine our remarks to the processes in man (Fig. 4-41), with only a few references to other species.

The innermost membrane, immediately surrounding the embryo, is the *amnion*. It is a fluid-filled, nonvascular membrane. It prevents the embryo from desiccation and acts as a shock-absorber. In the human

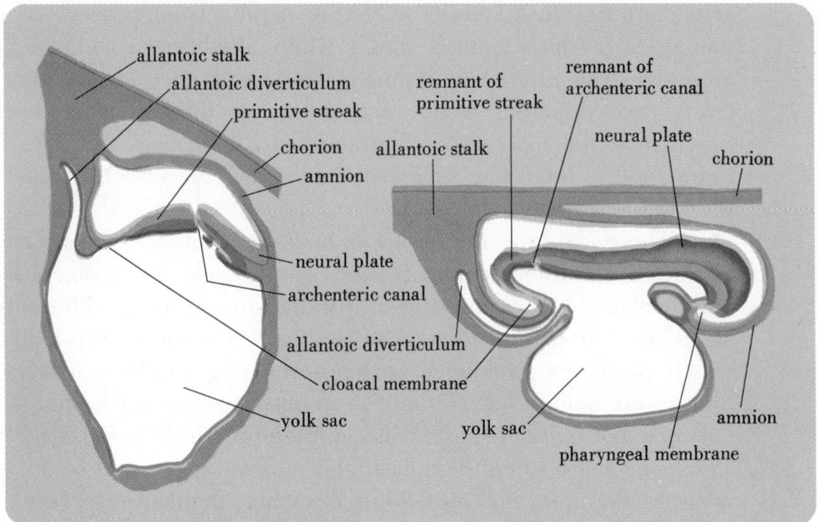


**Fig. 4-40** *Semidiagrammatic median sections of the human embryo in the primitive streak stage* (left) *and the neural plate stage* (right). *(Left, after Jones and Brewer, 1941, and Grosser, 1945; right, after Grosser, 1945.)*

embryo, the amniotic cavity is formed by a crevice appearing in the inner cell mass before gastrulation. This mode of formation of the amniotic cavity is known as *cavitation*. The amnion proper is a twin layer of epithelium.

The *primary yolk sac* is, in fact, the cavity of the blastocyst, filled initially with fluid, then gradually with extraembryonic mesenchyme, not with yolk, as it is in birds. A smaller, *secondary yolk sac* is formed by the hypoblast cells.

The vascularized *allantois* grows out of the hindgut. In birds (Fig. 4-42) and reptiles it functions as a urinary bladder. The insoluble end product of their nitrogen metabolism, uric acid, is stored in the allantois. In mammals, the soluble end product, urea, is passed from the embryo to the mother. In birds, a second function is respiration; the vascular *chorioallantois*, which is formed by the fusion of the protective *chorion* and the *allantois*, lies immediately beneath the shell membrane and functions in gaseous exchange. In mammals, the blood vessels of the allantois play a similar role, supplying the embryo with oxygen from the mother. In the human embryo, the allantois develops as a strand of mesenchyme leading from the blastodisk to the trophoblast. Only the mesodermal component of the allantois develops, giving rise to the allantoic circulation. The endodermal allantoic vesicle remains rudimentary; it does not serve as a bladder.

As the demands by the rapidly growing embryo for nourishment

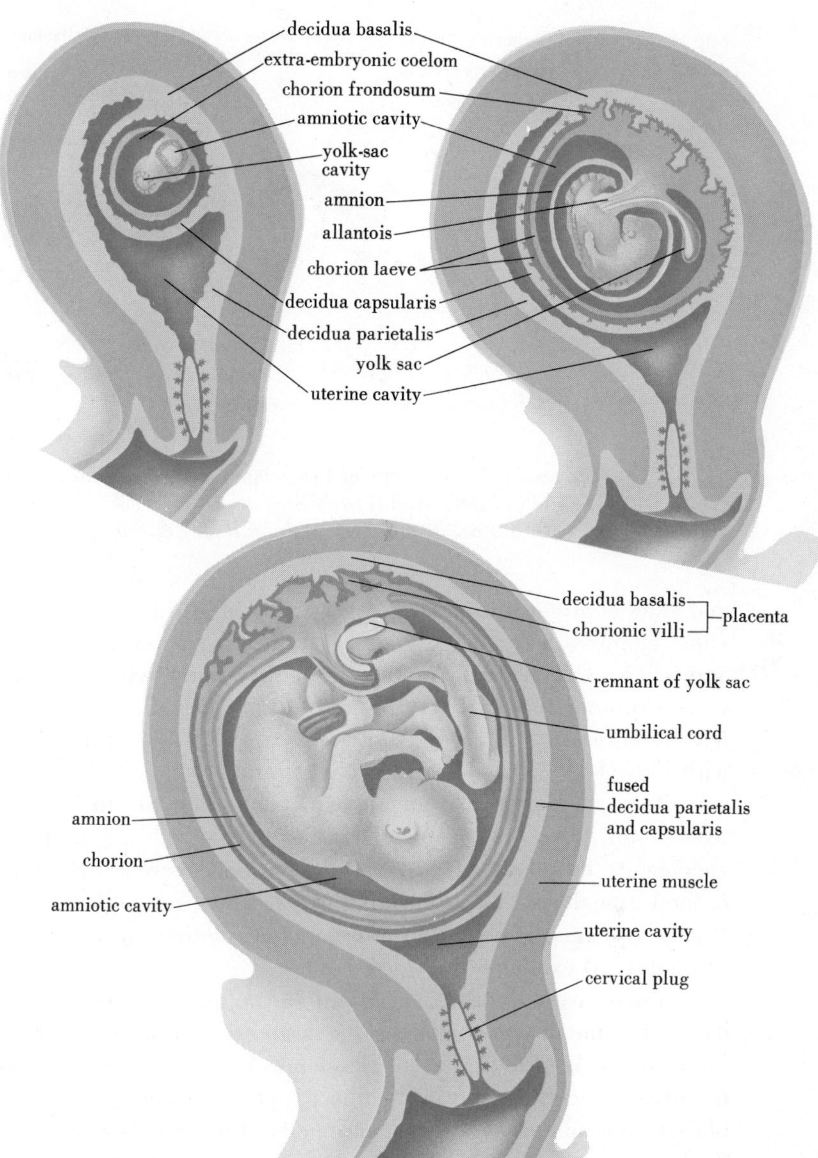


***Fig. 4-41*** *Relations of human embryo and uterus.* (a) *Early;* (b) *intermediate;* (c) *advanced.* (a *and* b, *from W. J. Hamilton, J. D. Boyd, and H. Mossman,* Human Embryology, *by courtesy of W. H. Heffer and Sons, Ltd.* c, *from T. W. Torrey,* Morphogenesis of the Vertebrates, *by courtesy of John Wiley & Sons, Inc.)*

and oxygen increase, the trophoblast increases the surface area it presents to the maternal blood. Fine, fingerlike processes, or *villi*, bud forth from the trophoblast of the human embryo when it has developed for about 12 days. About two days later, the villi begin to branch,

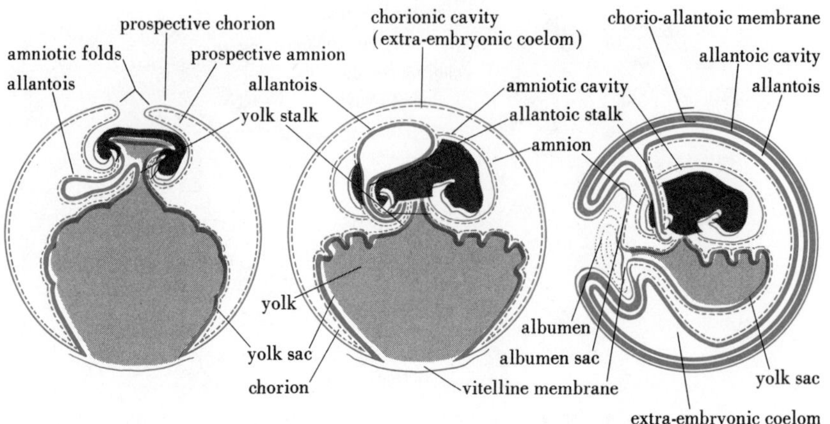


**Fig. 4-42** *Stages in development of extraembryonic membranes of the chick.* (Left) *Early;* (middle) *later;* (right) *fully mature. (After T. W. Torrey,* Morphogenesis of the Vertebrates, *by courtesy of John Wiley & Sons, Inc.)*

and they continue through several generations of branching until each villus acquires a treelike structure. The resulting surface is enormous and as the villi grow out and branch they carry with them a core of vessel-forming tissue that promptly establishes the pathways for the *fetal circulation* of the *placenta.* (Compare Figs. 4-41A and B and 4-43 with Figs. 4-41C and 4-44.)

Thus, a *placenta* is an apposition of trophoblastic and maternal tissues for purposes of physiological exchange, and in the human species its basic plan is a fetal circulation covered with trophoblast bathed in maternal blood circulating through a trophoblast-lined space. It joins two genetically different organisms, mother and fetus, providing physiological exchange between them.

There are numerous variations in the interactions of the egg and its environment: variations in the state of readiness of the uterine lining to receive the egg, in mechanisms of invasion, in the larger architecture of placentas, and in the fine structural relations between trophoblastic and maternal tissues and cells. But this is a large story in itself, the first chapters of which are told in George W. Corner's timeless books, *The Hormones in Human Reproduction* and *Ourselves Unborn.*

## DEVELOPMENT OF BEHAVIOR

In a brief but meaningful section entitled "The essence of a living animal" in *Animal Structure and Function,* in this series, Griffin and Novick tell us that many cells of the nervous system exhibit spontaneous activity even when they are isolated from excitation. "There is thus no dearth of excitation within a brain; the question is, how does it all add up to a system that controls the animal's activities in an appropriately coordinated fashion. Of all the organ systems the central nervous system contributes the most to making an animal the uniquely organized entity it is."

What can we say about the development of behavior when we cannot "explain how a brain really works, not even the nerve net of coelenterates"? We can make a beginning, and perhaps as a result of our doing so, one day one of our readers will find rewards in exploring the domain of brain and behavior through an embryologist's eyes, for it has been said that the best way of understanding the adult is by exploring the secrets of his development.

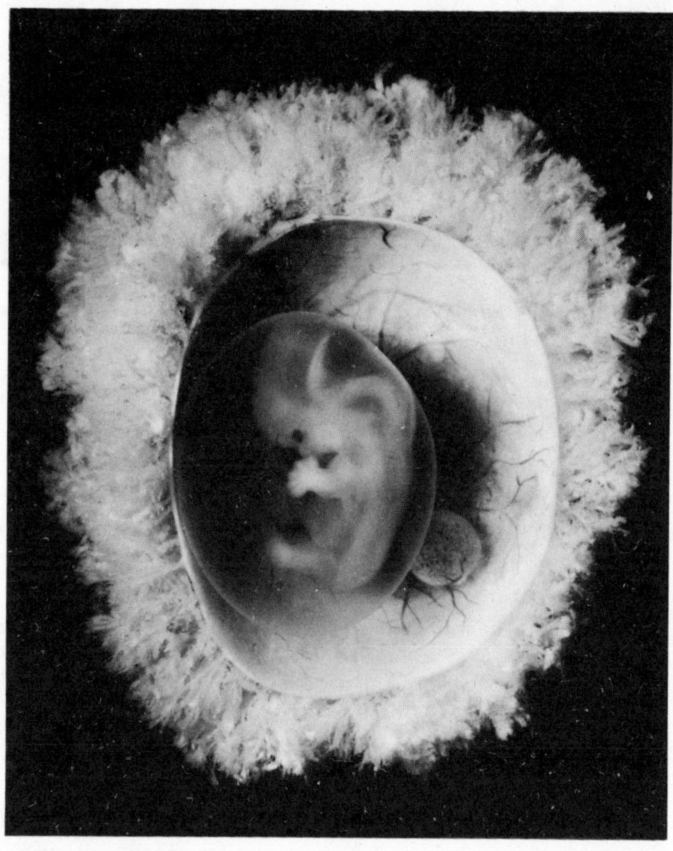

**Fig. 4-43** *A 40-day-old human embryo. × 1.5. (By courtesy of Carnegie Institution of Washington.)*

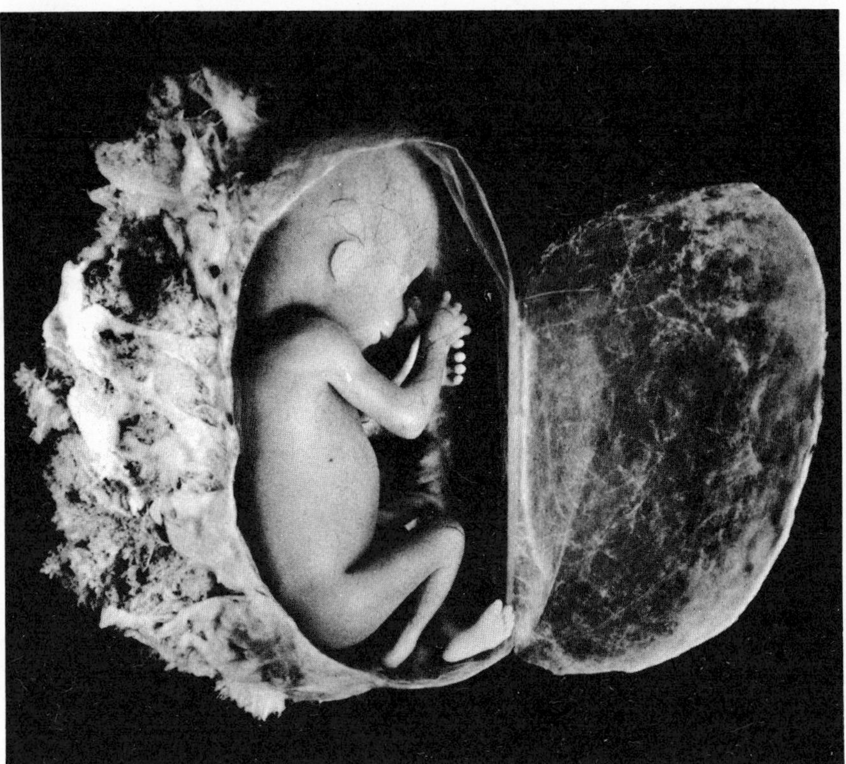

*Fig. 4-44 A human embryo at the beginning of the fifth month. × 0.75. (By courtesy of Carnegie Institution of Washington.)*

We may begin by emphasizing that the development of behavior is underlain by the development of neural connections. The human brain and spinal cord consist of many billions of cells, all arranged in an orderly manner with precise interconnections, both structural and functional. Communication among nerve cells depends, of course, on the passage of nerve impulses. These arise in individual cells and are effectively transferred from cell to cell either by one cell directly exciting another or by an intermediate step involving a chemical transmitter.

We have already observed that in the early embryo the young neurons develop sharp affinities and disaffinities for one another and for peripheral tissues. They reveal these properties both by entering selectively into specific cellular associations that become the interknit pathways and centers of the adult nervous system, and by selectively reestablishing some of these associations during nerve regeneration.

There is increasing evidence that the developing nervous system plays an essential role in the development of a large number of organs.

On the one hand, this role may be essentially "trophic," or nourishing, as in the maintenance of the body musculature. Cutting off the nerve supply to a skeletal muscle will cause it to wither. On the other hand, the influence may be more specific. Subtle details of muscle-cell structure are apparently determined by the particular nerves that enter the muscle. The functional properties of sensory endings, such as taste buds, are modified according to the specific neurons that innervate them.

So far as is known, such influences on peripheral tissues begin as soon as outgrowing nerve fibers reach them early in development. The question arises whether the periphery, in turn, exerts any influence in reverse on the nerve fibers that reach it. We have already seen how changes at the periphery affect the central nervous system quantitatively. Are there qualitative effects as well? Are the nerve fibers that grow into a particular muscle already different from nerve fibers entering other muscles? Do they become different only after they are in the muscle? Do axons destined to carry sensory impulses from the skin "know" in advance the local area of skin they must innervate? Or do they enter the skin more or less at random, then in effect "learn" where they are? Efforts to answer such questions have shown that peripheral tissues exert highly specific influences on their nerve fibers. As a consequence, previously indifferent nerve fibers acquire properties reflecting the precise zones in which they terminate. The fibers become specialized, each acquiring its own particular "local sign."

In support of these general statements, consider the consequences of anatomically disarranging the normal interrelations between sensory nerve fibers and the skin. If this is carried out in a suitable animal—for example, the frog—at a sufficiently early stage of larval life so that local sign properties have not already been stamped permanently on the neurons, the latter will acquire new properties in accordance with their new connections. A simple but instructive example is produced by rotating a large patch of flank skin in a frog tadpole. Turned through 180° and then allowed to heal in their new orientation, such skin patches become innervated again by the regeneration of sensory axons severed during the operation. But nerve fibers that once entered the flank skin near the back (dorsal) are now necessarily led into flank skin regions nearer the belly (ventral), and vice versa. Thus, residence in their new environments apparently imposes new properties on the regenerated neurons. Originally ventral axons now innervating dorsal skin acquire dorsal properties and presumably realign their interconnections with other neurons of the spinal cord. This may be demonstrated in the newly metamorphosed frog by irritating the dorsal part of the rotated piece (now of course, lying ventrally). The frog's characteristic wiping response to this offending stimulus will usually be misdirected; it will be aimed at the back even though the stimulus was delivered to an anatomically ventral site.

A large body of evidence supports the conclusion that muscles confer unique qualities on motor-nerve fibers, and sensory endings similarly influence sensory-nerve fibers. A surprising degree of specificity appears to be involved, extending to the level of the individual neuron in some cases.

Evidence to illustrate the extraordinary degree of this specificity in the case of central neurons may be drawn from the visual system of frogs and salamanders. Well after embryonic development is complete in these amphibians, and the neurons that subserve their vision have long since acquired specific local properties, some of the nerve cells can nevertheless regenerate if their elongate axons are cut. When they do so, they go back to their previous end station. The capacity to regenerate is especially apparent in the neurons of the retina. The axons of these cells run through the optic nerve to the midbrain, where, through contacts made with other neurons, impulses arising in response to stimulation by light are passed along to other parts of the brain. If the optic nerve is cut, its nerve fibers regenerate such that retinal neurons become reconnected with the midbrain. In so doing, small areas of the retina no larger than a few cells become selectively reassociated with equally restricted zones of the midbrain.

Strong support for this kind of statement comes from observations of the behavior of frogs or salamanders whose eyes have been surgically rotated through 180° at the same time their optic nerves were cut. When vision is recovered following nerve regeneration, the animal responds normally to the presentation of visual lures in all respects but one: it behaves as though every part of its visual field were upside down. Even more direct evidence that the regenerating axons actually reach their original end stations comes from neurophysiological studies. These show that tiny spots of light used to stimulate small areas of the retina produce recordable electrical changes in discrete zones of the midbrain. A map of this functional projection is the same after regeneration as it is in a normal, intact animal.

Taken all together, then, the evidence supports the view that neurogenesis depends heavily on those properties of neurons that enable them to enter selectively into functionally critical groupings and interconnections. The developmental events are flexible and dynamic rather than static, but a high degree of fixed structural order is attained.

### The Beginnings of Coordinated Movements

Observations and experiments on the chick embryo demonstrate the stepwise emergence of two basic components in the development of behavior. As soon as the first neuromuscular connections are established an *autonomous motor action system* begins to function. Overt, spontaneous motility can be observed as early as 3½ days of incubation, beginning with a flexion of the head. It spreads caudally along the trunk, and later wings and legs, beak, tongue, and eyelids begin spontaneous motility. In contrast to the heart, in which the early pulsations are *myogenic*, the spontaneous motility of the muscles we are discussing appears to be *neurogenic*. At first this motor action system performs in motility cyles of regular periodicity, but after about 13 days of incubation it performs almost continuously.

The *reflex apparatus* appears to develop independently, attaining functional maturity three to four days *after* the onset of spontaneous motility. At 6 to 6½ days, reflex arcs are completed; shortly thereafter, exteroceptive reflexes can be demonstrated.

The emergence of behavior patterns in teleosts is remarkably similar to that in the chick embryo. Observations on amphibians and mammals are inconclusive, but there is sufficient evidence to warrant the working hypothesis that autonomous motility is the foundation on which integrated behavior patterns are built. Reflex motility remains latent in the embryo owing to the absence of adequate stimuli; therefore, it probably does not contribute to the molding of behavior patterns before birth.

Is there a pattern to embryonic motility? We have noted that, in the chick embryo, motility starts with a flexion of the neck muscles, and that for a time a pattern of integrated cephalocaudal waves prevails. Until the 17th day of incubation, however, integrated behavior patterns cannot be recognized. There is no continuity of integration of behavior. The random activities suggest spontaneous, autonomous nerve "firings." Sensory stimulation appears to be absent as a patterning device.

On the other hand, the behavior pattern of amphibians appears to be more integrated almost from the beginning. If you have watched living frog or salamander embryos, you will have seen their first muscular contractions, at a time when their bodies are barely formed. At first only the neck region responds to touch. Later in development, one side of the body responds; next both sides of the body become involved; ultimately, fishlike swimming movements are produced.

Does this pattern develop autonomously, or is it learned, through "practice"? The questions may be answered by a simple experiment. Salamander embryos reared in an anesthetic drug develop normally, although somewhat more slowly than normal. They are paralyzed and do not exhibit motility. When control larvae of identical age have started swimming and feeding, the anesthetized animals are asleep. However, when returned to fresh water, without the drug, they recover, and at once behave like the normal controls.

In the chick embryo, coordinated movements emerge between 17 and 19 days. In these prehatching movements, the beak is lifted out of the yolk sac, and is brought into the hatching position. Hatching requires the coordination of all parts of the body. How is the transition made from unintegrated motility to integrated activities? How do exteroceptive stimuli, which appear ineffective in the chick embryo before the 17th day, take control of activity patterns?

## FURTHER READING

Balinsky, B. I., *An Introduction to Embryology*, 2d ed. Philadelphia: Saunders, 1965.

Blackler, A. W., "Transfer of Primordial Germ Cells between Two Subspecies of *Xenopus laevis*," *Journal of Embryology and Experimental Morphology*, vol. 10 (1962), p. 641.

Böving, B. G., "Anatomy of Reproduction," in *Obstetrics*, 13th ed., J. P. Greenhill, ed. Philadelphia: Saunders, 1965, p. 3.

Corner, G. W., *Ourselves Unborn*. New Haven, Conn.: Yale University Press, 1944.

———, 1963. *The Hormones in Human Reproduction* (reprint). New York: Atheneum, 1963.

Coulombre, A. J., "The Eye," in *Organogenesis*, R. L. DeHaan and H. Ursprung, eds. New York: Holt, Rinehart and Winston, 1965, p. 219.

DeHaan, R. L., "Emergence of Form and Function in the Embroyonic Heart," in *The Emergence of Order in Developing Systems*, M. Locke, ed. New York, Academic Press, 1968, p. 208.

Ebert, J. D., "The First Heartbeats," *Scientific American*, March 1959, p. 87.

Hamburger, V., "Emergence of Nervous Coordination. Origins of Integrated Behavior," in *The Emergence of Order in Developing Systems*, M. Locke, ed. New York, Academic Press, 1968, p. 251.

Harrison, R. G., "The Living Developing Nerve Fiber" (1907), reprinted in *Foundations of Experimental Embryology*, B. H. Willier and J. M. Oppenheimer, eds. Englewood Cliffs, N.J.: Prentice-Hall, 1964, p. 99.

Hörstadius, S., 1950. *The Neural Crest*. New York: Oxford University Press.

Jacobson, M., "Development of Specific Neuronal Connections," *Science*, vol. 163 (1969), p. 543.

Källén, B., "Early Morphogenesis and Pattern Formation in the Central Nervous System," in *Organogenesis*, R. L. DeHaan and H. Ursprung, eds. New York: Holt, Rinehart and Winston, 1965, p. 107.

Rawles, M. E., "Origin of Melanophores and Their Role in Development of Color Patterns in Vertebrates," *Physiological Reviews*, vol. 28 (1948), p. 383.

Saunders, J. W., Jr., and M. T. Gasseling, "Trans-filter Propagation of Apical Ectoderm Maintenance Factor in the Chick Embryo Wing Bud," *Developmental Biology*, vol. 7 (1963), p. 64.

Spemann, H., and H. Mangold, "Induction of Embryonic Primordia by Implantation of Organizers from a Different Species" (1924), reprinted in *Foundations of Experimental Embryology*, B. H. Willier, and J. M. Oppenheimer, eds. Englewood Cliffs, N.J.: Prentice-Hall, 1964, p. 144.

Torrey, T. W., *Morphogenesis of the Vertebrates*, 2d ed. New York: Wiley, 1967.

Twitty, V. C., *Of Scientists and Salamanders*. San Francisco: Freeman, 1966.

Van der Kloot, W. G., *Behavior*. New York: Holt, Rinehart and Winston, 1968.

Weiss, P., *Principles of Development*. New York: Holt, Rinehart and Winston, 1939.

———, "Nervous System (Neurogenesis)," in *Analysis of Development*, B. H. Willier, P. Weiss, and V. Hamburger, eds. Philadelphia: Saunders, 1955, p. 346.

Weston, J., "A Radioautographic Analysis of the Migration and Localization of Trunk Neural Crest Cells in the Chick," *Developmental Biology*, vol. 6 (1963), p. 279.

Witschi, E., "Migration of the Germ Cells of Human Embryos from the Yolk Sac to the Primitive Gonadal Folds," Carnegie Institution of Washington publication 575, *Contributions to Embryology*, vol. 32 (1948), p. 67.

Zwilling, E., "Limb Morphogenesis," *Advances in Morphogenesis*, vol. 1 (1961), p. 301.

*chapter* **5**

# *The Plant Embryo and Its Environment*

The cell and tissue movements described in the two preceding chapters and the new groupings and associations that they bring about are clearly essential for orderly development of the animal embryo. When they are experimentally prevented, or when they occur abnormally, development is abnormal. How, then, does embryonic development and morphogenesis occur in plants where cell movements do not occur? Plant cells are bounded by rigid walls, and the cells of the tissue are firmly bonded together by extracellular cementing substances. Cell separation is thus prevented, and although cell shape may change during growth, these changes are not reversible and do not result in movement. The presence of the walls also prevents the protoplast surfaces of adjacent cells from coming into contact, although the cells are interconnected by submicroscopic channels, the *plasmodesmata*, which traverse the walls. However, the close and elaborate contacts which the surfaces of animal cells make with one another are not possible in plants. Nonetheless, plant development involves interactions between cells and tissues. In this chapter we shall examine how embryonic development occurs in the absence of cell movement, and what kinds of interactions are involved in the development of the plant embryo.

## *PATTERNS OF EMBRYO DEVELOPMENT*

One of the most intensively studied plant embryos is that of a common, weedy flowering plant named shepherd's purse (*Capsella bursa-pastoris*). This embryo, like the embryo of all flowering plants, develops within an ovule deeply embedded in the female parts of the flower. The structure and relationships of the different tissues in the reproductive organs of higher plants have been described very clearly in *Plant Diversification*, in this series, and here we need only consider those aspects that have developmental significance for the embryo. Because the embryo is relatively inaccessible, at least when it is very small, it is difficult to make direct visual observations on it. Therefore, to follow the development of the embryo it is necessary to resort to preparation and examination of serial sections cut in various planes through embryos that have been killed and specially stained at successive stages of development. From light and electron microscopic examination of these it is possible to reconstruct three-dimensional interpretations of what the living embryo must have been like.

The gametes of *Capsella*, in common with those of all flowering plants, are structurally unspecialized at fertilization. The sperm consists simply of a haploid nucleus enclosed in a cytoplasmic sheath, and is transported through the style tissue in the pollen tube to the vicinity of the egg. The egg is a small, pear-shaped cell contained within the embryo sac in the ovule (Fig. 5-1). It consists of a haploid nucleus situated at the broad end of the cell and surrounded by cytoplasm, which, because of the scarcity of aggregated ribosomes, Golgi apparatus, and endoplasmic reticulum, appears to be metabolically inactive; it does not contain reserve nutrients.

Soon after fertilization, extensive changes take place in the cytoplasmic organization of the zygote. New ribosomes begin to be synthesized; these, and the preexisting maternal ribosomes, become aggregated into polysomes. The endoplasmic reticulum and the Golgi system become more abundant. The wall, which up to this time had been very thin, becomes increasingly thick, and many of the plasmodes-

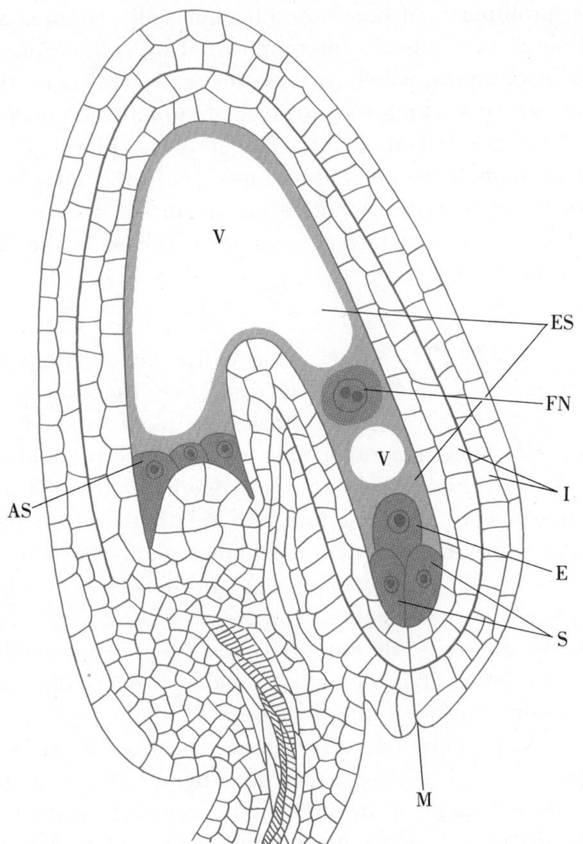


*Fig. 5-1* *Ovule of* Capsella *showing the embryo sac just before fertilization. Two integuments* (I) *form the outer cell layers of the ovule. The tips of the integuments come together to form a micropyle* (M) *through which the pollen tube will grow as it enters the ovule. The embryo sac* (ES) *consists of eight cells or nuclei. At the end nearest the micropyle are two synergid cells* (S) *and the larger egg cell, the female gamete* (E). *In the center of the embryo sac is the fusion nucleus* (FN) *which is formed by fusion of two haploid nuclei. This nucleus will fuse with one of the male nuclei during fertilization to initiate the endosperm development. At the basal end of the curved embryo sac are three vegetative antipodal cells* (AS). *The cytoplasm of the embryo sac is vacuolated and several vacuoles* (V) *are visible in this section.*

mata become blocked off, thereby increasing the physiological isolation of the protoplast from the surrounding cells. Following the first nuclear division the embryo is separated into two unequal cells: a small, terminal cell at the rounded end of the embryo and a larger, vacuolated cell at the tapered basal end (Fig. 5-2). These two cells are structurally differentiated from one another and have different developmental fates. The basal cell divides transversely and forms a linear *suspensor*,

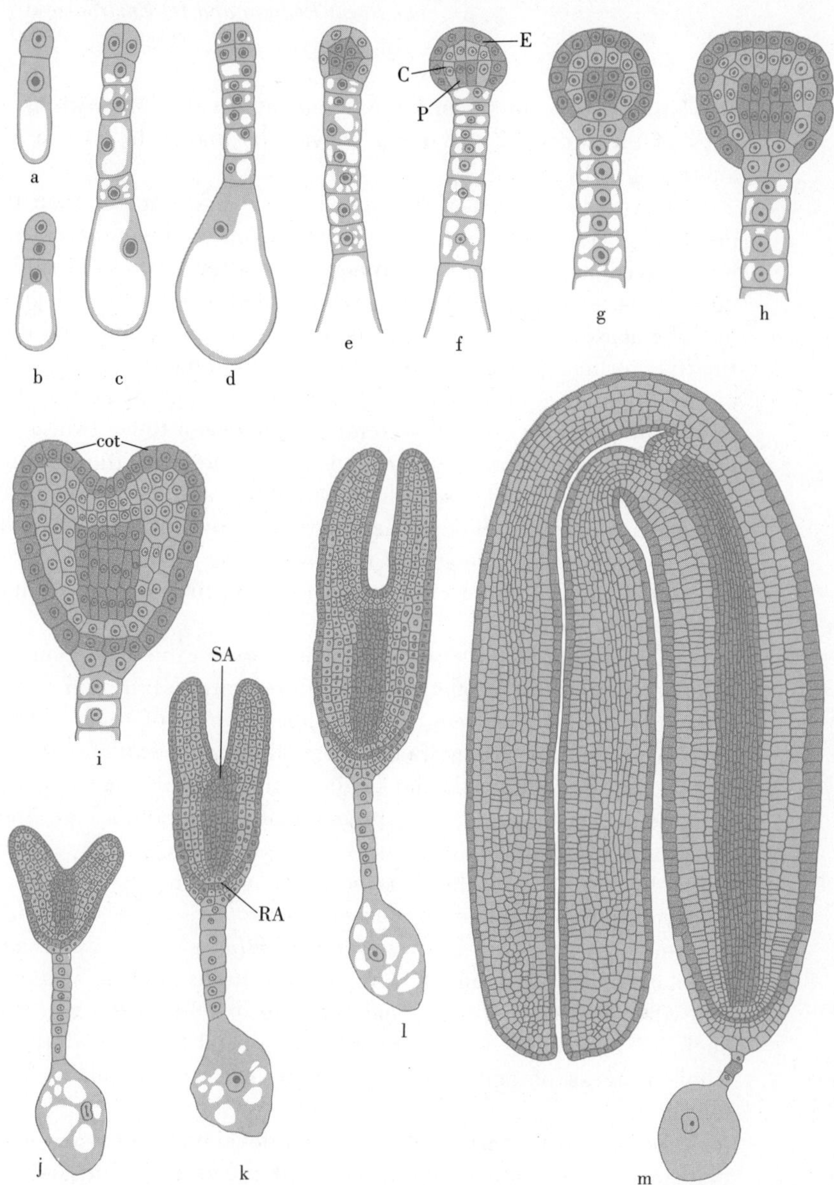


*Fig. 5-2 Embryo development in* Capsella. *The first division divides the zygote into two unequal cells* (a). *The lower cell forms the uniseriate suspensor* (b–d) *while the terminal cell forms the globular part of the embryo* (c–f). *In the globular embryo the three primary tissue systems are differentiated: a surface epidermis* (E), *procambium* (P), *and intervening cortex* (C) (f–h). *Subsequently the cotyledons* (cot) *are differentiated producing the heart-stage embryo* (i). *In later stages of development the terminal meristems are delineated: the shoot apex* (SA), *and the root apex* (RA) (j–l). *The mature, dormant embryo is curved to fit into the curvature of the ovule* (m). (a–i *from A. W. Haupt,* Plant Morphology, *by courtesy of the author and the McGraw-Hill Book Co.;* j–m *from M. Schaffner,* The Ohio Naturalist, *7:6, by courtesy of the editors of* The Ohio Journal of Science.)

consisting of five to seven large, vacuolated cells. Meanwhile, derivatives of the terminal cell divide in various planes to form a globular mass of small cells.

The globular part of the embryo is at first smaller than the suspensor, but it becomes much larger and forms all the embryonic organs. Early divisions in it are synchronous, but after a time the rate of cell division varies in different parts. When about 50 cells have been formed in the *globular stage embryo*, three tissues begin to differentiate: a distinctive surface layer of cells, which will produce the *epidermal tissue* of the plant; a central core of elongated *procambial cells*, which are precursors of the vascular system; and, between these two, a cylinder of cells that will form the *cortex*. A short time later, the terminal part of the embryo loses its radial symmetry and becomes bilaterally symmetrical as two hemispherical mounds grow out from the end opposite the suspensor. These are the *cotyledons*, or embryo leaves, and the shape they impart to the embryo at this time has led to its being called the *heart stage*.

At the pole of the embryonic axis between the cotyledons a group of cells remains in a relatively undifferentiated state and these form the *shoot apical meristem*. At the opposite pole of the axis, near the suspensor, another group of cells forms the *root apical meristem*. These two meristems, which do not begin to function until germination, are situated at the two ends of the procambial system and are progressively separated from one another by continued growth of the intervening embryonic axis, called the *hypocotyl*. Because of its shape the elongated embryo has been called the *torpedo stage*. The ovule cavity in which the embryo is growing is curved in *Capsella*; as the embryo continues to enlarge, the cotyledons become curved to fit into this space. By this time the seed is maturing, and the rate of embryo growth gradually declines, and finally is terminated as the seed becomes dormant. The whole process of embryo development in *Capsella* takes only about ten days.

Before investigating plant embryo development further we should first compare the embryo of *Capsella* with the animal embryos described in previous chapters. There is nothing comparable to a fertilization membrane formed in the plant zygote cytoplasm, so multiple fertilization must be prevented in other ways, although how this is done is not known. The changing shape of the plant embryo results entirely from the formation and growth of cells and there is no cell migration. In the absence of a stage of development comparable to gastrulation, the cells develop in the place where they are formed. This severely limits the kinds of interactions in which plant cells can participate. The fully developed embryo of *Capsella* does not possess all the organs that will be present in the adult plant. The shoot system and the root system are represented only by terminal meristems and there are no vegetative leaves, lateral shoots or roots, or any of the reproductive structures. These will be initiated only after germination. Finally, there is no stage comparable to cleavage, because the plant embryo begins to enlarge and differentiate soon after fertilization. However, the zygote is the largest cell of the plant embryo, and all later-formed cells are smaller because the rate of division is greater than the rate of cell growth. The effect is to partition the embryo into a relatively large number of small cells, a situation not unlike the blastula of animal embryos.

How general is the pattern of plant embryo development that we have just investigated? The sequence of early cell divisions is very regular in some species, and variable in others. The size of the globular embryo and the number of cells formed before organs and tissues begin to differentiate differs, and the extent of embryo development before the seed becomes dormant is quite variable. However, in the development of all plant embryos there are features that suggest that similar developmental processes are involved.

In the gymnosperms, where the embryo also develops enclosed in an ovule, the first nuclear divisions are not accompanied by wall formation; therefore, the young embryo is coenocytic, much like the early stages of insect embryos. Several thousand free nuclei may be formed, but in pine the coenocytic stage is relatively brief, and wall formation begins in the embryo when there are four nuclei (Fig. 5-3). The first four cells are located at one end of the egg cell, where two further synchronous divisions occur, producing 16 cells in four equal tiers. At this time the embryo begins to enlarge because the four cells of the second or suspensor tier expand and push the terminal cells into the surrounding tissues of the ovule. The terminal cells form a globular cell mass within which distinctive tissue systems and organs similar to those of *Capsella* differentiate. Thus, here also there is early differentiation of the embryo into a terminal small-celled organogenetic region and a region of larger suspensor cells.

In mosses and ferns the embryo develops in a free-living gametophyte and not in a nutrient-rich ovule, and perhaps the patterns of embryo development we have studied in the seed plants are adaptations to the abundant supply of nutrients surrounding the embryo. However, the development of embryos in mosses and ferns is remarkably like that in seed plants, the embryos soon becoming differentiated into two regions: one of enlarged vacuolated cells, called the *foot*, which resembles the suspensor, and which is thought to function in the transfer of nutrients into the embryo; and a small-celled region, within which the embryonic organs are differentiated (Fig. 5-4). In the moss embryo the only organ to be formed is the capsule, but the fern embryo

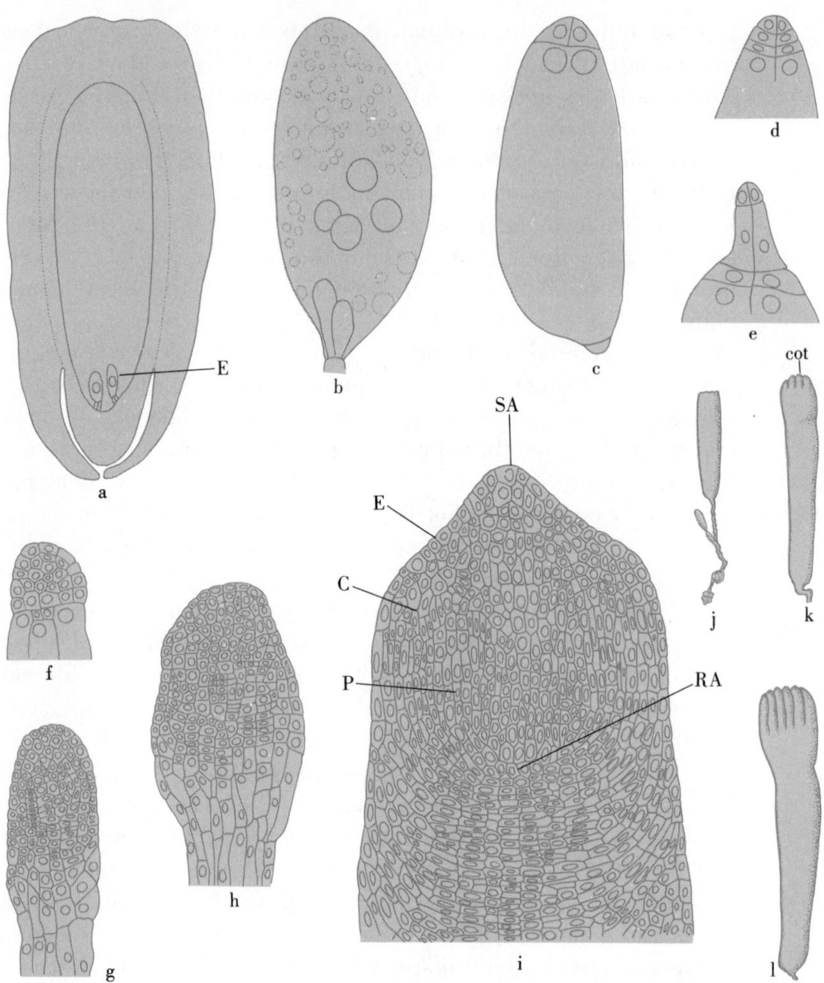


**Fig. 5-3** *Embryo development in* Pinus. (a) *The ovule just before fertilization containing two female gametes* (E), *the eggs. The first divisions of the embryo nuclei occur without wall formation* (b). *The nuclei subsequently move to the end of the cell opposite the micropyle and wall formation begins* (c–d). *The four suspensor cells start to elongate, pushing the terminal cells deeper into the nutritive tissues of the ovule* (e). *The terminal cells divide to form a globular embryo* (f–h) *in which a surface epidermis* (E), *underlying cortex* (C), *and procambium* (P) *are differentiated. Terminal meristems of the shoot* (SA) *and root* (RA) *are also delimited* (i). *Cotyledons* (cot) *are differentiated in the enlarging embryo* (j–l). (b–e *from J. M. Coulter and C. J. Chamberlain,* Morphology of Gymnosperms, *1917, by courtesy of the University of Chicago Press;* f–h, j–l *from R. N. Konar and S. Ramchandani,* Phytomorphology, *8:328, by courtesy of the authors and the International Society of Plant Morphologists;* i *from A. R. Spurr,* The American Journal of Botany, *36:634, by courtesy of the author and the Botanical Society of America.)*

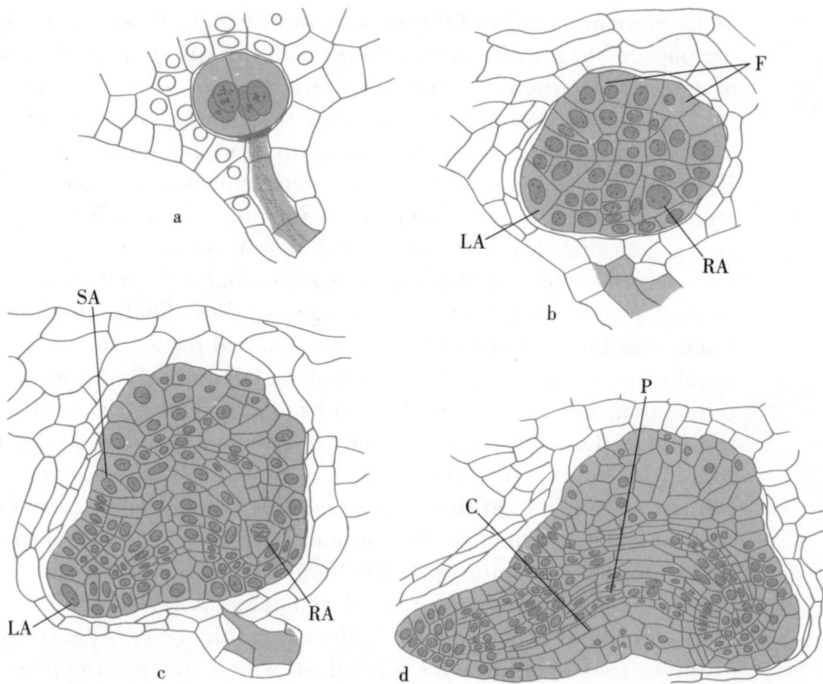


*Fig. 5-4 Embryo development in* Gymnogramme. *The female sex organ, the archegonium, contains a four-cell embryo* (a). *One of the walls is in the plane of the page and, therefore, does not show in the illustration. In the embryo the foot* (F) *is differentiated first, and subsequently the apical cells of the first leaf* (LA) *and the root* (RA) *enlarge* (b). *The apical cell of the shoot* (SA) *develops later and differentiation of the internal tissues, the cortex* (C) *and procambium* (P) *becomes more pronounced* (c). *As the embryo enlarges it breaks out of the surrounding archegonial wall and its growth continues without a period of dormancy* (d). *(From M. A. Vladesco,* Revue Genérale de Botanique, *47:422, by courtesy of the editors.)*

contains the terminal meristems of shoot and root, the first leaf, and the same three tissue systems that were differentiated in embryos of seed plants. These embryos are distinctive in that their growth is not interrupted by a period of dormancy, but development is continuous into postembryonic stages.

## EMBRYONIC POLARITY

How does the fertilized egg develop into the bipolar embryo possessing in plants a shoot pole and a root pole, and in animals a head end and a tail end? This question has occupied the time of many developmental biologists, and still we are unsure of the answer. Increasingly, however, it is becoming possible to provide partial answers to this question, and each piece of evidence sharpens the questions that remain.

A purely descriptive approach to polarity has limitations and answers are most likely to come from well-planned experiments. However, it is not easy to conduct experiments on the zygotes of higher plants, which are so deeply embedded in other tissues. What is needed is a zygote that develops outside the parent organism. Zygotes of this type, which occur in amphibia and also in some littoral brown seaweeds such as *Fucus*, have been used in many studies of polarity. Gametes of *Fucus* can be collected in large numbers, and fertilization and embryo development can easily be followed in laboratory cultures. Prior to fertilization the egg is spherical and radially symmetrical, with a centrally placed nucleus and uniformly distributed cytoplasmic organelles. About 15 hours after fertilization a tubular *rhizoidal outgrowth* is formed on one side of the zygote, and this is followed in another eight or nine hours by nuclear division and formation of a wall perpendicular to the rhizoidal outgrowth (Fig. 5-5). The zygote is thus cut into two unequal cells having different developmental fates, a situation remarkably like that in *Capsella*. Both cells continue to divide. Derivatives of the rhizoidal cell produce a holdfast, which attaches the plant to the substrate, and those of the terminal cell form the frondlike body. The developmental axis is formed before the first nuclear division occurs, and is independent of division. The rhizoid forms normally in zygotes treated with colchicine, which blocks mitosis by preventing the assembly of spindle fiber microtubules. Polarity cannot, therefore, depend on nuclear differences in different cells, but must result from localized changes within the cytoplasm of individual cells. How are such changes induced in the cytoplasm of a *Fucus* zygote?

When *Fucus* embryos develop in darkness the rhizoids arise at random positions, but if the embryos are very close together the rhizoids all develop on the side toward the center of the group. The position of the rhizoid, and therefore the axis of polarity, apparently can be influenced by factors external to the embryo. A number of environmental factors are known to interact with the zygote affecting the position at which the rhizoid emerges, and what is important is that these factors are effective only when they are present as a gradient across the zygote. Thus the rhizoid is formed on the warm side in a temperature gradient, on the more acid side in a pH gradient, and on the shaded side in a gradient of white light. It is not necessary to continue the gradients indefinitely; once the cell has become visibly asymmetric in a light gradient, for example, its original polarity will be preserved if it is then placed in darkness or is illuminated by light coming from a new direction.

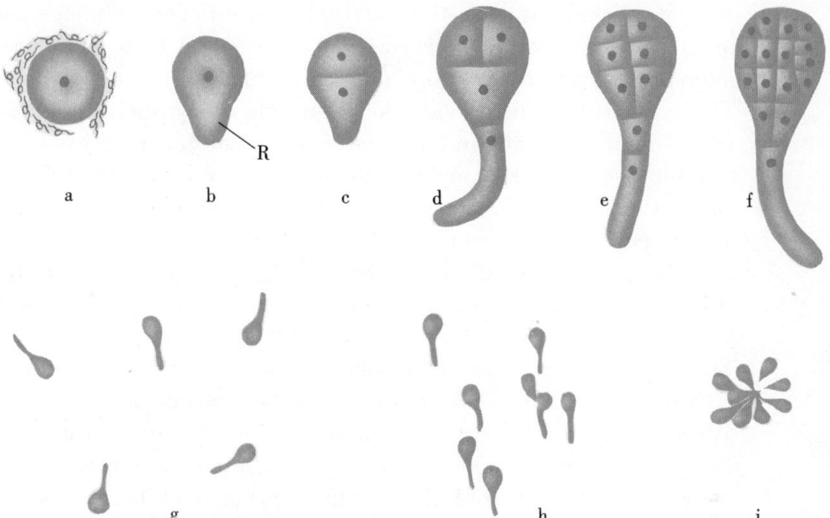


**Fig. 5-5** *Development and growth responses of the embryo of* Fucus. (a) *The nonmotile radially symmetrical egg is shown surrounded by numerous small, motile sperm at the time of fertilization. Early developmental change in the zygote is seen in the outgrowth of the rhizoid* (R) *before the nucleus has divided* (b). *After division the two cells undergo numerous divisions so that newly formed cells are each smaller in size than the zygote* (c–f). *When embryos develop in darkness the rhizoids in a population are randomly oriented* (g), *but if illuminated from one side (here from the direction of the top of the page) the rhizoids develop predominantly from the shaded side* (h). *When embryos develop in closely spaced groups the rhizoids tend to develop toward the interior of the group (the so-called "group effect") regardless of other conditions* (i). (a–f *from G. M. Smith,* Cryptogamic Botany, *1955, volume 1, by courtesy of The McGraw-Hill Book Co.)*

The environmental determinants seem to act as developmental triggers activating some intracellular mechanism which is insensitive to further environmental modification. What is the nature of the changes that occur in the cytoplasm of a polarized *Fucus* cell? It is here that the experimental evidence becomes less conclusive. There is some evidence that cytoplasmic vesicles accumulate in the region which will grow out as the rhizoid; and at this time an intracellular electrical gradient coinciding with the axis of polarity is established across the zygote. It is tempting to speculate that this gradient results in movement of vesicles through the cytoplasm, and that the accumulation of vesicles at a particular site is involved in the outgrowth of the rhizoid, but more evidence will be required to confirm this. However, the conclusion that polarity in *Fucus* zygotes is environmentally determined is well-established. Can this example provide us with a model for the determination of polarity in other plant embryos?

It is unlikely that the external environment regulates development of the zygote of flowering plants, and the embryo seems to be influenced more by the surrounding tissues of the developing seed. These tissues, then, provide the environment in which the embryo develops, and although we have previously analyzed patterns of embryo development without consideration of the surrounding tissues, we must now pay attention to the interactions that occur between the embryo and its cellular environment.

Even before fertilization the egg of *Caspella* has a polarized distribution of cytoplasmic organelles, which becomes accentuated after fertilization as organelles accumulate in different parts of the cell. The origin of polarity in higher plant embryos is unknown, but it is quite possibly caused by hormonal gradients present in the developing ovule and seed. Pollination triggers hormone production in the ovary. The *endosperm* surrounding the embryo is also an intense center of hormone synthesis, and three different kinds of hormones—*auxins*, *gibberellins*, and *cytokinins*, are known to occur in it. Because of the very small size of the young ovule it has not been possible to test it for hormonal gradients, which might be effective in determining polarity in the egg or the zygote; however, other kinds of experiments have shown conclusively that internal gradients of hormones are important in determining the bipolar pattern of development. One of the clearest of these is an experiment involving the regeneration of new shoot and root meristems in detached plant organs.

In an isolated piece of horseradish root, new shoots are formed at the original shoot end and roots form at the other end. The positions of these organs are independent of gravity or other external factors, and are related to internal hormonal gradients, shoots being formed at the end at which the concentration of cytokinin is highest and roots forming at the end at which the auxin concentration is highest.

It has been shown that the chemical environment in which organ initiation occurs is actually more complicated than the foregoing discussion might imply. In experiments with tobacco in Skoog's laboratory, pieces of pith were removed from the stem and grown in sterile culture on media containing different concentrations of *indoleacetic acid*, an auxin, and *kinetin*, a cytokinin, in addition to the usual inorganic salts, sucrose, and vitamins. Shoots were formed when the auxin-kinetin balance favored the kinetin, and roots formed when the balance favored the auxin (Fig. 5-6). When the concentration of kinetin was low, raising the auxin concentration increased the number of roots that were formed. As the concentration of kinetin was raised to intermediate levels, this same concentration of auxin resulted in the growth of undifferentiated tissue; at the highest concentration of kinetin tested, the same auxin concentration resulted in formation of shoots. These

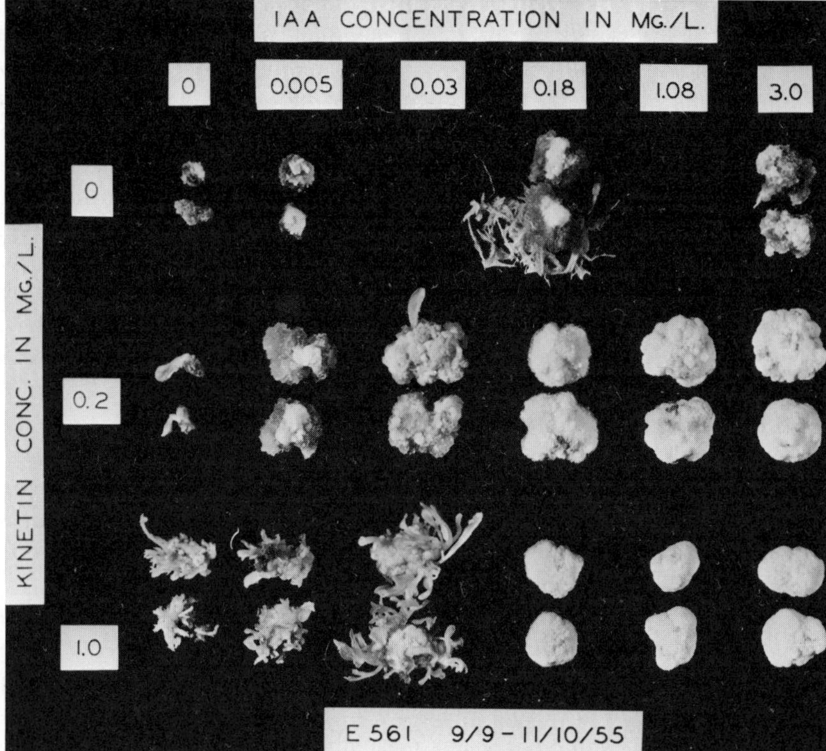


*Fig. 5-6 The effect of altering the concentration of auxin and kinetin on growth and organ formation of tobacco callus in sterile culture. Roots have been developed from tissue grown in culture media containing high concentrations of auxin and no kinetin. Shoots have been developed from tissue grown in low concentrations of auxin and high concentrations of kinetin. When the relative concentrations of the two substances are approximately balanced organs have not formed but the tissue has proliferated vigorously. (From F. Skoog and C. O. Miller,* Symposium of the Society for Experimental Biology, *11:118, 1957, by permission of the authors and the Society for Experimental Biology.)*

results indicate not only that these two substances are necessary for organ formation, but that their relative concentrations are important in determining what kind of organ is differentiated or indeed, whether organs are formed at all.

Certainly these experiments do not tell us that hormonal gradients are the determinants of the shoot and root meristems at opposite poles of the developing embryo, but they provide us with model systems in which hormonal gradients and balances are shown to be effective, and thus suggest further experimental approaches to embryo polarity.

***EMBRYO CULTURE*** There is direct evidence that the tissues surrounding the embryo influence its development and may have a determining effect on it. In flowering plants most of the nutrients used by the embryo for its growth accumulate in the endosperm, a tissue that completely surrounds the embryo. As the embryo grows, it digests the endosperm and enlarges to fill the cavity it creates. The endosperm is initiated after fertilization when a second sperm nucleus carried in the pollen tube enters the embryo sac and fuses with one or more centrally placed nuclei. The early development of this tissue is much more rapid than that of the embryo, and the endosperm usually consists of many cells before the zygote divides. In some species endosperm is cellular from the start, but in others, including *Capsella*, the first nuclear divisions are not followed by wall formation and a coenocytic liquid endosperm is formed. When wall formation does begin it may be complete—as happens in corn when the kernels pass from the milk stage to the starchy stage. Alternatively, walls may be formed in only part of the endosperm, leaving part as a fluid—as happens in the coconut, where the milk is actually liquid endosperm.

The critical importance of endosperm for the development of the embryo was first shown in crosses between different species that yield nonviable seeds. Although there are many causes of interspecific infertility it was sometimes found that fertilization occurred and the embryo began to develop normally but then died. In these crosses endosperm development was abnormal. In some it developed very slowly; in others it failed to accumulate nutrients; and in many the endosperm cells collapsed and died. In all cases developmental abnormality in the endosperm preceded that in the embryo, and there was close correlation between the extent of endosperm development and the stage at which the embryo died.

If death of the embryo is due simply to an inadequate supply of endosperm-contained nutrients it should be possible to remove the immature hybrid embryos from seeds and grow them to maturity on an adequate nutrient supply. This *embryo culture* technique has been used successfully to obtain hybrid plants from many otherwise infertile crosses. More importantly for our purposes it has provided a method by which the effect of nutrients and hormones contained in the endosperm on development of the embryo can be analyzed.

Fully developed embryos removed from dormant seeds and grown in embryo culture in the light on a nutrient medium containing only inorganic salts develop directly into seedlings (Fig. 5-7). Like adult green plants these embryos are autotrophic being able to synthesize organic molecules from the simple inorganic constituents of their environment. Younger embryos fail to grow on this simple medium and

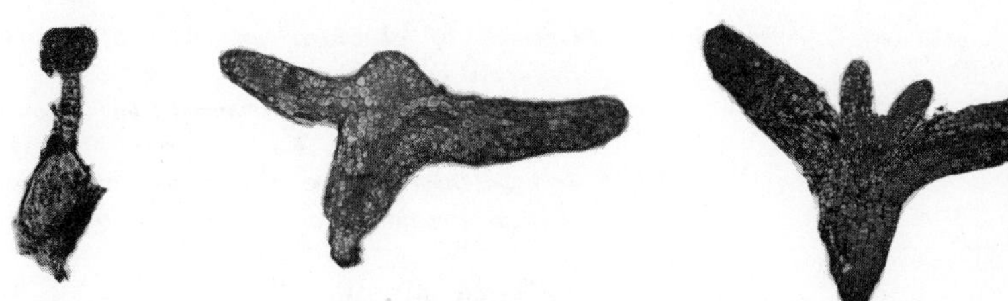

**Fig. 5-7** *Embryo culture of* Capsella. *Embryos were removed from the ovule at the early heart stage and grown on sterile nutrient culture media where they differentiated normal tissues and organs. The proportions of the parts of cultured embryos differ from those of embryos which develop completely within the ovule (compare these embryos with those in Fig. 5-2).* (Left) *An embryo which has just been removed from the ovule prior to transfer to a nutrient culture medium.* (Center) *An embryo after five weeks of growth.* (Right) *The first vegetative leaves have differentiated around the shoot apex and development as a seedling is commencing.* × *65. (From V. Raghavan and J. G. Torrey,* The American Journal of Botany, *50:540, 1963, by courtesy of the authors and the Botanical Society of America.)*

require various supplements before they will develop. In experiments with *Datura* embryos it has been found that the optimal sugar requirement changes during development. Globular embryos require 8–12 percent sucrose, heart stage embryos 4 percent, and torpedo stage embryos 0.1 percent. In addition, the medium must be supplemented with several vitamins and amino acids for torpedo-stage embryos to grow optimally, and heart-stage embryos require the further addition of coconut milk, which was selected because it was assumed to contain many embryo-growth-promoting factors.

What kinds of growth promoting factors are present in coconut milk? Chemical and biological analyses have revealed a wide range of substances which affect plant growth. These include sugars and sugar alcohols, organic acids, vitamins, many nitrogenous compounds, and auxins, cytokinins, and gibberellins. That some of these latter are important in the regulation of embryo growth was shown in experiments by Raghavan and Torrey. They grew very young globular embryos of *Capsella*, consisting of 16–32 cells and showing no evidence of internal tissue differentiation or formation of meristems or organs when they were removed from the seed, on culture media containing indoleacetic acid, kinetin, and adenine (Table 5-1). All of these substances were required for optimal growth, and all had previously been shown to affect meristem initiation in tobacco tissue cultures.

How do these experiments relate to normal embryo development? The youngest embryos seem to be completely heterotrophic, depending on the endosperm not only for organic nutrients but also for hormonal substances that determine the bipolar developmental pattern. As the

*Table 5-1* ***Growth of globular embryos of* Capsella *in different culture media.***
*(When the embryos were removed from the ovule their average length was 54 μ. The results indicate that each additive results in improvement of growth or survival over that obtained in the preceding media.)*

| *Culture medium* | *% size increase over initial length after 10 days* | *No. of embryos that grew* | *No. of embryos cultured* |
|---|---|---|---|
| Basal medium | 75.0 | 1 | 24 |
| Basal + indoleacetic acid 0.1 mg/1 | 25.0 | 2 | 8 |
| Basal + IAA 0.1 mg/1. + kinetin 0.001 mg/1 | 23.0 | 5 | 8 |
| Basal + IAA 0.1 mg/1. + K 0.001 mg/1. + adenine sulfate 0.001 mg/1 | 154.2 | 13 | 14 |

Data from V. Raghavan and J. G. Torrey, *American Journal of Botany, 50:540, 1963.*

embryo develops morphologically it also develops biochemically, and thus is able to synthesize a progressively greater array of organic molecules. Or to put this another way, during embryo development there is progressive activation of the metabolic machinery to synthesize the hormones, vitamins, and carbohydrates required for development. The conclusion is strengthened by another experiment on globular *Capsella* embryos. Although these embryos had been found to be dependent on externally supplied auxin, kinetin, and adenine, this requirement was completely abolished if the concentration of inorganic salts or the osmotic concentration of the nutrient medium was raised. From experiments with other plant tissues growing in sterile culture it is known that increasing the salt concentration of the culture medium activates metabolic pathways that are otherwise inactive, and possibly in the young embryos a similar effect occurs. The osmotic concentration and salt content of endosperm is high, and it is an interesting possibility that in addition to its role in supplying nutrients and hormones to the developing embryo the endosperm activates metabolic pathways that make the embryo increasingly autotrophic in its later developmental stages.

So far we have limited this analysis of embryo development to the flowering plants. The relationship of embryo to surrounding tissues is very different in mosses and ferns. Here embryonic development is

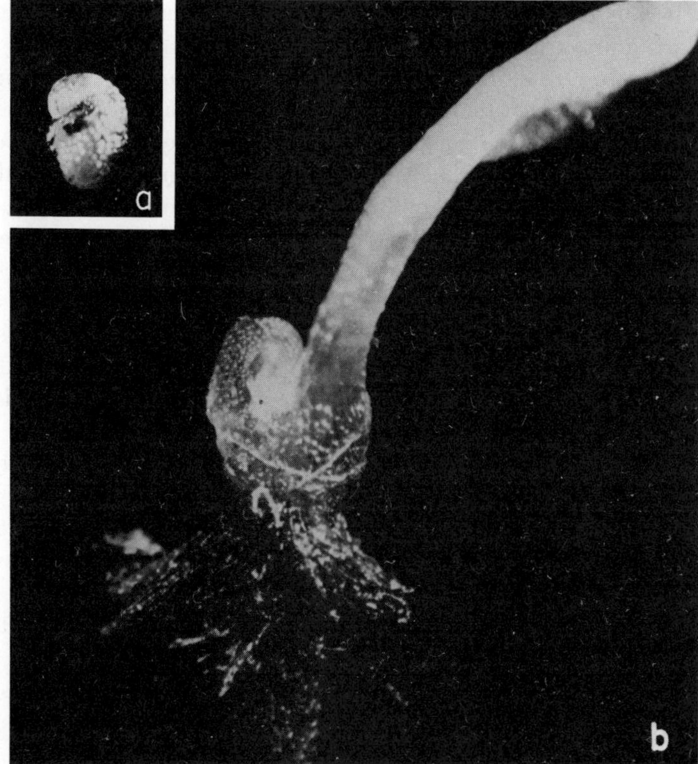


*Fig. 5-8* *Embryo culture of* Todea. *The embryo in* (a) *was removed from the archegonium 17 days after fertilization, at which time only the foot was differentiated, and cultured on a simple medium for one month. Its growth has been relatively poor. The embryo in* (b) *was removed at 20 days after fertilization, when organ differentiation has been initiated, and grown for one month on the simple culture medium. It has grown much more than the 17-day embryo. × 26. (From A. E. DeMaggio and R. H. Wetmore,* The American Journal of Botany, *48:551, 1961, by courtesy of the authors and the Botanical Society of America.)*

completed entirely within the female sex organ, the archegonium, on an independent, autotrophic gametophyte that does not accumulate nutrients in a specialized tissue. Do interactions between the embryo and the surrounding cells in these plants regulate development also? Embryo culture experiments have shown that the biochemical development of the embryo of the fern *Todea* is surprisingly like that of flowering plants. Embryos removed from the archegonium when all the organs were formed grow rapidly on very simple culture media, whereas those removed when only the foot is differentiated require coconut milk for continued development (Fig. 5-8). The young embryo is thus initially dependent on the surrounding gametophytic tissues for a supply of regulatory substances.

In fern embryo development the surrounding tissues seem to play a further role, which is apparently unimportant in the flowering plants, where the very young embryo lies freely suspended in liquid endosperm. The cells of the archegonial wall surrounding the fern embryo divide and the wall becomes multilayered. As the embryo enlarges, it displaces this multilayered wall and the adjacent gametophyte cells without digesting them. The embryo is, therefore, probably under considerable and increasing physical constraint as it enlarges. Does this force play any part in regulating development of the embryo? This has been studied by Wetmore and his students. If the archegonial walls are carefully cut away to relieve some of the pressure, development is abnormal. The embryo grows as a number of nodular or cylindrical outgrowths and does not form organs at the usual time. Organ formation occurs later at the ends of the outgrowths, and in this way a single embryo can give rise to several plants (Fig. 5-9). An even greater alteration of the developmental pattern occurs when *Todea* zygotes are removed from the archegonium and grown in culture. They do not form embryos at all but develop as flattened two-dimensional structures that resemble

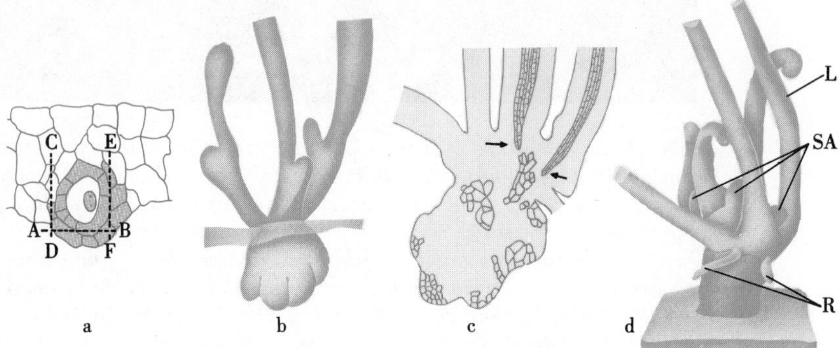


*Fig. 5-9 Development of fern embryos under conditions of reduced physical restraint. The experimental technique involves making a series of incisions in the gametophyte tissues which closely surround the young developing embryo. These incisions are shown in* (a). *The incision A–B removes the surface of the archegonium. Incisions in the planes C–D and E–F cut into the sides of the archegonium. In* (b) *and* (c) *are shown surface view and internal tissue section respectively of the embryo of* Phlebodium *after removal of tissue by an incision in the plane A–B. The embryo has formed several lobes from which outgrowths have developed.* (d) *A comparable experiment on the embryo of* Thelypteris *has resulted in the formation of numerous shoot apices instead of the one which develops in normal embryos* (SA). *Numerous leaves* (L) *and roots* (R) *have also been formed on the abnormal embryo.* (a–c *from M. Ward and R. H. Wetmore,* The American Journal of Botany, *41:428; 1954, by courtesy of the authors and the Botanical Society of America;* d *from R. D. E. Jayasekera and P. R. Bell,* Planta, *54:1, 1959, by courtesy of the authors and Springer–Verlag.)*

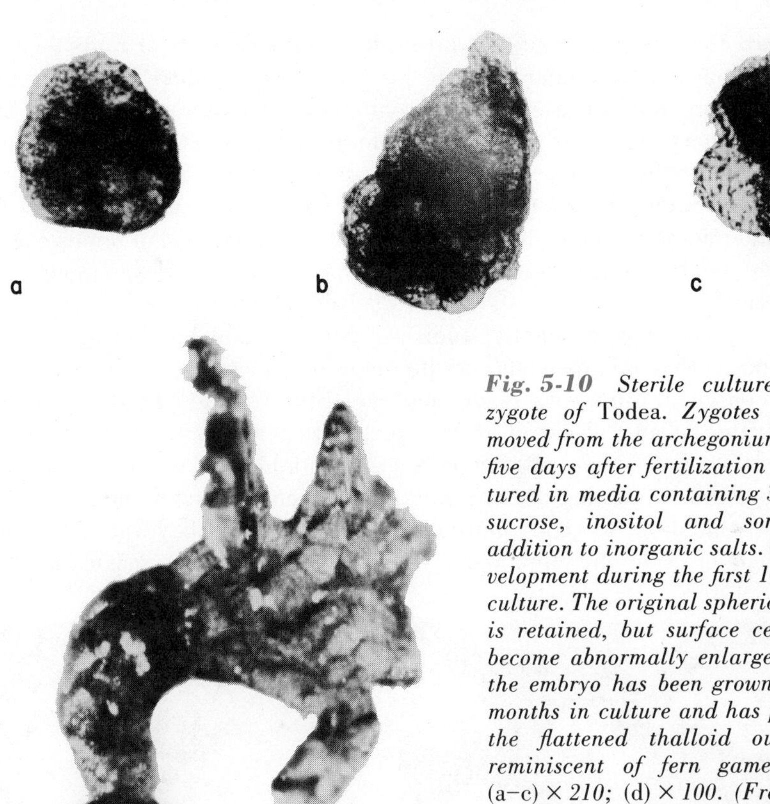


*Fig. 5-10 Sterile culture of the zygote of* Todea. *Zygotes were removed from the archegonium four or five days after fertilization and cultured in media containing 3 percent sucrose, inositol and sorbitol in addition to inorganic salts.* (a–c) *Development during the first 14 days of culture. The original spherical shape is retained, but surface cells often become abnormally enlarged. In* (d) *the embryo has been grown for four months in culture and has produced the flattened thalloid outgrowths reminiscent of fern gametophytes.* (a–c) × *210;* (d) × *100. (From A. E. DeMaggio and R. H. Wetmore,* The American Journal of Botany, *48:551, 1961, by courtesy of the authors and the Botanical Society of America.)*

immature gametophytes (Fig. 5-10). It is probable that these experimental treatments alter the nutritional and hormonal relationships between the embryo and the surrounding tissues; it is not known how much of the effect should be attributed to these changes and how much to altered physical pressure. However, the experiments extend the conclusion that development of the plant embryo involves interactions between it and the surrounding cells.

## VEGETATIVE EMBRYOIDS

If endosperm provides an optimal environment for development of embryos, would natural endosperms or synthetic substitutes act on differentiated, mature vegetative cells of the organism directing them to develop as the zygote does? Previously we saw that tobacco tissue cultures initiate either shoots or roots depending on the particular balance of hormones supplied in the culture medium, but will a single differentiated cell produce both shoot and root meristems and thus give rise to an entire plant? If differentiation does not involve irreversible changes in the DNA this should be possible. Such an experiment would provide important information for understanding how the developmental potentialities of mature cells are held in check so that orderly and integrated development can occur.

The morphogenetic studies of Steward and his colleagues have shown that mature cells can be made to repeat stages in embryo development, often with surprising regularity. They grew tissue from cultivated strains of carrot or from the wild carrot—both of which belong to the same species, *Daucus carota*—in rotating flasks in a liquid culture medium containing coconut milk. As the cells divided and grew some became detached and continued to develop while they were freely suspended in the medium. When the suspension, which consisted of some

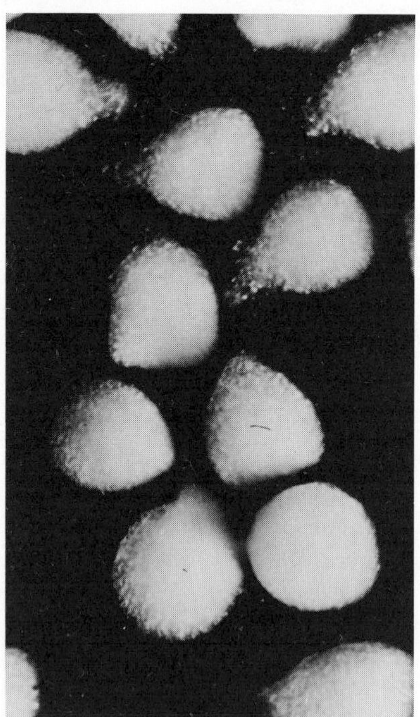

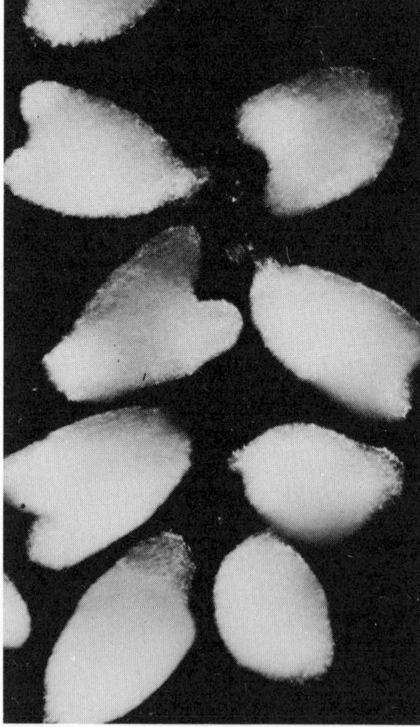

**Fig. 5-11** *Vegetative embryoids of carrot.* (Left) *Globular stage embryoids.* (Right) *Heart stage embryoids. (From W. Halperin,* The American Journal of Botany, *53:443, 1966, by courtesy of the author and the Botanical Society of America.)*

single cells and some cell clusters, was spread on the surface of an agar medium almost all of the cellular units gave rise to embryolike structures, which progressed from globular to heart and torpedo stages (Fig. 5-11). Because of the similarity between these and normal embryos they have been called *embryoids.* Embryoids were formed most abundantly from suspensions made from wild carrot embryos, but mature cells of the petiole and root also formed them (Fig. 5-12). Several other species of flowering plants and gymnosperms have now been found to produce embryoids, and haploid embryoids have even been produced from pollen of tobacco and *Datura.*

It will be recalled that during the early development of *Capsella* the embryo became increasingly isolated from surrounding cells by blockage of the plasmodesmata, and it was thought that the physical isolation of cells in carrot suspension cultures may have been a necessary condition for embryoid initiation. However, some embryoids seem to arise from groups of cells, and in suspension cultures the development of individual cells cannot be followed because they are grown

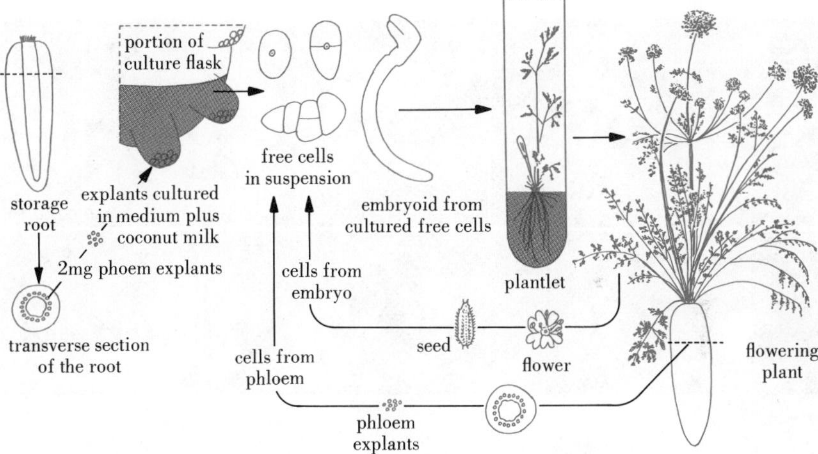


*Fig. 5-12 Production of embryoids and whole plants from tissue cultures of carrot. On the left side of the illustration a carrot root has been used to obtain explants of phloem tissue which are then cultured in a liquid medium in slowly rotating flasks. Here the tissues proliferate and single cells and aggregates of cells separate and are released into the medium. Some of these continue to develop producing roots, shoots, and embryoids which can be transferred individually into test-tubes and then into soil where they grow as typical carrot plants producing enlarged tap-roots and flowers. On the right side of the illustration such a plant obtained from tissue culture is shown as the source of an embryo which was placed in sterile culture to proliferate and to repeat the cycle of embryoid production to whole flowering plant. (From F. C. Steward et al.* Science, *143:20-27, no. 3601, 1964, copyright 1964 by the American Association for the Advancement of Science, by courtesy of F. C. Steward and the American Association for the Advancement of Science.)*

under conditions wherein microscopic observation is not possible. In tobacco, however, there is conclusive proof that entire plants arise from single cells (Fig. 5-13). Single tobacco cells from a suspension were transferred to a drop of nutrient medium on a microscope slide and covered to exclude microorganism contaminants. The slide was then maintained on the stage of a microscope and photographed periodically. Under these conditions many of the cells were seen to divide, producing small tissue masses, which were subsequently transferred to larger culture vessels containing media with both auxin and kinetin. Roots and shoots were initiated on the tissues and developed into whole plants, which since were transferred to soil, where they completed their life cycle by flowering and setting viable seed.

***SEED DORMANCY*** Embryos grown in embryo culture, as well as vegetative embryoids, develop directly into seedlings. Their growth is not interrupted by dormancy, which

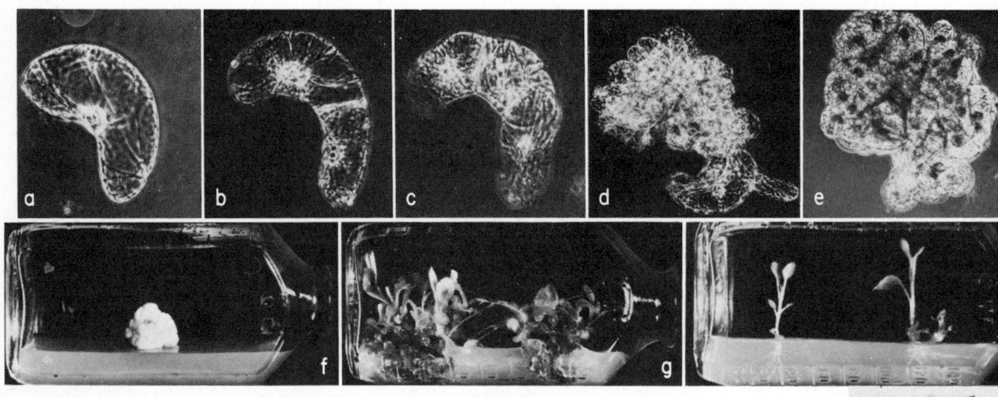

***Fig. 5-13*** *Development of tobacco plants from single tissue culture cells. In this technique single cells are removed aseptically from liquid suspension cultures and grown in microcultures where they can be continuously observed and photographed. In microculture the cells undergo division* (a–e) *forming small callus masses which can be transferred to the surface of an agar culture medium* (f). *After further growth on a medium containing both auxin and kinetin shoots and roots are differentiated* (g). *Young plants are then transferred to fresh culture medium* (h) *and when large enough are transplanted into soil where they continue to grow vegetatively and later flower* (i). *(From V. Vasil and A. C. Hildebrandt,* Science, *150:889–892, no. 3698, 1965, copyright 1965 by the American Association for the Advancement of Science, by courtesy of the authors and the American Association for the Advancement of Science.)*

terminates development of the embryo in the seed. In this respect their development is like that of moss and fern embryos. What is the cause of this difference of embryo behavior?

Growth of the embryo in the seed follows a typical sigmoid curve, as does that of most of the other seed tissues (Fig. 5-14). Embryo growth rate increases during the period of most rapid endosperm growth and

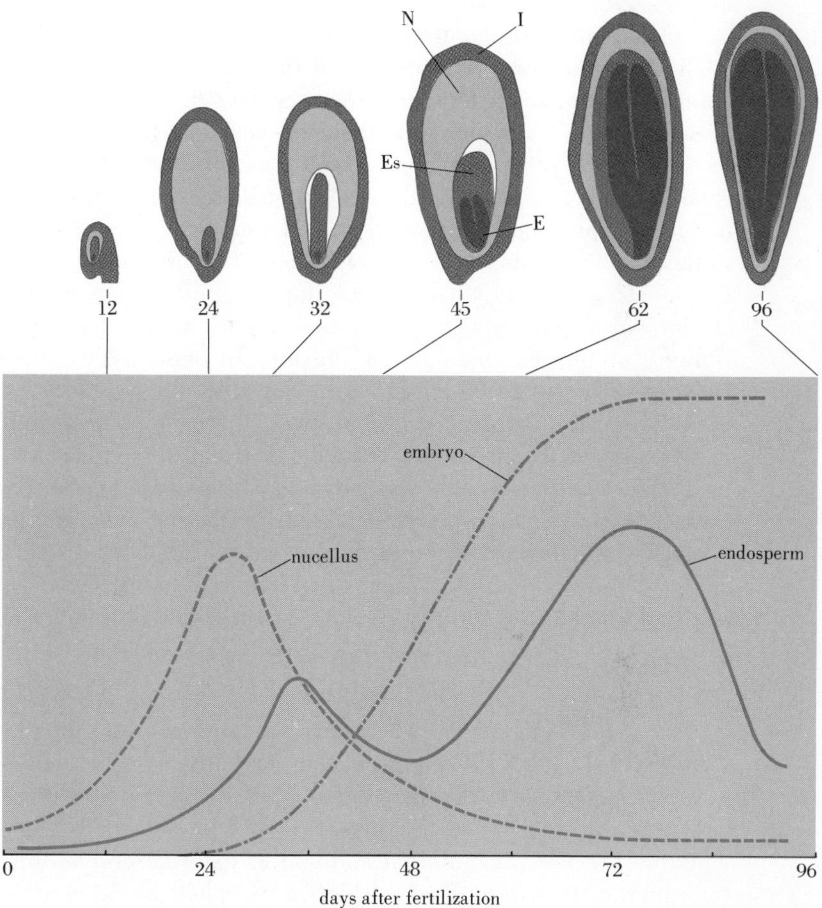


*Fig. 5-14 Development of the internal tissues of the apple seed. The diagrams show longitudinal sections of the seed at successive times after fertilization, and the graph shows the relative rate and extent of development of the various tissues. Tissues are labeled* I: *integument;* N: *nucellus;* ES: *endosperm;* E: *embryo. In the seed the nucellus starts to develop before fertilization, and develops rapidly soon after fertilization. It is subsequently almost entirely absorbed, and its nutrients transferred into the endosperm. The endosperm, after a period of rapid growth, decreases in volume as its nutrients are absorbed by the enlarging embryo. (Diagrams of seeds from L. C. Luckwill,* Journal of Horticultural Science, *24:32, by courtesy of the author and the editors; graph from L.C. Luckwill,* Symposium of the Society for Developmental Biology, *17:223, by courtesy of the author and the Society for Developmental Biology.)*

declines as the outer tissues of the seed develop most extensively. By the time the embryo is fully developed, all or most of the endosperm has been digested and its nutrients transferred to the cotyledons of the embryo. In any case, cessation of embryo growth cannot result from lack of nutrients at this time, and it seems that the cause is to be sought in the chemical and physical changes that the more superficial tissues of the seed are undergoing.

The concentration of gibberellin in the tissues around the embryo declines as the seed matures, and if the embryo is still dependent on externally supplied gibberellin there will be a decline in its rate of growth. However, in the superficial tissues that are differentiating as the seed coat, other changes—which may be even more important in retarding embryo growth—are occurring. Differentiation of these cells includes deposition of pigments; development of thick, hardened cell walls; and increasing dehydration. These have the effect of reducing the level of light in the seed, of imposing increasing physical constraint on the enlarging embryo, and of altering the internal concentrations of oxygen and carbon dioxide. In some seeds this simply reduces the rate of embryo growth to zero. In other species, however, growth-retarding substances accumulate in the seed coat during the final stages of development. The chemical nature of these retardants is diverse, but they include abscisic acid, coumarin, and several organic acids. As the balance between growth promoting and retarding substances shifts toward the retardants, embryo growth is progressively arrested and the embryo becomes inactive, remaining in this condition in the dormant seed until the balance is reversed during germination.

## *THE PLANT EMBRYO AND ITS ENVIRONMENT*

The embryo consists of diploid cells, which contain both maternal and paternal genomes. It is surrounded in the seed by endosperm, which is usually triploid and contains two sets of maternal and one of paternal chromosomes, and by superficial ovular tissues, which are diploid but contain only the maternal genome. The genetic relationships between embryo and surrounding cells is simpler in mosses and ferns, but in all cases the embryo develops in a complex cellular environment of changing nutritional levels, hormonal balance, and physical forces.

How are we to interpret the complex series of events by which a single-celled zygote becomes a multicellular organism? The zygote must possess the genetic capacity for development, but we have seen that this is possessed also by differentiated cells of the plant body. Therefore, embryonic development must be expressed in response to an environment that provides the cues necessary for development. This kind of environment occurs in the embryo sac or the archegonium and it can be duplicated in sterile culture. As the embryo develops morphologically and biochemically the same environment will probably affect it differently, but the environment itself changes as the embryo alters levels of nutrients and hormones in it. Therefore, there is likely to be continual flux, and changes in the embryo and the environment are probably concurrent throughout development. Thus different environmental factors act on the embryo at different stages of development, establishing polarity, directing the differentiation of meristems and specific tissues, activating additional biochemical pathways, and finally retarding embryo growth as dormancy occurs.

## *FURTHER READING*


Amen, R. A., "A Model of Seed Dormancy," *The Botanical Review*, vol. 34 (1968), p. 1.

DeMaggio, A. E., "Morphogenetic Studies on the Fern *Todea barbara* (L.) Moore. II. Development of the Embryo," *Phytomorphology*, vol. 11 (1961), p. 64.

———, and R. H. Wetmore, "Morphogenetic Studies on the Fern *Todea barbara*. III. Experimental Embryology," *The American Journal of Botany*, vol. 48 (1961), p. 551.

Miller, H. A., and R. H. Wetmore, "Studies on the Developmental Anatomy of *Phlox drummondii* Hook. I. The Embryo," *The American Journal of Botany*, vol. 32 (1945), p. 588.

Pollock, E. G., and W. A. Jensen, "Cell Development during Early Embryogenesis in *Capsella* and *Gossypium*," *The American Journal of Botany*, vol. 51 (1964), p. 915.

Quatrano, R. S., "Rhizoid Formation in *Fucus* Zygotes: Dependence on Protein and Ribonucleic Acid Synthesis," *Science*, vol. 162 (1968), p. 468.

Raghavan, V., "Nutrition, Growth and Morphogenesis of Plant Embryos," *Biological Reviews*, vol. 41 (1966), p. 1.

———, and J. G. Torrey, "Growth and Morphogenesis of Globular and Older Embryos of *Capsella* in Culture," *The American Journal of Botany*, vol. 50 (1963), p. 540.

Skoog F., and C. O. Miller, "Chemical Regulation of Growth and Organ Formation in Plant Tissues Cultured *in vitro*," *Symposium of the Society for Experimental Biology*, vol. 11 (1957), p. 118.

Steward, F. C., M. O. Mapes, A. E. Kent, and R. D. Holsten, "Growth and Development of Cultured Plant Cells," *Science*, vol. 143 (1964), p. 1.

Vasil, V., and A. C. Hildebrandt, "Differentiation of Tobacco Plants from Single Isolated Cells in Microcultures," *Science*, vol. 150 (1965), p. 889.

Ward, M., and R. H. Wetmore, "Experimental Control of Development in the Embryo of the Fern *Phlebodium aureum*," *The American Journal of Botany*, vol. 41 (1954), p. 428.

Wardlaw, C. W., *Embryogenesis in Plants*. New York: Wiley and Sons, 1955.

*chapter* **6**

# *Interactions of Nucleus and Cytoplasm*

Two principal elements underlie our modern theory of development. We know, first, that genes determine the specific nature of many molecules—and thus, ultimately, cell types and structural configurations. But, second, as we stated earlier, each of the different regions of the egg cytoplasm has its own characteristic specificity. Usually regional differences in the embryo can be traced back to regional cytoplasmic differences in the egg and its cellular descendants. Thus, in defining the nature of developmental processes we must explore the interactions of genes with the cytoplasm. This concept requires at the outset a differential response of genes, which during cleavage come to lie in different cytoplasmic surroundings. At first, then, topographic diversity leads to differential gene action; later, during morphogenesis, cell-to-cell interactions become more complex, leading to changes in the cytoplasm, which again may be manifested in differential expression of genes.

But we are running ahead of our story. Is it not possible that the cytoplasm is merely "building material," to be fashioned under the influence of genes, which are "sorted out" and distributed differentially during cleavage? This idea was embodied in a theory advanced by August Weismann in 1892. His theory required that, after cleavage, not only must the cells of the blastula (and their descendants) contain different "determinants," but that each cell be limited to a specific position in the embryo.

We now know that this theory is untenable, and that all nuclei of the early embryo are equivalent in their capacity to interact with cytoplasm leading to normal development. This generalization is sound for embryos as advanced as the early gastrula, and probably beyond.

## EQUIVALENCE OF CLEAVAGE NUCLEI

In both the frog and sea urchin egg, the first two cleavage planes are *vertical*, passing through the animal and vegetal poles, and the third division is *horizontal*, or *equatorial*, dividing the embryo into four upper and four lower blastomeres. If eggs are mounted between glass plates, animal pole upward, and gently compressed, the direction of the third cleavage plane is altered; it is now vertical. If the pressure is released at the eight-cell stage, the fourth cleavage plane will be horizontal. This procedure, first performed by Driesch, results in a complete reshuffling of the nuclei. Nuclei that normally would have been located in dorsal organs are now located ventrally (Fig. 6-1). If the hypothesis of unequal

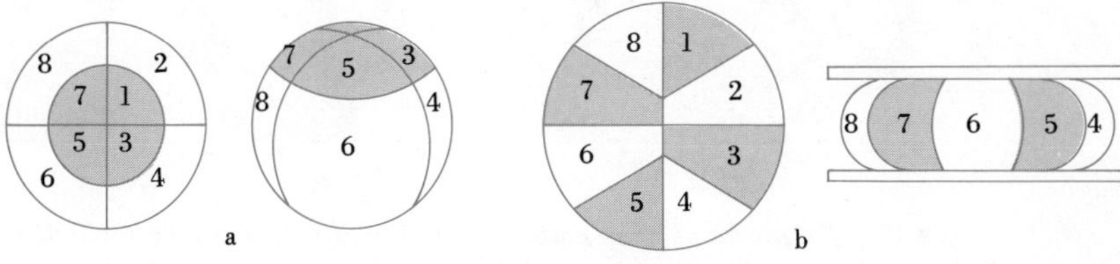


*Fig. 6-1 Diagram showing position of blastomeres in normal cleavage* (a), *and in eggs compressed during the first three cleavages* (b). *Although nuclei are not shown in the diagram, the daughter nuclei of eggs cleaving under pressure will be positioned in areas of cytoplasm different from those in which they normally are located. (From Huxley and DeBeer,* The Elements of Experimental Embryology, *by courtesy of Cambridge University Press.)*

nuclear division were correct, a disorganized embryo would result; normal embryos develop, however.

Further proof is offered by another ingenious experiment, first performed by Spemann using the fertilized egg of the newt. The zygote is constricted into two halves, one containing the fused egg and sperm nuclei, and the other lacking nuclei altogether. When the egg is constricted, just a small protoplasmic bridge is left (Fig. 6-2). The nucleated part cleaves normally; the sister, nonnucleated half does not. Eventually, often as late as the 16-cell stage, a nucleus from the cleaving half slips across the narrow bridge. Now both halves may develop, and the result is the formation of two whole embryos. Thus the nucleus of a 16-cell stage was found to be equivalent to the original nucleus.

### Nuclear Transplantation

Taken together, the evidence thus far suggests that somatic cells contain identical nuclei. However, we have considered proof of this statement only for early cleavage stages.

In an effort to detect changes in the properties of nuclei during development, Briggs and King perfected a method for transferring nuclei of embryonic cells into enucleated eggs of the frog. The transplantation is carried out in two main steps (Fig. 6-3). First, the recipient eggs are activated parthenogenetically,with a glass needle, and subsequently enucleated. Second, donor cells are isolated by disaggregating an older embryo. A given cell is drawn up into the tip of a micropipette, the inner diameter of which is smaller than that of the cell. The cell surface is

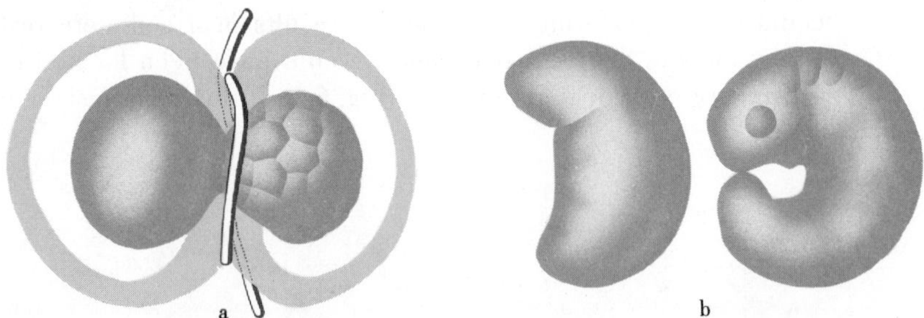


**Fig. 6-2** *Twinning of newt embryos combined with delayed nucleation of one half.* (a) *Constriction of the egg along the advancing first furrow leading to the 2-cell stage has left the copulation nucleus in the right half. This half has cleaved. At the stage illustrated the nucleus of the blastomere nearest to the "bottleneck" escapes into the unsegmented left half, which presently begins to cleave.* (b) *Twin embryos developed from* (a). *The lag of the left partner owing to the initial delay still is appreciable. (After H. Spemann,* Embryonic Development and Induction, *by courtesy of Yale University Press.)*

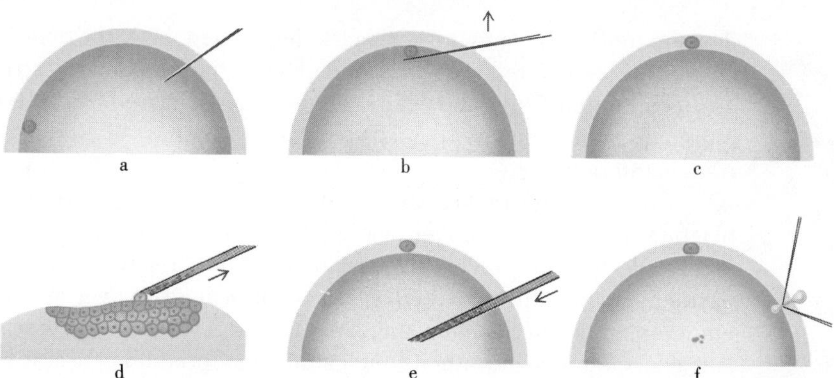


*Fig. 6-3 Diagram illustrating method for transplanting nuclei from embryonic cells into enucleated eggs* (Rana pipiens). (Upper) *Steps in preparing enucleated eggs. The egg is activated* (a), *and enucleated* (b) *with a clean glass needle. The enucleation operation results in the formation of an exovate containing the egg nucleus, shown trapped in the jelly surrounding the egg* (c). (Lower) *Procedure for transplanting nuclei. The donor cell is isolated on the surface of an intact blastula (which serves as an operating platform) and is then drawn up into a micropipette* (d). *Since the diameter of the pipette is somewhat smaller than that of the cell, the cell surface is broken. When the broken cell is injected, as shown in* (e) *the nucleus is liberated undamaged in the egg cytoplasm. As the pipette is withdrawn it tends to draw the surface coat of the egg with it, forming a small canal to the exterior which must be severed, as shown in* (f), *to prevent leakage. (From R. Briggs and T. J. King, in* Biological Specificity and Growth, *E. G. Butler, ed., by courtesy of Princeton University Press.)*

broken but not dispersed; thus, the nucleus is protected by its own cytoplasm until the pipette is inserted into the recipient egg. One possible objection to the technique, the inclusion of a small volume of donor cytoplasm in the transfer, is at least partly met by control experiments in which cytoplasm injected alone does not support development.

It was found that late blastula nuclei can be transplanted in undamaged condition; they are equivalent to the nucleus at the beginning of development, for the majority of eggs that receive nuclei from late blastulae develop normally. Similar results were obtained by Fischberg and his colleagues in experiments with the South African clawed toad, *Xenopus*. Thus far our conclusion is clear: stable nuclear changes affecting the capacity to support development have not been observed through the blastula stage.

However, when nuclei of cells from older embryos are transferred, the outcome of the experiment is different. For example, when nuclei taken from late gastrula endoderm cells are transplanted, they frequently, *but not invariably*, are unable to support normal development. Defective embryos are produced; although the endodermal structures develop normally (and to a remarkable extent also the mesodermal structures associated with the endoderm), the ectodermal structures are usually deficient. Thus some change has occurred in these nuclei. Moreover, this change persists under conditions of serial nuclear transplantation.

Compare the records of serial transplantation experiments with blastula and gastrula nuclei in Figs. 6-4 and 6-5. In the first step (Fig. 6-4) single blastula nuclei are transplanted into a series of enucleated, parthenogenetically activated eggs, as described above. When these embryos reach the blastula stage, nuclei are removed from one of them and transplanted into other enucleated eggs, producing a first blastula generation. From one of these blastulae, nuclei are again transferred to enucleated eggs, producing a second blastula generation. Serial transplantation of blastula nuclei confirms the original proposition that blastula nuclei are equivalent. When the experiment is repeated with only one difference, namely that the original donor nuclei are derived from the prospective midgut of a late gastrula (Fig. 6-5), each *clone* (or group of eggs activated by nuclei from the single donor) perpetuates the pattern of abnormalities—the restricted developmental capacity—exhibited by the original recipients of late gastrula endoderm nuclei.

What are these nuclear changes? Are these results meaningful in relation to our theory of differentiation? What other facts must be taken into account? It is possible that the defects result from deleterious effects of the operative methods on the nuclei. As cells get older, they may become more "sensitive," more subject to damage in the course of the experiments. There is some evidence to support this contention. On close examination, many of the abnormal embryos produced by transplantation of late gastrula nuclei are found to have chromosomal abnormalities. These abnormalities, too, are perpetuated

**Fig. 6-4** *A record of serial transplantation experiments with nuclei of undifferentiated blastula. In this experiment a nucleus (and a small amount of adherent cytoplasm) is removed from a blastula cell* (top) *and injected into an anucleate egg activated to parthenogenetic development. The host egg then proceeds to develop normally. In serial transplantation experiments a nucleus is removed from the blastula and transferred, as in the first cage, to an anucleate egg. When this egg reaches the blastula stage, nuclei are removed and transplanted into other anucleate eggs to produce a first blastula generation. From one of these, blastulae nuclei are again transplanted into anucleate eggs to produce a second blastula generation. In this way it is possible to test the effects of development upon the nucleus and its changes during development. The record of the results of serial transplantations of three blastula donors A, B, C* (bottom). *Each of the blastula donors developed from nuclei of a single original blastula donor. (From C. W. Bodemer,* Modern Embryology, *by courtesy of Holt, Rinehart and Winston, Inc.)*

original blastula
nuclear donor

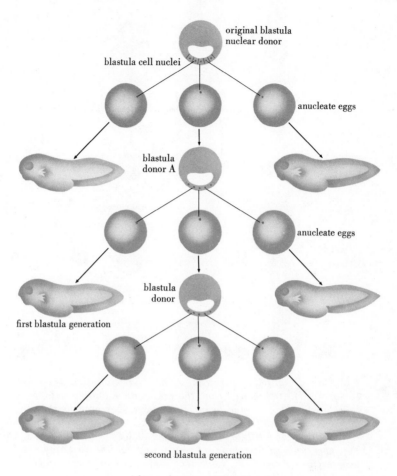

blastula cell nuclei

anucleate eggs

blastula
donor A

anucleate eggs

blastula
donor

first blastula generation

second blastula generation

serial transplantation of blastula nuclei

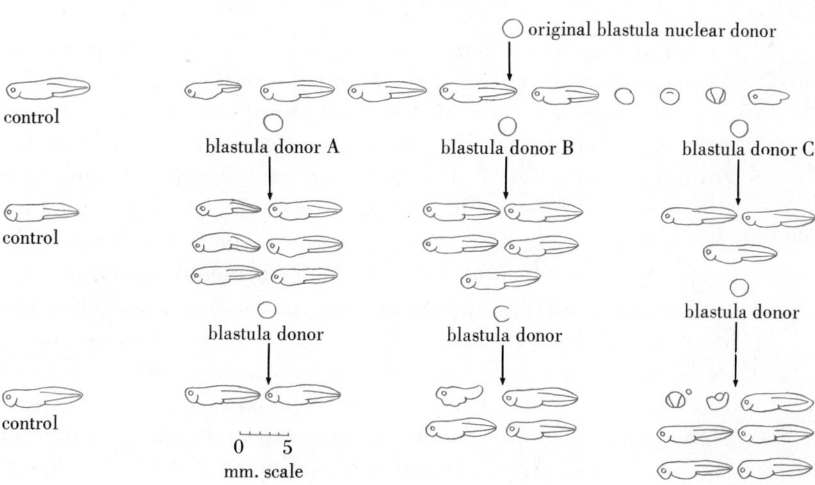

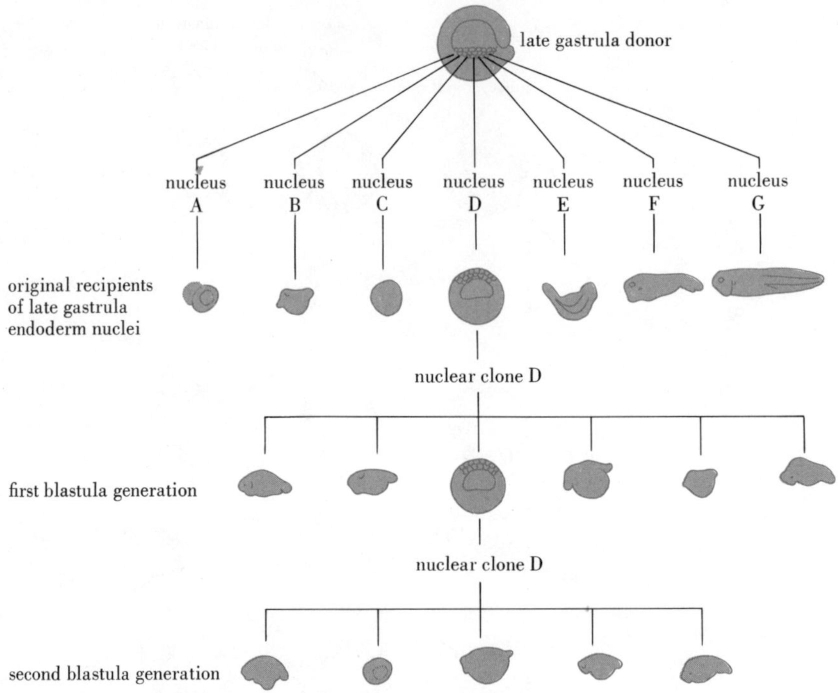


**Fig. 6-5** *Diagram illustrating serial transplantation of endoderm nuclei. Donor nuclei are removed from the presumptive anterior midgut region of the late gastrula. Transferred to anucleate eggs, they promote the various types of development shown for "original recipients" in the diagram. One of the original recipients at the blastula stage provides nuclei for a single clone, which shows the more uniform development illustrated for the first and second blastula generations. (From C.W. Bodemer,* Modern Embryology, *by courtesy of Holt, Rinehart and Winston, Inc.)*

in serial transplantation experiments, but they can hardly be taken as evidence of progressive nuclear differentiation.

We must also consider another line of evidence: as Gurdon and his colleagues have shown in experiments carried out on *Xenopus*, nuclei from the majority of differentiating endoderm cells and from fully differentiated cells of the intestine can support the formation of nerve and muscle cells after transplantation to enucleated eggs. Moreover, a number of fertile adult frogs have now been obtained from eggs whose only active nuclei were transplanted intestine nuclei. The evidence is by now convincing; in *Xenopus*, at least, nuclei of some differentiated cells can support the whole range of normal development. Do the differences in findings in *Rana* and *Xenopus* simply reflect the fact that *Rana* cells are more sensitive to experimental manipulation than *Xenopus* cells? Are *Xenopus* cells simply more "hardy" than those of the frog? Or should it be argued that the time of onset of observable stable change is later in *Xenopus?*

Our questions must remain unanswered for the present. We are confronted with two large possibilities. *First*, it is possible that stable nuclear changes affecting the capacity to support development do occur. They may be cell-type specific, although fully convincing evidence is lacking thus far. For reasons that are not entirely clear, such changes are better demonstrated in *Rana* than in *Xenopus*. *Second*, it is possible that the changes so far observed in *Rana* are not meaningful to our theory of differentiation in their present context. We would emphasize the finding that *some* nuclei of swimming tadpoles support normal development. Therefore, it can be argued that the observed deficiencies are the consequence of the experiment and that they do not reflect fundamental restrictions in the capacity of the nucleus to support development. It is possible, for example, that the "deficiency" of the older nucleus lies in its inability to enter the mitotic cycle characteristic of the zygote.

It would be misleading to select one or the other of these points of view from the sketchy data just presented. However, there is a store of information, yet to be taken into account, which shows that many kinds of cells retain some ability to transform into other cell types throughout life. Both the invertebrates and vertebrates possess remarkable capacities for regeneration of lost parts; some cells from specialized tissues of mature higher plants are capable, when isolated, of giving rise to whole plants. Not all of the evidence is this dramatic, but many specialized cells are capable, under appropriate conditions, of extending their repertoires, of making new products and assuming new roles. Thus it is difficult to accept the idea of irreversible, stable changes in the capacity of all cell nuclei to support development.

Therefore, while keeping an open mind pending the outcome of further experiments, we shall emphasize the regulation of genetic and other nuclear functions, which depend on continuing interactions with the cytoplasm, rather than their irreversible restriction.

## *EVIDENCE FOR REGIONAL DIFFERENCES IN EGGS AND EARLY EMBRYOS*

What is the evidence for regional cytoplasmic differences in eggs? Let us state our conclusion first: the amount of regional differentiation varies with the species. At one end of the scale—in eggs of ascidians, annelids, and mollusks for example—the undivided egg is highly organized; from the very beginning of cleavage, the daughter cells are specialized; the restriction of developmental capacity occurs early.

Thus, in these forms, nuclei come to lie in regionally differentiated cytoplasm from the outset of development. These regional differences are shown most graphically by defect experiments in annelids and mollusks. In the eggs of some marine snails and mussels immediately before the first cleavage, part of the egg protrudes as a "polar lobe"; at division it remains attached to one blastomere into which it is withdrawn, only to reappear during the next division. This lobe makes an attractive "target" (Fig. 6-6). When the lobe is amputated, the embryo develops, but imperfectly, the resulting larvae lacking mesodermal structures. Thus, this localized region contains materials essential to formation of mesodermal structures. In another example, at the fourth cleavage in the eggs of one of the ctenophores (comb jellies), 16 cells are produced: eight large *macromeres* and eight small *micromeres.* Direct observations show that the micromeres contain materials that are originally localized as a cortical layer in the fertilized egg. Then, gradually during cleavage, these materials are concentrated in the micromeres (Fig. 6-7). If one or more micromeres are removed, specific defects are observed later in development. The normal ctenophore possesses eight rows of comblike swimming plates. It has been found that the number of swimming plates present is proportional to the number of micromeres present. Removal of micromeres results in a corresponding deficit in swimming plates.

Such experiments, and the number might be multiplied many times, show that in the eggs of many species regional cytoplasmic differences are expressed at the outset of development. Consequently each blastomere, or each group of blastomeres, in isolation can perform only a restricted role.

Let us carry out the same kind of analysis in the egg of another form, the sea urchin. Here, as Driesch showed in 1892, at the two- or four-cell stage each isolated blastomere can give rise to a complete

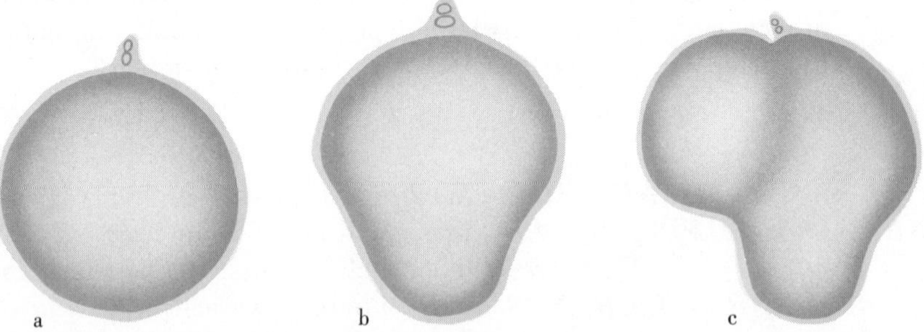


**Fig. 6-6** *Early development of the mussel,* Mytilus edulis. (a) *Formation of polar bodies.* (b) *Formation of polar lobe.* (c) *First cleavage. (From R. M. Eakin,* Vertebrate Embryology, *after I. A. Field, 1922, U.S. Bureau of Fisheries, 38:127–159, by courtesy of Dr. Eakin.)*

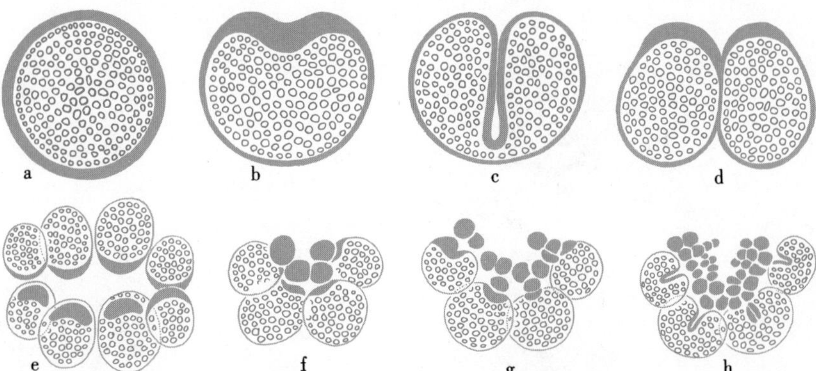


*Fig. 6-7 Redistribution of cortical (micromere-forming) protoplasm during the cleavage of the ctenophore* Beroë ovata, *as seen in the dark field. (After Spek.) All views lateral, except* (e) *which is a polar view.* (a–d) *First cleavage division.* (e) *8-cell stage with accumulation of micromere material near the upper pole.* (f) *Pinching off of first micromeres; transition from 8-cell to 16-cell stage (only one half of the embryo is shown).* (g–h) *Continued micromere formation (only one half of embryo shown). If the 16-cell stage is transected into two unequal parts, one retaining five, and the other three, micromeres, each fragment develops into a larva with just as many rows of swimming plates as it had contained micromeres. Both fragments add up to the normal number of rows, which is eight. (From P. Weiss,* Principles of Development, *by courtesy of Holt, Rinehart and Winston, Inc.)*

larva, which is normal and proportionate, though small. Thus, at the outset of development it would *appear* that there were no regional cytoplasmic differences. In another experiment, performed two decades later, two whole sea urchin eggs were fused. A single larva—giant, but otherwise normal—developed.

But we have omitted an important point. When the embryos are fused, their axes must coincide; thus, there is an implication of some organization in the fertilized egg, to which we now must return. Instead of allowing the egg to cleave normally and then separating the blastomeres, let us cut the unfertilized egg in half, and fertilize each half (Fig. 6-8). If we make the cut vertically (as the first cleavage does), two normal larvae result; but if the egg is cut along the equator between the animal and vegetal hemispheres, the animal half forms a ball of ciliated cells and the vegetal half forms an incomplete embryo. Returning for a moment to the normal development of the sea urchin embryo, we recall that the first two cleavages are vertical, but that the third is horizontal. What would you expect to find if you separated the animal and vegetal halves at the eight-cell stage? If you said the results would be similar to those obtained by cutting the fertilized egg horizontally, you would be correct.

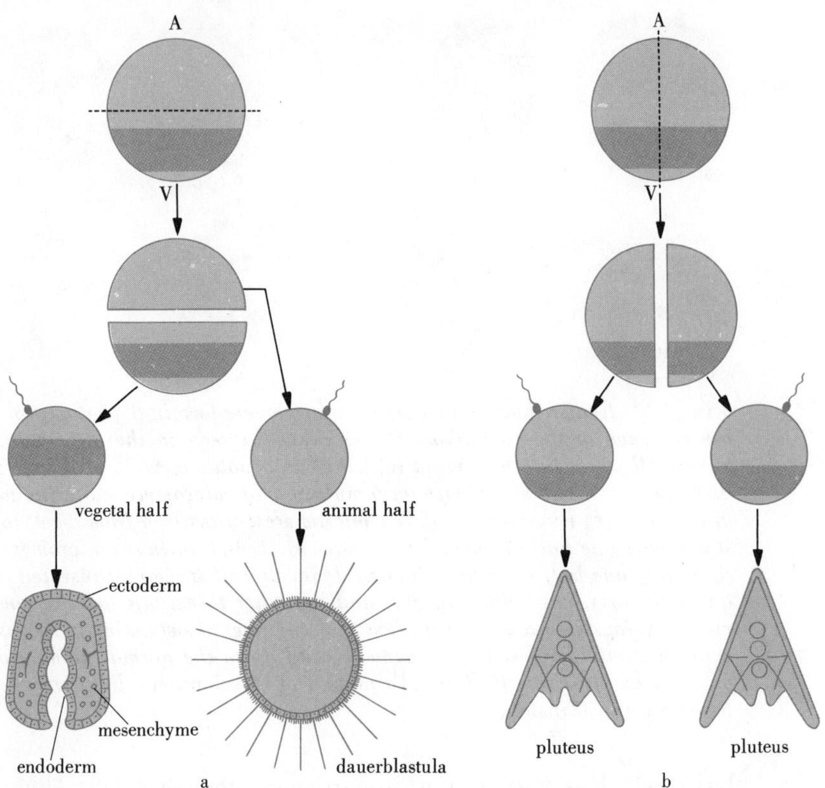


***Fig. 6-8*** *Development of halves of the sea urchin egg.* (a) *Animal and vegetal: the egg is cut in two through the equator by means of a glass needle. The animal half* (right) *is fertilized and develops only into a ciliated blastula (dauerblastula) failing to form endoderm. The vegetal half* (left) *after fertilization forms an incomplete embryo.* (b) *The egg is cut in two longitudinally, A–V. Each half is fertilized, and each half develops into a normal embryo of half size. (From L. G. Barth,* Embryology, *by courtesy of Holt, Rinehart and Winston, Inc.)*

Therefore, even the sea urchin egg shows some evidence of regional differences early in development, differing from our other examples in that the differences appear to be distributed along the animal-vegetal axis. Numerous other experiments prove the point: the animal and vegetal hemispheres differ significantly, and it is only through their interaction that normal development results. The removal of cells from one pole or the other, or any treatment that produces an imbalance between the two hemispheres, results in abnormal larvae. For example, if the vegetal hemisphere is allowed to develop in the absence of cells from the animal pole, so that endoderm exceeds ectoderm, abnormal larvae result. For reasons we do not yet understand, eggs allowed to develop in a solution of lithium chloride behave much the same way as those without animal cells. This substance inhibits cells at the animal pole. We speak of embryos in which the cells at the vegetal pole—the endoderm—predominate as *vegetalized*, and lithium chloride as a *vegetalizing agent*.

Other agents, including sodium thiocyanate, depress endodermal cells, leading to a disproportionate array of ectodermal structures. Such embryos are said to be *animalized*.

Nor is the sea urchin egg an isolated example. We have already alluded to the observation that the division of newt eggs at the two-cell stage could produce twins. The same may be said for the frog's egg; but again, the plane of division is critical. It must be vertical, but it also must divide the egg in such a way that each half of the embryo receives part of the *gray crescent* (Fig. 6-9).

When each of the first two blastomeres contains part of the gray crescent, two normal embryos result. If cleavage occurs so that one cell

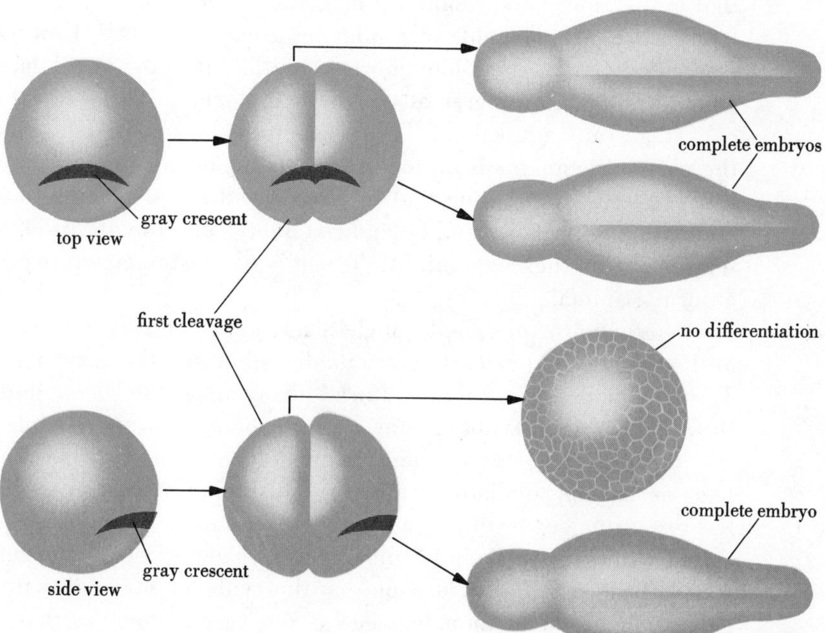


*Fig. 6-9 The significance of the gray crescent as revealed by separating the cells at the two-cell stage.* (Left) *The gray crescent is shown in top view. When the egg divides, one half of the gray crescent passes into each of the two cells. If these cells are separated from each other, each will form a complete embryo.* (Right) *A side view of the gray crescent, and a first cleavage in which one cell contains all of the crescent, whereas the other cell lacks any part of it. If these cells are separated, the one lacking the crescent fails to develop. (From L.G. Barth,* Embryology, *Holt, Rinehart and Winston, Inc.)*

contains all of the crescent and the other cell lacks any part of it, then when the cells are separated the one lacking the crescent fails to develop normally. Mapping and transplantation experiments have shown that this area marks the future dorsal lip of the blastopore; hence, in these species there is an early regional differentiation (page 52). You already know the critical role of the cells at the dorsal lip of the blastopore; after invagination they interact with the overlying ectoderm inducing the neural plate. Thus, you see why half an egg that did not contain some of the gray crescent region could not develop normally.

Finally, you may ask, do not the inescapable facts of identical twinning in man and other mammals tell us that if the first two blastomeres of mammalian embryos could be separated experimentally, each would give rise to a normal embryo? We knew of identical twins—"experiments of nature"—long before the experiment was performed in the laboratory.

The following evidence is available. Early experiments showed that in the mouse and rabbit, a single blastomere from either the two- or four-cell stage is capable of producing a normal animal. However, the evidence is not as complete as we would like it to be. We do not know whether *both* blastomeres at the two-cell stage, or *all four* at the four-cell stage, are equivalent. These experiments consisted of removing the cleaving embryos from the uterus, killing one or three blastomeres, respectively, by pricking with a glass needle, and placing the mass, containing both intact and killed blastomeres into the uterus of another female. Under these conditions, the surviving blastomere often produces a normal animal.

In further experiments single blastomeres were actually separated at the two- to eight-cell stage by mild digestion with the enzyme pronase. The blastomeres were then cultured in a complex, sterile medium until they reached the blastocyst stage, after which they were transplanted to the uterus of a foster mother. Again the results show only that *some* became normal, miniature embryos. This experimental procedure is a far cry from separating sea urchin blastomeres by mild shaking in calcium-free seawater and returning the separated cells to normal seawater. Thus it is not surprising that the evidence for lability in the organization of the mammalian egg is not as complete as that for sea urchins and frogs. We can carry the argument one step further, however. Two or more cleaving mouse eggs may be fused in culture to form a normal morula or blastocyst, which when transferred to the uterus of a foster mother produces a normal mouse. Embryos fused as late as the eight-cell stage routinely develop normally; even somewhat later stages can be fused successfully, but it has been reported that with late morulae (about 32 cells) there is some decline in success.

These experiments are important to our present discussion, for they contribute to our understanding of the labile organization of the mammalian egg. They also create new avenues for exploration. These techniques make it possible to create new kinds of adult mice in which cells with two or more different, rather than identical, genotypes would be included. Cleaving embryos with distinct genotypes, bearing a wide range of genetic markers, may be fused and reared to adulthood (Figs. 6-10 and 6-11). These mice are called *allophenic* mice, that is, individuals with a simultaneous, orderly expression of two or more allelic cellular phenotypes or allophenes, each with a distinctive genetic basis (*alleles* being alternative genes at the same locus).

Thus far, nothing we have said is inconsistent with a theory of differentiation which requires the continuing interaction of nucleus and cytoplasm, *the action of genes in a nucleus in any given tissue being constantly regulated by their changing chemical environment.*

## THE NEED OF GENIC MATERIALS FOR DEVELOPMENT

Let us begin to examine the evidence for this theory in more detail, reentering the cycle at the nucleus.

Both nucleus and cytoplasm are indispensable for development. The egg will not develop if the nucleus is removed. There are scattered observations of abortive cleavages of enucleated salamander and sea urchin eggs. As we shall see, however, there is a ready explanation for such partial cleavages; put most simply, they result from residual activities initiated during oögenesis.

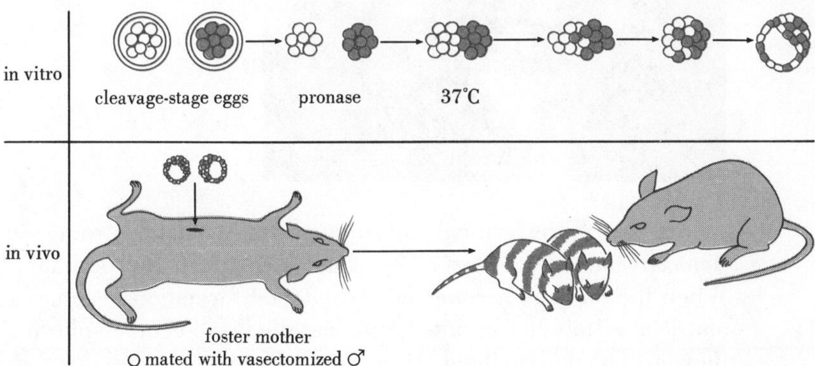


**Fig. 6-10** *Diagram of the experimental procedures for producing allophenic mice from aggregated eggs. (From B. Mintz,* Proceedings of the National Academy of Sciences, *vol. 58, 1967, p. 345, by courtesy of Dr. Mintz and the Academy.)*

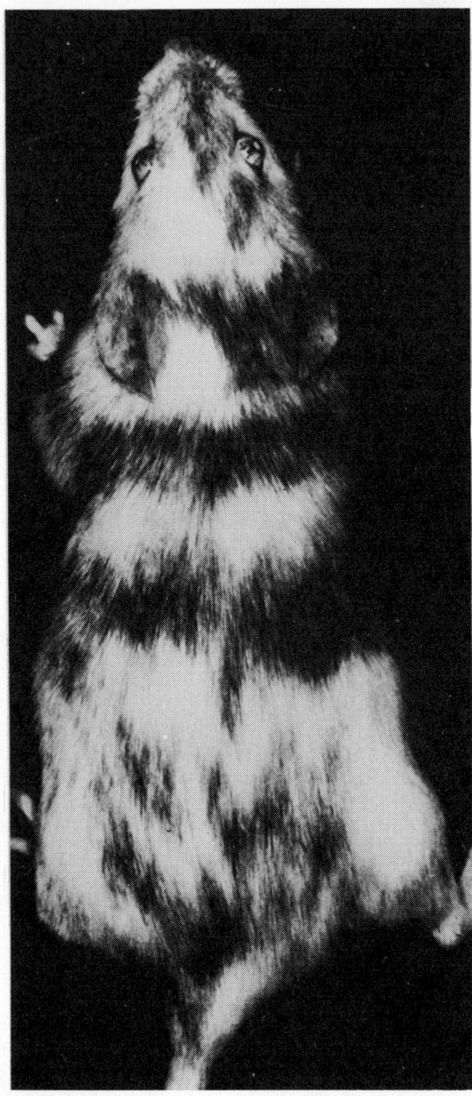

*Fig. 6-11* *Adult allophenic mouse containing both homozygous black (CC) and albino (cc) melanocytes from C57BL6, and ICR genotypes, respectively. (From B. Mintz,* Proceedings of the National Academy of Sciences, *vol. 58, 1967, p. 348, by courtesy of Dr. Mintz and the Academy.)*

Even more critical information is available from studies of *Amoeba* and *Acetabularia*. We shall begin with the familiar *Amoeba*. When the nucleus is removed, pseudopod formation becomes sporadic and loss of motility results. Anucleate halves become spherical. Since they are unable to feed, they gradually decline, although they may survive as long as two weeks. For a time, the changes produced by enucleation are reversible, for if a nucleus is reintroduced within two or three days, normal activity is restored. If one waits six days before returning a nucleus, the amoeba no longer responds.

The classic experiments of Hämmerling on the unicellular alga, *Acetabularia*, have been described repeatedly, but they deserve retelling. During most of its life this green alga is a stalk, containing chloroplasts, and bearing rootlike *rhizoids* (Fig. 6-12). Its single nucleus is located in one of the rhizoids. The tip of the stalk forms a cap or umbrella. When the umbrella is almost completely formed, the nucleus breaks down and "daughter nuclei" spread through the stalk and cap, where cysts are formed. As a result of germination of cysts, flagellated gametes are formed; following mating, the zygote again differentiates into rhizoid and stalk.

Anucleate stalks survive for a long time—as long as several months. Moreover, an anucleate stalk is able to complete its normal ontogeny by forming a cap and to regenerate one new cap, provided a new growing tip has formed before the nucleus was removed. However, the new cap—in fact, the entire cell—is incapable of further regeneration. On the other hand, cells containing a nucleus can regenerate a new cap repeatedly. Again, then, we conclude that the limited capacity for development in the absence of the nucleus results from residual products synthesized earlier by the nucleus or under its control.

Moreover, the nuclear products are species specific, as shown by combining parts of two species (Fig. 6-13). A capless stalk of one species (*A. crenulata*) is enucleated by the removal of the rhizoid containing the nucleus. In its place is grafted a rhizoid containing a nucleus of a second species, *A. mediterranea.* (The design of the cap is species

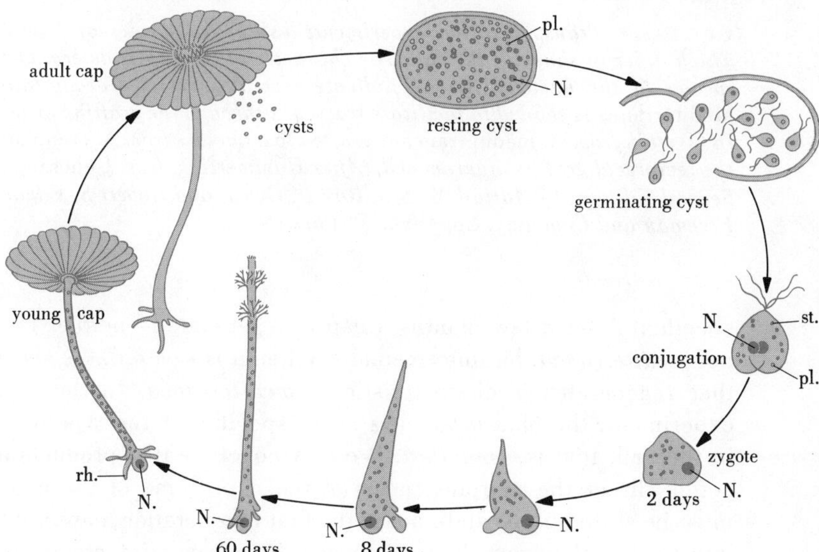


**Fig. 6-12** *Life cycle of* Acetabularia mediterranea. *(After J. Brachet,* Biochemical Cytology, *by courtesy of Academic Press.)*

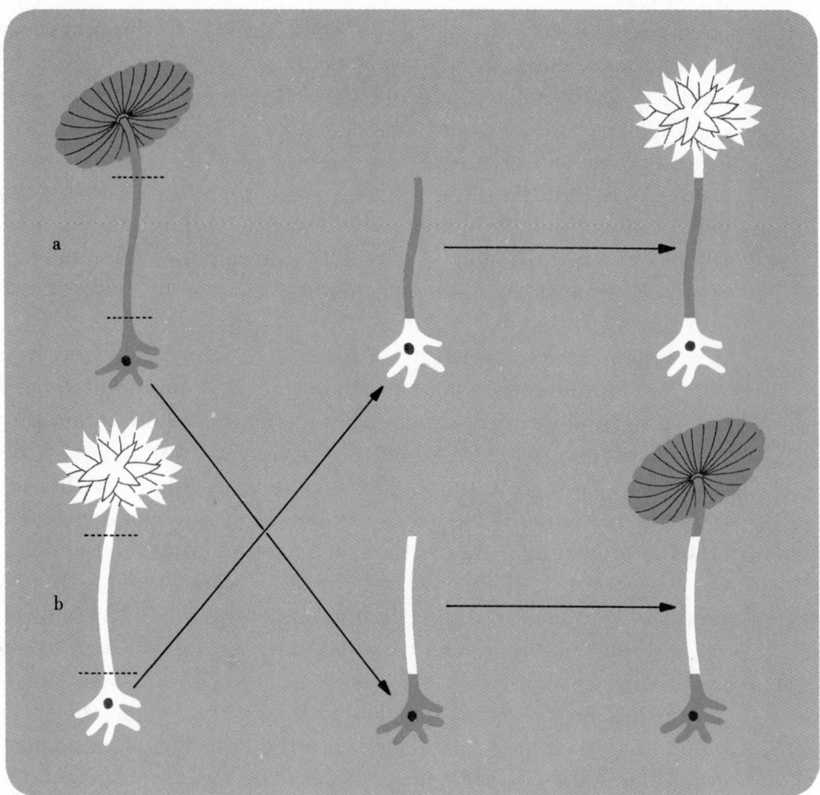


*Fig. 6-13 Transplantation experiments using two species of* Acetabularia, *which differ in shape of the cap. The character of the cap is determined by the species of the nucleus. Arrows indicate several successive decapitations.* (a) Mediterranea *is shaded to facilitate tracing it through the grafting experiments. In* (a) *stalk from* A. mediterranea *is grafted to a nucleus from* A. crenulata. *In* (b) *the reciprocal graft is diagrammed. (After Hämmerling, from* General Genetics, *Second Edition, by Adrian M. Srb, Ray D. Owen, and Robert S. Edgar. W. H. Freeman and Company. Copyright © 1965.)*

specific.) After a few months a cap is regenerated—neither *crenulata* nor *mediterranea*, but intermediate. When it is severed, the second cap that regenerates is characteristic of *mediterranea*. In the *reciprocal* experiment, the final product is again specific for the type of nucleus in the cell, that is, *crenulata*. We may conclude that products of gene action direct the morphogenesis of the cap. Some of these products must be stored in the stalk, for in the first regeneration, caps are formed intermediate between the two species. Other experiments suggest that there is a gradient of these products, for their activity is higher in the apex than in the base of the stalk.

Although the nucleus is essential, is it essential that there be a full haploid complement of chromosomes? Again there is ample evidence for an affirmative answer. Boveri showed that sea urchin eggs in which certain chromosomes were lacking, failed to develop properly. In that classic object of study, *Drosophila melanogaster*, the necessity of having each of the four kinds of chromosomes represented in the fertilized egg has been proved. For example, one X chromosome must be present; without an X chromosome, although there are a few cleavages, development stops; differentiation and morphogenesis fail to occur.

Muller's hypothesis of *dosage compensation*, formulated in 1932, held that the expression of genes on the X chromosome is the same in females and males although there are twice as many loci in the female. This hypothesis required essentially that one X chromosome in the *female* be inactivated or suppressed.

In 1949, Barr observed that in female mammals, including the human species, the two X chromosomes differ markedly in appearance. One is a "typical" chromosome; the other is *heteropycnotic*—that is, condensed and deeply staining. In the interphase nucleus, when the typical, fully extended chromosome cannot be seen by light microscopy, the heteropycnotic chromosome appears as a clearly defined chromocenter, which is now called the *sex chromatin* body. Thus, cytologic evidence first suggested that one X chromosome in the female may be inactive, a conclusion later confirmed by autoradiographic methods which reveal it to be inactive in nucleic acid synthesis. At the outset of development, however, both chromosomes are active. In rabbits no X-derived chromatin body can be seen in two-day embryos, but it is observed at 4½ to 5 days. In the hamster embryo both X chromosomes are still active at the eight-cell stage. Hence one X chromosome is modified during early development, becoming inactive before implantation. Is the same X chromosome destined to be inactivated in each cell? Or is it a matter of chance?

Evidence from genetic studies on mammals, first brought together effectively by Lyon, favors the latter possibility. We know that certain traits are sex-linked. For example, in the mouse there are many sex-linked coat color mutants. Females that are heterozygous for these genes show *variegation*, their coats exhibiting patches of both mutant and wild-type coloration. In interpreting these observations, Lyon proposed that the genetic inactivation of one X chromosome must occur early in development paralleling the cytologic changes, and that the inactivation must be random. After inactivation occurs in a cell, all its progeny would bear the same inactive loci. Thus, if it arises from several cells present at the time of first appearance of sex chromatin, a tissue should show variegation. In fact analysis of X-linked markers should make it possible to estimate the number of progenitor cells from which a given cell type is derived. For example, use has been made of X inactivation in cell-lineage studies on populations of red blood cells. A number of human females heterozygous for the X-linked gene for the enzyme glucose-6-phosphate dehydrogenase have been found to contain hemizygous red cells; that is, their pooled red cells have enzyme levels no higher than those of deficient males (having only one X chromosome that contains the mutant gene for this enzyme). It has been estimated that more than 1 percent of all heterozygous females have red cells that are hemizygous. Assuming a random inactivation of X chromosomes, it has been estimated that eight or fewer cells were *determined* to become red cells when X inactivation took place. Experiments were carried out to decide whether the inactivation occurred at a time when there were only eight cells in the embryo or eight red-cell progenitors. The latter alternative was shown to be more likely by demonstrating that other tissues such as skin and subcutaneous fat had enzyme levels characteristic of heterozygotes in the same individuals who had low (hemizygous) levels of enzyme in their red cells.

Studies of this kind complement those using allophenic mice. Thus the combination of a pigmented and an albino embryo produced the coat-color pattern shown in Figs. 6-10 and 6-11. Mintz has interpreted each band on each side as a clone descended from one cell determined to be a melanoblast. Thus the melanocytes that provide coat color in the mouse are derived from 17 cells on each side (34 in all), which are determined as melanoblasts at an early stage.

In allophenic mice 50 percent of the fused animals are expected to contain a mosaic population of XX and XY cells and therefore could be expected to be hermaphrodites. However, hermaphrodites have been found in only a very low percentage. The reason for the low incidence of hermaphrodites is unknown but it may be related to selection of one kind of gamete during the proliferative phase. However, the presence of mosaic populations of germ cells in allophenic mice can easily be demonstrated by including autosomal markers in the embryos used in the original fusion. Allophenic mice were produced by fusing an embryo carrying a dominant coat-color gene with one carrying a recessive gene. The allophenic mouse was then crossed to a normal mouse with the recessive coat-color gene. Both color types were found in the progeny, showing the allophenic mouse to be mosaic in its germ cells. This finding shows that there must be two or more cells that are independently determined to be germ cells in the mouse. When germ cells can first be cytologically identified in the mouse embryo they are ten in number. These two kinds of information show the initial number of primordial germ cells to be between two and ten.

***Chromosome imbalance*** It is a curious fact that although one X chromosome appears to be inactive, abnormalities are found when one chromosome is missing or when there is an imbalance in the diploid condition. One of the better known examples is found in the human species. When there is a deletion of an entire sex chromosome—the X chromosome being unaccompanied by either another X or Y chromosome (XO), so that the individual has only 45 chromosomes (Fig. 6-14)—the gonads fail to develop. The individual is female, but the rudimentary gonads consist only of connective tissue. The theoretically possible YO condition has never been observed in man, since this combination is probably incompatible with life.

What effects are produced by the presence of one or more *extra* chromosomes? Again, several good examples have been described, one being the combination XXY in man. In these individuals, prenatal and postnatal development is normal until puberty, except for a reduction in spermatogonia. Then the seminiferous tubules atrophy, the testes remain small, and the individual becomes eunuchoid.

Another classic example is found in *Langdon Down's syndrome* or *trisomy 21* (formerly inappropriately called *mongolism*). It is now known that most individuals who are in this condition of mental retardation have one supplementary autosome, their total chromosome number thus being 47. This discovery was a milestone in the search for the origin of the defect. The earlier evidence had suggested a hereditary basis; if one of identical twins is affected, the other is also; in contrast, only very rarely would both dizygotic (nonidentical) twins be affected.

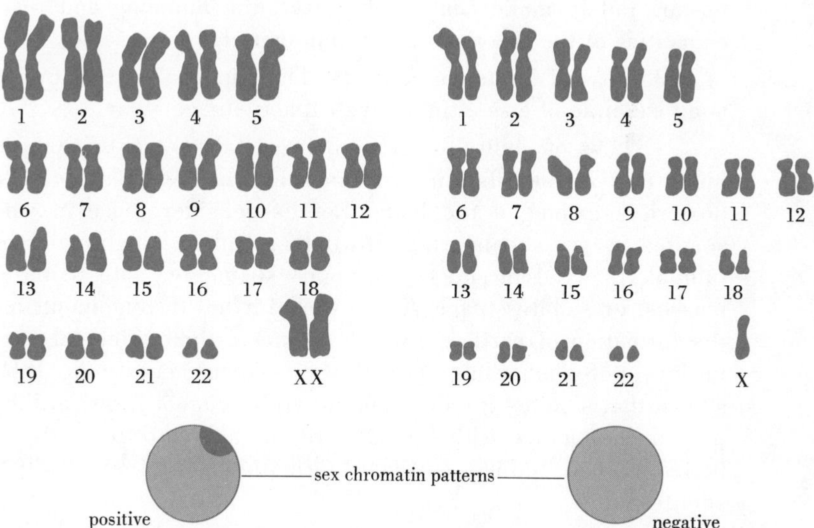


*Fig. 6-14 Chromosome complements and sex chromatin patterns in humans.* (Left) *Normal female.* (Right) *A patient with an XO sex chromosome complex (gonads remained rudimentary). (After M. Barr, in* Congenital Defects, *by courtesy of the author and International Medical Congress, Ltd.)*

**Defects Produced by Specific Mutations** Large aberrations of genetic material are not required to alter the course of development. Discrete point mutations and deficiencies of single genes, or translocations of small blocks of genes, may produce marked changes in development. Rarely do we know the initial molecular "lesion"—that is, the synthetic reaction first affected by the altered or absent gene. However, we have obtained some insight into the mechanisms by which defects are expressed morphologically. Let us begin with a familiar example. How might specific genetic defects be expressed in tissue interactions?

1. Distortion of the inductive pattern. *Polydactyly* may result from an atypical distribution of the factor produced by limb mesoderm and required for maintaining the apical ridge, it being well known that tissue interactions are required for normal development of the limb (page 80).

2. Failure of the interactants to make proper contact. An example of this type of failure is *anophthalmia* (eyelessness). The developmental events that precede eyelessness in certain strains of mice are the following. In the extreme expression, the optic vesicle forms but does not make contact with the overlying ectoderm in which the lens normally is induced (page 76). As a consequence, a lens does not form and the vesicle remains quite rudimentary. In some cases a reduced optic vesicle does make contact with the ectoderm and induces a small lens. It is presumed that eye development fails because the interacting tissues fail to make contact; however, the inducing and responding capacities of the two tissues are unimpaired.

3. Loss of inductive capacity. The wingless mutant in the chick is an example of loss of inductive capacity (page 79).

4. Loss of ability to respond. In certain *brachyuric* (short-tailed) mice, part of the offspring are abnormal, having gross defects in the posterior regions of the body. Let us consider those in which the somites are grossly abnormal. It will be recalled that normally the longitudinal, paraxial mesoderm becomes segmented into paired cubical masses, or somites (page 70). In their further development, including the formation of cartilage, which is one of their principal roles, cells in the somites must interact with the adjacent neural tube. It has been shown that somites from the mutant strain cannot form cartilage when placed in contact with spinal cord; spinal cord from the mutant, however, can induce cartilage formation from genetically normal somites.

We have already emphasized the importance of cell movements in morphogenesis, including those that encompass large territories of the embryos. One of our principal examples was the migration of the primordial germ cells (page 93). In one especially interesting mutant of the mouse, known as *W* (dominant white spotting), primordial germ cells appear in the yolk sac endoderm in the same numbers and at the same time as in the normal mouse. However, during their migration, mutant cells fail to increase in number, whereas in the normal animal, the number continues to increase by repeated mitoses.

This observation is interesting in itself, but the story is not yet complete. In this mutant, two other kinds of cells also are defective: pigment cells and blood cells. What do the progenitors of these cells have in common with germ cells? The development of two of the three types of cells definitely involves cell migration; and there is increasing evidence that migration occurs in the third. We have already reviewed the compelling evidence for the primordial germ cells (page 93) and pigment cells, derived from the neural crest (page 91). We shall take up the development of the blood cells in Chapter 14.

This mutant illustrates an important principle: a single mutation may have multiple consequences in development. Another example emerges in the study of the *dwarf* gene of the mouse, which affects a specific class of cells in the pituitary gland; as a consequence, these cells fail to produce the hormone that can control the endocrine activities of the thyroid and adrenals. Dwarfism and sterility result. The mouse pituitary dwarf also illustrates another principle: the malfunction of a tissue is not necessarily caused by malfunction of the genes in its own cells, but often by malfunction of genes of other cells. Dwarfism and sterility can be corrected in the mutant by hormones obtained from normal pituitaries. Such treatments do not "cure" the underlying disorders, that is, they do not repair the primary genetic lesions. They merely compensate for the consequence of the lesion. Thus the thyroid and adrenal are equipped to function normally provided they receive the proper stimulus from the pituitary.

Genetic dwarfing occurs also in plants. Some dwarf mutants of corn are fertile and at maturity are less than half the height of normal plants. The reduced height results from a single gene mutation and is caused by reduced synthesis of the hormone gibberellic acid required for stem elongation. When dwarf mutant seedlings are sprayed with gibberellic acid solutions they grow to normal heights and are phenotypically indistinguishable from nonmutant plants.

### Time of Genic Action

Perhaps you were struck a few pages ago by the fact that in XXY human males, development of the gonad was essentially normal until puberty despite the extra chromosome. Normally it is only at puberty that the mechanisms of spermatogenesis are activated, in response to hormonal stimuli (themselves the consequence of a series of gene-directed events). Only then in the abnormal male is the effect of the extra chromosome expressed. What does this fact tell us? In itself, less than we would like to know. It again suggests, but does not prove, that genic action may occur late in development, even postnatally. But has genic action immediately preceded puberty, or did it occur much earlier, thus setting the stage for the latter event? Until we understand all of the biochemical events underlying developmental defects, this kind of question will always plague us.

In one sense it is redundant to discuss the "time of genic action," inasmuch as all development reflects the orderly, sequential expression of gene action. Genes are activated during oögenesis; it is during this time that the "machinery" for the early stages of development is assembled. Thus early development, through cleavage, is dependent largely on the action of maternal genes (page 16). The period of cleavage itself is not a time of intense genic action, for it is during this period that the genome is being repeatedly replicated. It is probably somewhat of an oversimplification to say that "It is at gastrulation that genic action is resumed," for genes are not quiescent during cleavage. However, as cleavage draws to a close, as divisions become less rapid, there is clearly a shift in nuclear function. It is at this point, also, that in the normal course of events the biparental origins of the zygote are first expressed.

There is little merit in drawing up a catalog or chart of the mutations affecting embryonic processes at different stages of development. It is important, however, to understand why the authors have elected to omit it. The reason is this: most of the observations that are available do not tell us *when* a gene is presumed to have acted (as judged by its failure or modification in the mutant). They reveal only that the new end result—death, failure of organization, duplication of parts, deficiency or excess of a specific product—is observable at a given time. The action of the gene may have immediately preceded the observable event, or it may have occurred well in advance of it. Moreover, many sequential reactions may be initiated or primed by genic action, thereafter proceeding independently of the nucleus.

Finally, let us examine the nuances of the phrase "sequential expression." To some it may mean precisely what it says: that individual genes are expressed one after another throughout development. To others, struck by the fact that development proceeds through a series of discrete stages—each having its own biochemical as well as morphological characteristics—the phrase may suggest the progressive expression of genes in "blocks," one block functioning in oögenesis, another at gastrulation, and so on. Bear this subtle distinction in mind as we consider the synthesis of nucleic acids and proteins during embryogenesis and the possible mechanisms for regulating gene function.

## *FURTHER READING*

Barth, L. J., *Development: Selected Topics*. Reading, Mass.: Addison-Wesley, 1964.

Boveri, T., "On Multipolar Mitosis as a Means of Analysis of the Cell Nucleus" (1902), reprinted in *Foundations of Experimental Embryology*, B. H. Willier and J. M. Oppenheimer, eds. Englewood Cliffs, N.J.: Prentice-Hall, 1964, p. 74

Driesch, H., "The Potency of the First Two Cleavage Cells in Echinoderm Development. Experimental Production of Partial and Double Formations" (1892), reprinted in *Foundations in Experimental Embryology*, B. H. Willier and J. M. Oppenheimer, eds. Englewood Cliffs, N.J.: Prentice-Hall, 1964, p. 38.

Gurdon, J. B., "Nuclear Transplantation in Amphibia and the Importance of Stable Nuclear Changes in Promoting Cellular Differentiation," *Quarterly Review of Biology*, vol. 38 (1963), p. 54.

King, T. J., and R. Briggs, "Changes in the Nuclei of Differentiating Gastrula Cells as Demonstrated by Nuclear Transplantation," *Proceedings of the National Academy of Sciences*, vol. 41 (1955), p. 321.

Levine, P., *Genetics*, 2d ed. New York: Holt, Rinehart and Winston, 1968.

Lyon, M. F., "Gene Action in the X Chromosome of the Mouse, *Mus musculus* (L.)," *Nature*, vol. 190 (1961), p. 372.

Mintz, B., "Gene Control of Mammalian Pigmentary Differentiation. I. Clonal Origin of Melanocytes," *Proceedings of the National Academy of Sciences*, vol. 58 (1967), p. 344.

———, "Hermaphroditism, Sex Chromosomal Mosaicism and Germ Cell Selection in Allophenic Mice," *Journal of Animal Science*, vol. 27, suppl. 1 (1968), p. 51.

Roux, W., "Contributions to the Developmental Mechanisms of the Embryo. On the Artificial Production of Half-Embryos by Destruction of One of the First Two Blastomeres, and the Later Development (Postgeneration) of the Missing Half of the Body," reprinted in *Foundations of Experimental Embryology*, B. H. Willier and J. M. Oppenheimer, eds. Englewood Cliffs, N.J.: Prentice-Hall, 1964, p. 2.

Stern, C., "Two or Three Bristles," *American Scientist*, vol. 42 (1954), p. 212.

———, "Gene Action," in *Analysis of Development*, B. H. Willier, P. Weiss, and V. Hamburger, eds. Philadelphia: Saunders, 1955, p. 151.

Ursprung, H., "Genes and Development," in *Organogenesis*, R. L. DeHaan and H. Ursprung, eds. New York: Holt, Rinehart and Winston, 1965, p. 3.

Weiss, P., *Principles of Development*. New York: Holt, Rinehart and Winston, 1939.

*chapter* 7

# *The Molecular Basis of Gene Expression During Development*

Is it possible to detect signs of gene activity along the length of a chromosome? Specifically, if the ideas we are developing are correct, we should be able to demonstrate that different gene loci show activity in different tissues and at different developmental stages within a given tissue.

## *CHROMOSOMAL DIFFERENTIATION*

We have already learned that no single material is suited for the analysis of every problem; we must constantly be on the lookout for forms that are uniquely fitted to the study of specific questions. Certain organisms lend themselves particularly to the investigation of cell division or cell migration. In others specific types of metabolic processes are most readily revealed; still others are particularly suited for the study of heredity. But nowhere is this generalization more applicable than in the analysis of chromosome structure. The giant, banded chromosomes in the salivary glands and other tissues of several flies (*Diptera*), including *Drosophila*, *Rhynchosciara*, and *Chironomus*, provide exceptionally favorable material. During the development of these tissues, the chromosomes are replicated, but not the cells. The newly formed chromosomes do not separate, but remain together with the old ones, forming a functional unit or *polytene* chromosome. These giant chromosomes have banding patterns so distinct that an experienced observer can use them to identify a specific chromosome or even part of a chromosome. Moreover, in *Drosophila* the banding pattern has been correlated with the genetic map as revealed by breeding experiments.

When one examines these giant chromosomes carefully, he observes characteristic features along their length. At points along the chromosome, instead of a sharp band, there may be diffuse, "puffed" regions (Fig. 7-1). These puffs reflect a "loosening" of the chromosome structure, the many strands of DNA in these giant chromosomes having separated to form loops.

Studies of the pattern of puffing activity during development have been especially revealing. First, the locations of puffs are different in different tissues at any one time; second, they are different in the same tissues at different stages of development (Fig. 7-2); and finally at any given time, all the cells of any one type in any one tissue show the same pattern.

These observations led to the proposal, first in the early 1950s, that the puffs indicate regions where genes are especially active. It is now clear that puffs are visible manifestations of gene activity. The further evidence for this statement will be better appreciated after we have reviewed the molecular basis of gene expression.

## THE MESSENGER HYPOTHESIS

We are now prepared to consider the molecular basis of the interactions between nucleus and cytoplasm. Because it is so dramatic—one of the truly thrilling chapters in the history of science—it is tempting to retell the story of the cracking of the genetic code. An understanding of the nature of the genetic material and its role in protein synthesis is essential for what follows. But this great concept, important as it is, is just the beginning of our story. What we must do is ask how the existing facts fit, and contribute to, our theory of differentiation. We must also ask, what new information may the embryologist contribute?

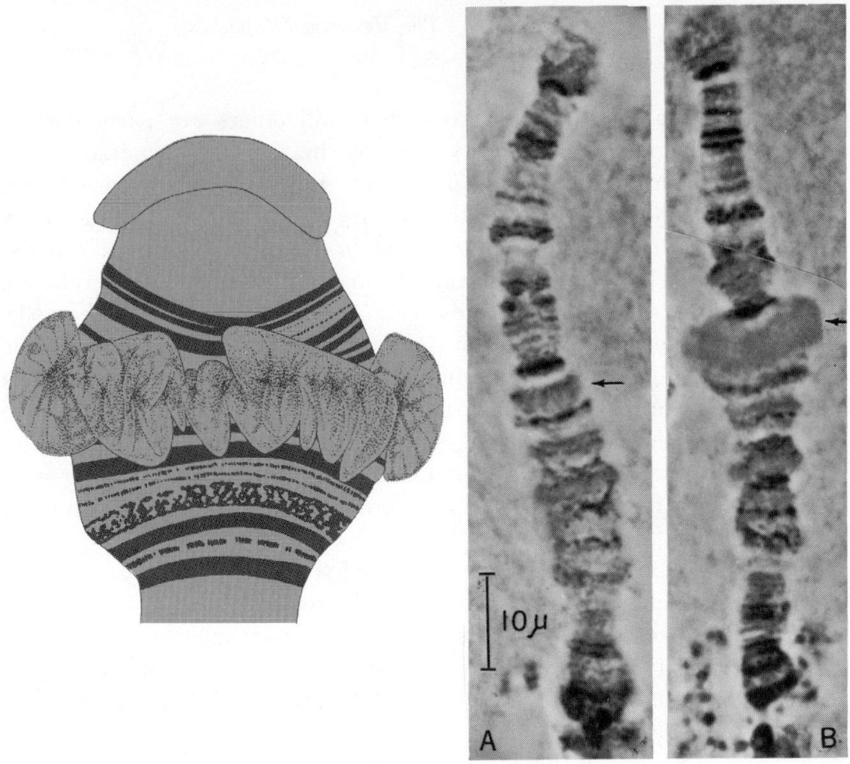

**Fig. 7-1** (Left) *A segment of a* Chironomus *salivary gland chromosome showing a moderately developed puff.* (Right) *Short arm of chromosome "C" at two stages of development of the fly,* Sarcophaga bullata. (A) *Late day 8 showing minimum puffing activity (beginning of puff at arrow);* (B) *At day 10 this same puff is much larger. (Left from Beermann, in* Chromosoma *(Berl.) 5:139-198 (1952); Right from Whitten, ibid., 26:215-244 (1969); Springer-Verlag, Berlin.)*

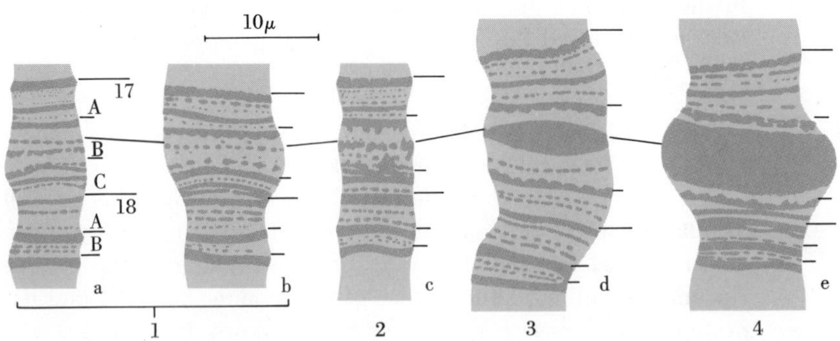


**Fig. 7-2** *Various degrees of puffing at band I-17-B in* Chironomus; (a–c) *From untreated control animals;* (d) *and* (e) *from larvae that had been injected with the hormone ecdysone. (From U. Clever, in* Chromosoma *(Berl.) 12:607-675 (1961), Springer-Verlag, Berlin, Göttingen, Heidelberg).*

To be sure of our intellectual footing, however, let us briefly, in barest outline, restate the fundamental principles on which the discussion will be based:

1. DNA is the essential material of heredity. This statement rests securely on over two decades of intensive research, the most convincing direct evidence being the demonstration that, in bacterial transformation, the *transforming principle* is DNA.

2. DNA directs the synthesis of proteins, a fact first demonstrated convincingly in now classical studies on the bread mold, *Neurospora.* Proteins are chains of amino acids (of which at least 20 are biologically important), and their functional characteristics are a product of their three-dimensional form, which is generated by folding of the linear array of amino acids. Genetic alterations lead to changes in this linear sequence, and consequently to changes in function.

3. DNA is a duplex structure, a double helix, or, in the words of Watson and Crick, "A pair of templates."

4. Each of the two strands of DNA is composed of nucleotides, of which there are four kinds, the order of nucleotides being the basis of the genetic code.

5. Thus both DNA and proteins are characterized by linear sequences of subunits, nucleotides and amino acids, respectively. There is convincing evidence that the coding unit, or *codon* for an amino acid is three nucleotides long. All but one of the amino acids (*tryptophan*) are specified by at least two different codons.

6. The code can be *duplicated* (or *replicated*) in DNA, thereby providing exact copies for hereditary transmission. It can also be *transcribed*, generating complementary copies of RNA. It is not known whether at transcription it is necessary for the nucleotide chains to uncoil and separate; it is known that one of the chains acts as a template for the synthesis of a single-stranded DNA-like, or messenger RNA (mRNA) (Fig. 7-3). Nucleotides are paired, for example, adenine (DNA) with uracil (RNA), and thymine (DNA) with adenine (RNA), and cytosine with guanine.

7. Double-stranded DNA can be broken into single strands experimentally; single strands of DNA can be recombined (hybridized) with DNA, or with RNA providing key evidence on the complementarity of DNA and RNA.

8. Once formed, messenger RNA usually leaves the nucleus and enters the cytoplasm where it is attached to ribosomes, structural units composed of at least three kinds of ribonucleic acids and a number of different proteins. Ordinarily each molecule of mRNA becomes associated with several ribosomes (the complex being known as a *poly-*

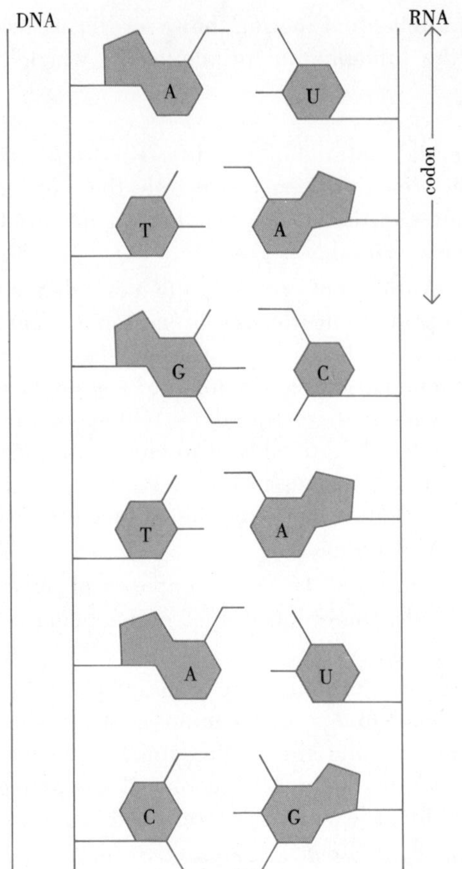


*Fig. 7-3 Nucleotide pairing in the synthesis of messenger RNA on the DNA template. (From H. Curtis,* Biology, *by courtesy of Worth Publishers, Inc.)*

*ribosome* or *polysome*) (Fig. 7-4). Now the information in mRNA must be *translated* into amino acid sequences in the protein.

9. Single, activated amino acids become associated with *transfer* or *soluble* RNA molecules (tRNA or sRNA).

The tRNA next attaches to mRNA, which in turn is complexed to part of the ribosome. The precise positioning mechanism is not known in detail: presumably a nucleotide triplet or *anticodon* in tRNA pairs with the mRNA *codon.* There is at least one transfer RNA specific for each codon—that is, one for proline, one for threonine, more than one for leucine, and so on. Moreover, there are at least 20 different enzymes, one for each combination of amino acid with transfer RNA. The alignment of the RNAs in turn lines up the amino acids in the proper sequence. The energy released when the bond between the tRNA and amino acid is broken is used in forming peptide bonds between amino acids. As now visualized, the evidence being far from complete, the

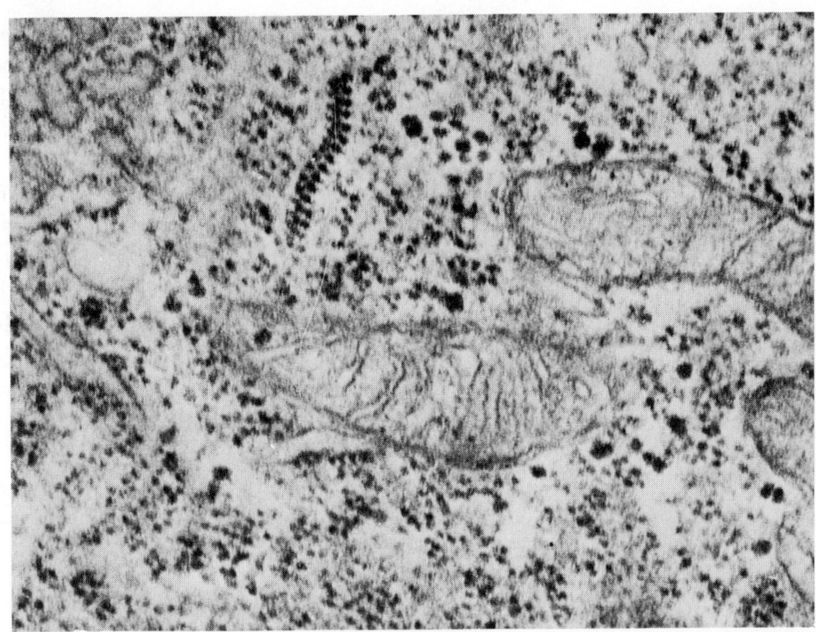

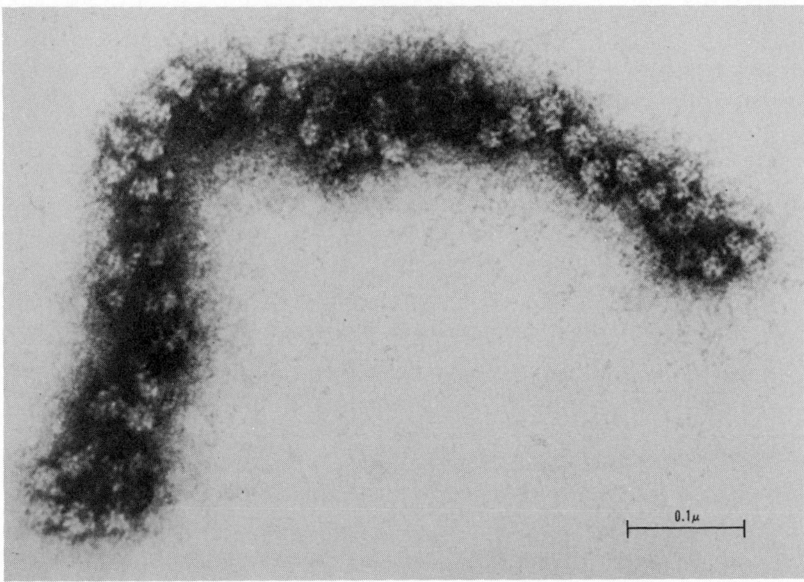


**Fig. 7-4** (Top) *Electron micrograph of cardiac muscle from ventricle of a chick embryo having 14 somites, showing helical polyribosome. Note size in comparison with mitochondria.* × *68,000. (By courtesy of F. J. Manasek.)* (Bottom) *Electron micrograph of large polyribosomes from embryonic chick thigh muscle. (From A. Rich, in* The Neurosciences. *G. C. Quarton, T. Melnechuk, and F. O. Schmitt, eds., by courtesy of the author and Rockefeller University Press.)*

polyribosome mechanism consists of a long strand of messenger RNA to which individual ribosomes attach temporarily. The ribosome "travels along" the thread, "reading" the information required for the synthesis of a polypeptide chain. Each ribosome makes a complete chain (Figs. 7-4 and 7-5).

What are the consequences of these facts, deductions, and far-reaching hypotheses for embryogenesis? Before examining this question we must ask: What kinds of evidence support the hypothesis? How was it derived? Does it apply equally to bacteria, mice, and men?

The idea that RNA is an intermediary between DNA and the synthetic machinery of the cytoplasm is an old one. The concept that the function of a gene can be equated with the formation of a protein, which came to full flower in the one gene—one enzyme hypothesis of Beadle and Tatum, appeared to require such an intermediary. Brachet and Caspersson, especially, believed that in the results of cytochemical and early biochemical studies, they found a constant relation between RNA content and protein synthesis. But the critical evidence for it is of recent vintage. The highlights are these:

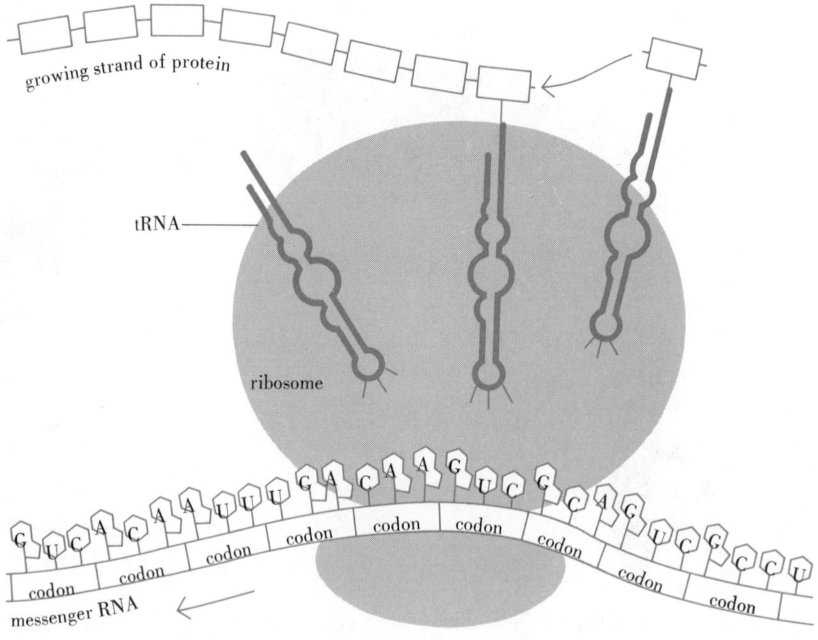


*Fig. 7-5 Possible mode of action of polyribosomes. Single ribosomes are attached temporarily to a long strand of DNA-like RNA. As each ribosome travels along the strand, it reads the information needed to synthesize a polypeptide chain. (From H. Curtis,* Biology, *by courtesy of Worth Publishers, Inc.)*

1. Bacteria and nuclei of mammalian cells were found to contain RNA-polymerases, enzymes for the synthesis of RNA that depend for their function on the presence of DNA.

2. RNA can carry genetic information; for example, certain viruses contain only RNA.

3. Following infection of bacterial cells by bacteriophage containing DNA, the infected cells synthesize RNA, which is like bacteriophage DNA, its base ratios mirroring those of the DNA.

4. In rapidly growing bacteria, ribosomes are relatively stable. The addition of substrates to the medium leads to a very rapid increase in the synthesis of specific proteins, and deletion of the same substrates bring about a comparably rapid decrease in specific protein synthesis. Therefore, it was predicted and found that a small fraction of the RNA of bacterial cells should be unstable—that is, turn over very rapidly—and that it should have a base composition mirroring the DNA of the cell. In fact, the average lifetime of an mRNA molecule in *E. coli* is two minutes.

Does the messenger hypothesis apply outside the world of microbes? Specifically, does it contribute to our understanding of the mechanisms of development? The answer, clearly, is "yes," as documented in the pages to follow. However, problems do remain. Let us consider only one. In microorganisms, messenger RNA is "defined" by (1) its size, (2) its complementarity to DNA, (3) its instability, and (4) its specificity in coding a polypeptide. In bacterial systems, all of these criteria can be met; however, as the methods have been adapted thus far for cells of frogs and chickens, protozoa, rabbits, and men, all too often only two criteria are accepted, usually the first two. The third criterion, the instability of the messenger, clearly does not always apply. The argument is best understood if we consider an experiment.

The oxygen-transporting protein of red blood cells of vertebrates is hemoglobin. In mammals it is produced in immature erythrocytes. In the course of their development, erythroblasts lose their nuclei. At this stage they are called *reticulocytes*. Thus both reticulocytes and mature erythrocytes are anucleate cells. They are highly specialized, hemoglobin being the only protein produced in bulk. More than 95 percent of the total protein is hemoglobin. Polyribosomes isolated from anucleate reticulocytes and provided with amino acids, appropriate energy sources and enzymes synthesize hemoglobin *in vitro*. An exogenous supply of mRNA is not required. In short, in the absence of the nucleus, the mRNA of the reticulocyte polyribosome is effective in coding for hemoglobin. Thus this messenger clearly must have a relatively long lifetime.

Of what earlier experimental systems does this observation remind you? In *Acetabularia*, anucleate cells are capable of regeneration; anucleate frog eggs cleave abortively. We argued that a product of the nucleus could remain in the cytoplasm long enough to accomplish a specific function or set of functions. Clearly that product could be a stable mRNA. How often is mRNA produced at one time and used at a later time? In our discussion of the time of genic action (page 157), we raised the question in another way when we asked whether the action of a gene immediately precedes an observable event in development. We may answer the question in a general way by saying that there is abundant evidence for mRNAs with a wide range of stabilities, from short- to long-lived. Thus as we proceed it will be necessary to inquire into the stability of the mRNA in each system we examine.

In bacteria the evidence shows that ribosomes are nonspecific protein-synthesizing machines, the specificity of the process being embodied in short-lived messengers. In animal and plant cells, however, the specificity of mRNAs has rarely been defined. Thus it is impossible to say, critically, whether the ribosomes of embryonic tissue are nonspecific. Is "a ribosome a ribosome a ribosome?" Are there "families" of ribosomes—kidney, liver, muscle? Are there specific ribosomes for each protein—actin and myosin of muscle, for example? Do ribosomal proteins and RNAs differ from tissue to tissue? If so, does a given ribosome continue to synthesize the same protein, always obtaining the same instructions from a corresponding gene? Or does a "muscle" ribosome synthesize now actin, now myosin, now tropomyosin, picking up messengers in sequence?

We are just beginning to dig into these questions. What little evidence is available suggests that "nonspecific" ribosomes may be reused for different synthetic functions. But the gaps in our knowledge are enormous.

And finally, we must ask, how do these synthesizing systems themselves differentiate?

### "Puffing" and the Messenger Hypothesis

On page 161 we concluded that the puffs of dipteran giant chromosomes represented active gene loci. If our earlier argument is correct, it should be possible to demonstrate the active synthesis of mRNA at the puffs. Such evidence is available. Experiments with labeled uridine, a derivative of uracil (that is incorporated into RNA but not into DNA), show that RNA is accumulated rapidly at the puffs. Biochemical studies of this RNA further suggest that it is messenger or DNA-like in its composition. Moreover, puffing can be prevented by *actinomycin D*, an antibiotic known to inhibit the synthesis of RNA.

Our story would be even more convincing if we had evidence that the mRNA actually conveyed a message, that is, if we could identify the final product it specifies. Unfortunately, that link in the chain of evidence has not yet been forged. However, there is evidence relating puffing, at one extreme, with the formation of a cytoplasmic product at the other. For example, two related species of the midge, *Chironomus, tentans* and *pallidivittatus* differ in one of the lobes of the salivary gland. In *C. pallidivittatus*, the lobe contains granules; in *C. tentans*, they are lacking. Crosses show that the trait is heritable. Moreover, the gene locus has been mapped to a thickened region near the end of a chromosome in *C. pallidivittatus*, whereas in *C. tentans* this puff is missing.

## *RNA SYNTHESIS DURING EARLY DEVELOPMENT*

The time when any gene begins to act can be most accurately defined by determining the time when DNA-dependent RNA synthesis begins. In all the species that have been studied, some of the genes of the new individual come into play at the very beginning of embryonic life, or very early therein. But do the genes for the several forms of RNA begin to function as a coordinated unit, or independently?

Studies of bacterial metabolism have shown that ribosomal RNA and soluble RNA are synthesized independently, since synthesis of the former correlates with the rate of protein synthesis, whereas synthesis of the latter is correlated with the rate of DNA duplication. Since the different stages of oögenesis and embryogenesis are so obviously different from each other with respect to rates of growth, cell division, and morphogenesis, variations in the relative synthesis of the different classes of RNA might be expected. It is now clear that the three different functional classes of RNA are synthesized independently of each other at rates determined by the stage of the embryo.

### ***DNA-like or Messenger RNA***

Throughout this section we shall use the expressions DNA-like RNA (dRNA) and messenger RNA (mRNA) interchangeably. We shall use dRNA from time to time as a device for emphasizing that most of the evidence does not permit us to speak with any degree of accuracy about mRNA in higher organisms. We know next to nothing about the informational content of these molecules, which are defined principally on the basis of their size and base composition. However, since most authors use the term "messenger," we shall fall in line, asking our readers to bear this reservation clearly in mind.

Reference has already been made (page 16) to the massive synthesis of RNA during the growth phase of oögenesis. This statement seems to be universally true, since oöcytes of all species carry on an intense RNA synthesis up to the time of meiosis and even beyond it. As we have already observed, most of this RNA is ribosomal. However, it is clear that unfertilized eggs of all animals studied (except, possibly, mammals) contain dRNA. We know more about dRNA in the sea urchin embryo than in any other form. The unfertilized egg appears to contain a complete "program" for early development, up to the onset of gastrulation, in the form of cytoplasmically stored messages.

Several kinds of experiments provide evidence for the existence of such stored or "masked" messages. First, the treatment of sea urchin eggs with actinomycin D inhibits their ability to incorporate labeled precursors into RNA, but permits protein synthesis. Second, when anucleate fragments of sea urchin eggs are activated parthenogenically, there is some increase in the incorporation of amino acids, recalling the fact that such anucleate eggs will cleave abortively. Both types of experiments suggest the presence of relatively stable messenger RNAs, produced during oögenesis and held, in some manner, in an inactive state prior to fertilization. There is also evidence of the synthesis of small amounts of dRNA during oögenesis and at ovulation in amphibians.

However, we know nothing about the specificity of these stored mRNAs nor about their precise location in the cytoplasm. Nor do we know how many such mRNAs must be stored to encode the proteins required for cleavage, for example. However, in principle, the existence of such "maternal messages," along with ribosomes and other components required for protein synthesis, helps explain the fact that the paternal genes do not need to act for development to begin.

At varying times after fertilization, depending on the species, embryonic genes are activated and new mRNAs are formed. There is clear evidence in both sea urchins and amphibians for the sequential activation of the genome.

In the sea urchin, some dRNA is made during cleavage. It is not entirely clear, however, just how much of the dRNA being made is translated promptly, for undoubtedly much of the protein synthesis during this period is encoded by "maternal mRNA." For the amphibians, we shall consider only one species, *Xenopus laevis.* Although small quantities of dRNA are formed in cleavage and blastula stages, a substantial synthesis of dRNAs occurs at gastrulation. The presence of those new DNA-like RNAs is revealed by studies of their size, nucleotide composition, and ability to hybridize with DNA. If double-stranded DNA is heated, the strands separate. If dRNA is present, as the mixture is slowly cooling, single strands of DNA will associate or hybridize with the single-stranded dRNA. If the dRNA is radioactively labeled, hybridization can be detected readily. This technique has been used to determine whether embryos at any given stage contain newly synthesized dRNA. They do, and their content increases at gastrulation. However, this method alone does not tell us whether the dRNA present in embryos at different stages is complementary to the same or different regions of the genome. For this next step of the investigation, *competition* experiments are required. First a fixed amount of a given radioactively labeled dRNA (the "reference" RNA) is incubated with a fixed amount of DNA. The amount of hybridization is observed. Next the procedure is repeated except that in addition to the labeled reference dRNA, a second nonradioactive, competitive dRNA is added to the mixture. The competitor will reduce the amount of hybridization between DNA and reference dRNA in proportion to the degree to which it and the reference dRNA share complementarity with the DNA.

Although the evidence is far from complete, the following tentative conclusions are possible:

1. In early embryos—that is, gastrulae and neurulae—genes are activated, function for a limited period, and become inactive. We may speculate that some of these genes direct the synthesis of products required for the formative movements of gastrulation and neurulation. From a biochemical standpoint, gastrulation deserves special emphasis. We have emphasized its fundamental nature in relation to future morphogenetic events. We have also referred to gastrulation as a critical, or sensitive period. In reviewing the fate of hybrid amphibian embryos, L. J. Barth concluded that only 42 percent of hybrid combinations that have been studied produce normal tadpoles or adults. Many are arrested at the onset of gastrulation. These observations and many others have fostered the idea that qualitative changes occur in the metabolism of the developing embryo at the beginning of gastrulation.

2. Some genes are activated at the blastula or early gastrula stage and probably remain active throughout the animal's lifetime.

3. In postneurula stages, some DNA sites are active in RNA synthesis for limited periods. The evidence does not permit us to say whether they are relatively short-lived genes, or whether they are genes that function intermittently, short periods of activity resulting in the production of relatively stable messengers.

### *Soluble Transfer RNA*

All of the embryos studied thus far make at least small quantities of sRNA during oögenesis. In sea urchins, the sRNA of unfertilized eggs is functional, hence it can be properly called tRNA. An egg of *Xenopus laevis* contains about one percent of its total RNA as sRNA. Although the content is high for a single cell, it is low relative to the egg's content of rRNA, representing less than one molecule for each ribosome in the egg. During ovulation there is no detectable synthesis of sRNA, although there is some incorporation of labeled precursors into preexisting molecules. This appears to reflect turnover in one of the ends of the sRNA molecule: the C:C:A (cytidylic acid:cytidylic acid:adenylic acid) terminus. The significance of this turnover is unknown. New synthesis of sRNA begins shortly after fertilization in sea urchins and in *Xenopus*, after which the embryo doubles its content with each doubling of cell number. By the time the swimming larva stage in *Xenopus* is reached, about 15 percent of the total RNA is sRNA.

**The Ribosomal RNAs** It cannot be said that we have a comprehensive knowledge of RNA synthesis in any species; however, more is known of the synthesis of the ribosomal RNAs (rRNAs) than of any of the other classes; and more is known of these processes in the amphibians than in any other forms.

Ribosomal RNA constitutes more than 80 percent of the RNA in all tissues of higher organisms studied thus far. At least three classes of rRNA are associated with animal and plant ribosomes. They are commonly designated by their sedimentation constants, which are measures of the velocity of the particles in a field of force in the ultracentrifuge. The values vary slightly from species to species; in the amphibians the three classes are designated as 28S, 18S, and 5S rRNAs. The corresponding molecular weights are estimated to be 1,600,000, 600,000, and 30,000 daltons, respectively (a dalton being $1.650 \times 10^{-24}$ gram). All three classes of rRNA have a higher content of guanylic plus cytidylic acids (G + C) than the overall DNA of the organism. There is also a clear difference in base composition between 28S and 18S rRNAs, suggesting that they are products of different genes.

In *Xenopus*, and in other amphibians, the pattern of synthesis of rRNAs is particularly striking. During oögenesis their synthesis is intense, greatly predominating over that of dRNA and sRNA. This synthesis occurs in immature oöcytes. Once the growth phase of oögenesis is terminated, no further synthesis of the rRNAs occurs until gastrulation, and then it is very low. In short, most if not all of the ribosomal RNA synthesized during oögenesis is conserved not only during the entire period of oögenesis, a period of many months and, in *Xenopus*, perhaps years, but during a large part of the period of embryogenesis as well.

During oögenesis, then, the "maternal genes" are highly active in the synthesis of the rRNAs. However, during fertilization, cleavage, and blastula stages, the embryonic genes for the rRNAs are "silent." The synthesis of ribosomal RNA, inactive since oögenesis, starts anew at the beginning of gastrulation. This synthesis is correlated with the first appearance of nucleoli in the embryonic cells, suggesting along with numerous other observations, a relation between nucleoli and ribosomal RNA synthesis.

Although rRNA synthesis begins at gastrulation, its pace is relatively slow in *Xenopus*. It is not until later, when the tail bud is forming (at about the time of onset of yolk utilization) that there is a significant increase in the *net* amount of the rRNAs and of ribosomes.

Except for minor differences in detail, this pattern holds good for many other forms, mammals and the worm *Ascaris* being among the exceptions. Biochemical evidence suggests that rRNA synthesis begins very early in mammalian embryogenesis, for example, following the second cleavage in the mouse. It is at this time that the nucleoli appear in the early blastomeres of the mouse embryo.

In thinking about this fundamental difference in the biochemistry of development in amphibians and mammals, it may be useful to reflect on the nutrition of the two kinds of embryos.

Biochemically speaking amphibian embryogenesis has several features that should be emphasized. The entire process of early development takes place in very nearly a "closed system." The early embryo is impermeable to most exogenous organic molecules, and the materials required for embryogenesis are derived from nutrients present in the unfertilized egg. The overall growth of the embryo is limited. Thus, cleavage is the opposite of oögenesis: it is a time of intense cell division and DNA duplication in the absence of growth.

Note that we said the absence of growth, not synthesis. There must be some synthesis apart from the synthesis of DNA; for example, the proteins of the spindle, as well as new cell surfaces, must be formed.

In mammalian embryos, however, growth and net increase of mass begin early, for the embryo develops from the outset in an environment fully and constantly supplied with nutrients. The early increase in cytoplasmic mass would require an early production of ribosomes if the ribosomal density in the cytoplasm were to be maintained.

An unusual pattern of RNA synthesis is observed in the development of *Ascaris lumbricoides*. After activation, development proceeds for from 12 to 24 hours as the eggs pass through the uterus. During this time large amounts of rRNA are accumulated, especially around the male pronucleus. It will be noted that earlier we used the word "activation" rather than fertilization; we had a good reason. During the first 50 to 60 hours of development, the male and female pronuclei do not fuse. The sperm nucleus remains separate, and the evidence indicates that the rRNA that is synthesized is made under the direction of the *paternal* genome. Thus the oöcyte produces yolk, but not ribosomal RNA. The sperm produces rRNA after activation while the egg nucleus completes meiosis. Thus the egg begins cleavage with stored rRNA—but of paternal, not maternal, origin.

## *EMBRYOGENESIS IN THE ABSENCE OF RIBOSOME SYNTHESIS*

The late appearance of new ribosomes suggested the possibility that the developing amphibian embryo could utilize the ribosomes synthesized during oögenesis and present in the unfertilized egg until the tailbud stage.

The opportunity to prove this point was provided by the existence of a mutant strain of *Xenopus*. The normal wild-type *Xenopus* has two nucleoli in the majority of its diploid cells (2-*nu*). The viable mutant has only one nucleolus in each cell (1-*nu*). The heterozygous condition (1-*nu*) has no apparent effect on the viability or reproductive capacity of the animal. When two heterozygotes (1-*nu*) are mated, the genotypes of the progeny give the Mendelian segregation pattern: 2 nucleoli (2-*nu*), 1 nucleolus (1-*nu*), and no nucleoli (0-*nu*) in a ratio of 1:2:1. The homozygous mutant (0-*nu*) develops normally past the hatching stage to an early swimming stage. At this time these anucleolate embryos stop growing and development is arrested; death follows several days later (Fig. 7-6). In view of the late onset of ribosome synthesis in normal embryos, it seemed likely that the molecular deficiency in these embryos might involve the synthesis of the ribosomal RNAs. This prediction was already supported in part by earlier studies, which had demonstrated that both nuclear and cytoplasmic RNA were lower in the 0-*nu* mutants after hatching than were the control embryos of the same stage.

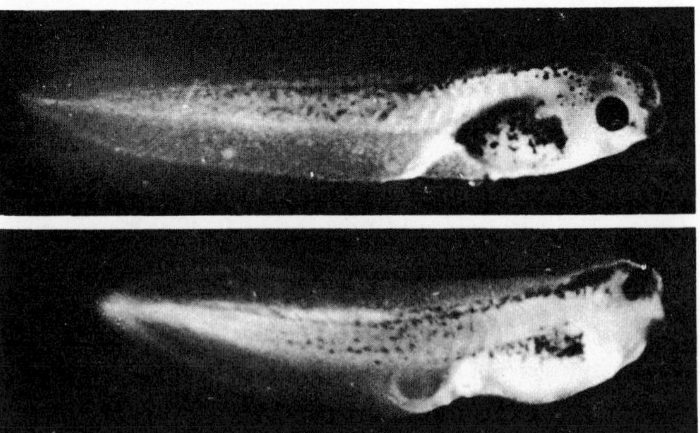

**Fig. 7-6** *Comparison of normal* (top) *and anucleolate* (bottom) *embryos of* Xenopus laevis. *These embryos are siblings that have developed for the same length of time. (From D. D. Brown and J. Gurdon, in* Proceedings of the National Academy of Sciences, *51 by courtesy of the authors and the Academy.)*

To prove that the homozygous mutants are incapable of synthesizing the ribosomal RNAs, Brown and Gurdon performed an experiment that distinguished newly synthesized RNAs from those synthesized during oögenesis, when the growing oöcyte was still heterozygous with respect to the genetic mechanism for nucleolus formation. The experiment was designed to subject control and mutant sibling embryos to a radioactive nucleic acid precursor ($C^{14}O_2$) at a time when wild-type embryos were capable of synthesizing new ribosomal RNAs but before any differences could yet be detected morphologically between the control and 0-*nu* embryos. Following incubation with the labeled precursor, the embryos developed for two days in a nonradioactive medium, and then were assorted according to genotype by determining the number of nucleoli in the cells of each individual by phase-contrast microscopy. The anucleolate (0-*nu*) embryos were sorted from the wild-type (2-*nu*) and heterozygote (1-*nu*) embryos, and the RNAs separated. Only the RNA that was synthesized between neurula and tailbud stages is radioactive (Fig. 7-7). The 0-*nu* embryos

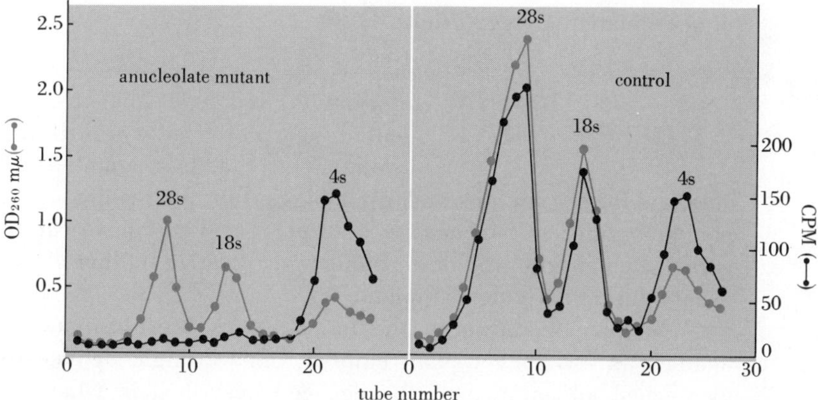


*Fig. 7-7 Sucrose density gradient centrifugation of total RNA isolated from 0-*nu *and control embryos. Two heterozygote (1-*nu*) adults were mated and the embryos allowed to develop to neurulation when they were incubated in a closed serum bottle at pH 6.0 with about 0.2 μc $C^{14}O_2$ for 20 hours at 18°C with mild shaking. By the end of this incubation period, development had proceeded to the stage of muscular response, and the mutant embryos were still indistinguishable grossly from the control. The medium was changed, and the embryos continued development in nonradioactive tap water for 48 hours at 20°C. The anucleolate mutants were recognized by examination of their tail tips with a phase contrast microscope and separated from the two control genotypes (1-*nu *and 2-*nu*). Both groups were then washed with distilled water and frozen at −20°C before chemical analysis. The bulk of the RNA (•——•) is represented by optical density measurements at 260 mμ. The RNA synthesized between neurula and muscular response stage is represented by the radioactive measurements (•——•). (From D. D. Brown and J. Gurdon, in *Proceedings of the National Academy of Sciences*, 51 by courtesy of the authors and the Academy.)*

contain all classes of the ribosomal RNAs and soluble RNA. However, only the soluble RNA is radioactive, in contrast to the control embryos, in which the rRNAs as well as sRNA are radioactive. Since the total amount of ribosomal RNA in the mutant is about the same as that found in the unfertilized egg, it is concluded that, during development, the mutant conserves the RNA synthesized during oögenesis and present in the unfertilized egg but is incapable of synthesizing new ribosomal RNA during embryogenesis. The defect prevents the synthesis of all classes of ribosomal RNA without affecting the synthesis of sRNA, DNA, or dRNA. Bear in mind, however, that although we have said that the mutant affects the synthesis of 28S, 18S, and 5S rRNAs, we have not said *how*.

In the anucleolate mutant the structural genes for 28S and 18S rRNAs are deleted. However, 5S rRNA is coded by genes distinct from those for the 28S and 18S rRNAs and not necessarily contiguous with them. Although the two sets of genes are independently located, they must somehow interact. In the next chapter we take up the evidence for these statements, for they lead us to a consideration of mechanisms of coordination or *regulation*.

## RNA IN PLANT DEVELOPMENT

Information on the changes in RNA content of plant embryos during their development is far less complete than that obtained for sea urchins and amphibians. But by drawing on studies of plants as different as *Fucus*, wheat, peas, and cotton we can piece together enough information to indicate a tentative outline of molecular events during early development.

Nothing is known of the immediate postfertilization molecular changes in any higher plant embryo, but we might deduce from the cytological appearance of the egg cell described in Chapter 5, and from the immediate start of growth after fertilization that, like the mammalian embryo, there is early synthesis of RNA. Rhizoid formation in the zygote of *Fucus* depends on RNA synthesized after fertilization, because zygotes cultured continuously in actinomycin D fail to form a rhizoid. However, if actinomycin treatment is delayed until five hours after fertilization rhizoid formation occurs normally, suggesting that the RNA is synthesized during this five-hour period.

More is known of RNA changes associated with germination of the seed. Like amphibian embryos, seeds are essentially "closed systems" containing nutrients and requiring to be activated for further development. Although RNA synthesis begins early in germinating seeds this RNA is apparently not essential for the early developmental stages, and protein synthesis and development proceed at rates comparable to untreated control seeds in actinomycin D treated seeds. Recently experiments have shown that polysomes are present in the dormant embryo and that mRNA present in the dormant embryo is translated during the first two days of germination. There is, therefore, evidence of a masked form of mRNA in the dormant embryo.

## FURTHER READING


Britten, R. J., and D. E. Kohne, "Repeated Sequences in DNA," *Science*, vol. 161 (1968), p. 529.

Brown, D. D., "The Genes for Ribosomal RNA and Their Transcription during Amphibian Development." in *Current Topics in Developmental Biology*, vol. 2, A. Monroy and A. A. Moscona, eds. New York: Academic Press, 1967, p. 47.

———, and J. Gurdon, "Absence of Ribosomal RNA Synthesis in the Anucleolate Mutant of *Xenopus laevis*," *Proceedings of National Academy of Sciences*, vol. 51 (1964), p. 139.

Gall, J., "Chromosomes and Cytodifferentiation," in *Cytodifferentiation and Macromolecular Synthesis*, M. Locke, ed. New York: Academic Press, 1963, p. 119

Gross, P., "Biochemistry of Differentiation," *Annual Review of Biochemistry*, vol. 37 (1968), p. 631.

Kaulenas, M. S., and D. Fairbairn, "RNA Metabolism of Fertilized *Ascaris lumbricoides* Eggs during Uterine Development," *Experimental Cell Research*, vol. 52 (1968), p. 233.

Levine, R. P., *Genetics*, 2d ed. New York: Holt, Rinehart and Winston, 1968.

Loewy, A., and P. Siekevitz, *Cell Structure and Function*, 2d ed. New York: Holt, Rinehart and Winston, 1969.

Spiegelman, S., "Hybrid Nucleic Acids," *Scientific American*, May 1964, p. 48.

Ursprung, H., "Genes and Development," in *Organogenesis*, R. L. DeHaan and H. Ursprung, eds. New York: Holt, Rinehart and Winston, 1965, p. 3.

Woodland, H. R., and C. F. Graham, "RNA Synthesis during Early Development of the Mouse," *Nature*, vol. 221 (1969), p. 327.

*chapter* **8**

# The Regulation of Gene Expression: Levels of Control

In Chapter 6 we discussed the interactions of nucleus and cytoplasm. In Chapter 7, however, we narrowed our field of vision to the genes and the synthesis of their immediate products, the ribonucleic acids. We now return to our main theme, nucleocytoplasmic interactions, this time at the molecular level.

One of the most striking examples of the cytoplasmic control of nuclear activity is found in the following experiments. It will be recalled that the synthesis of rRNA that predominates during oögenesis in *Xenopus* stops at maturation, not to be resumed until gastrulation, and not to be resumed in large amounts until substantially later. Nuclei from late embryos or young tadpoles were transplanted into eggs. Following transplantation, the nucleoli which were prominent in the embryonic nuclei disappeared within 40 minutes, and rRNA synthesis was not detected until gastrulation. In other words, the genes controlling rRNA synthesis are active in cytoplasm of cells at the tadpole stage, but become inactive when placed in cytoplasm of cleaving eggs.

It may be helpful first to consider some of the kinds of questions with which we shall be dealing. We raised one question at the close of the preceding chapter: within a given genome, how are the activities of genes at different loci, even on different chromosomes, coordinated? If only a fraction of the genes in a given cell are acting at any moment, how are the neighboring genes held in check? Once the expression of a gene has resulted in the formation of a stable message (as in hemoglobin synthesis in the mammalian reticulocyte, where, having acted, the gene is not just "silent," but lost), what regulates protein synthesis? In short, we must go to the very root of our subject to ask: how are these intracellular reactions coordinated and regulated in differentiating cells?

Gene action may be regulated at four levels, at least. First, the number of copies of a given gene may differ from cell to cell or from time to time in the history of a given cell, as a consequence of a *differential replication* or *amplification* of part of the genome. We shall consider conclusive evidence for the control of ribosomal RNA synthesis by *differential gene content.*

Second, the frequency with which a given gene is transcribed into RNA copies may vary. Such controls at the *level of transcription* appear to be commonplace. We refer to the third level of control as *intranuclear processing.* By this we mean that the immediate products of transcription—for example, the ribosomal RNAs—may not be identical with the final functional product we identify in the cytoplasm. The immediate product may have to be "processed" before it becomes functional.

Finally, transcribed copies of genes in the form of mRNAs may or may not be translated into products; that is, regulation may occur at the *level of translation.*

## DIFFERENTIAL GENE REPLICATION (AMPLIFICATION) IN OÖCYTES

As we have seen, at least three kinds of ribosomal RNA are made; they are designated according to their size. Oöcytes and embryos of *Xenopus* synthesize rRNA at different rates depending upon the stage of development of the egg or embryo. Ribosomal RNA is formed in large quantities during oögenesis, and ribosomes are assembled during that period. When the egg becomes mature, however, rRNA synthesis ceases, and resumes only well after development is initiated. Thus, the activity of genes for rRNA must be controlled in a sensitive way during oögenesis and embryogenesis. The control of this activity involves both the availability of genes for transcription and frequency with which available genes are transcribed.

Evidence from many laboratories shows that the nucleolus is the site at which the 28S and 18S ribosomal RNAs are made. The most convincing demonstration of this was obtained in the *anucleolate* mutant of *Xenopus*, in which the nucleolus is completely absent.

Embryos which are homozygous for the mutation do not synthesize ribosomal RNA. The embryo contains enough rRNA made previously during oögenesis to support early development, but when in later stages the embryo fails to make its own new rRNA, it dies. The nucleolus originates from a specific chromosomal site, the "nucleolar organizer," which is visible in most animal cells as a secondary constriction on one or more metaphase chromosomes. In *Xenopus*, each set of haploid chromosomes contains one autosome with a single nucleolar organizer. In the homozygous lethal anucleolate mutant, both nucleolar organizers are absent. Moreover, molecular hybridization studies show that at least 99 percent of the DNA coding for the 28 and 18S ribosomal RNAs is also missing. In other words, the anucleolate mutant is, in fact, a *deletion* of that part of the chromosome coding for these two rRNAs.

The occurrence of this deletion, and the development of refined fractionation and molecular hybridization techniques have made possible the "molecular mapping" of these genes. We now know that each haploid set in normal *Xenopus* cells contains about 450 genes for 28S and about 450 genes for 18S rRNA. These genes are clustered on one part of one of the 16 chromosomes in the set, probably along with some "spacer" DNA which differs from that controlling the two rRNAs. Moreover, the evidence suggests that the 28S and 18S genes are alternating. It is not known whether spacer DNA separates the 28S and 18S genes, but it seems unlikely.

There are over 20,000 genes for 5S rRNA in the normal set, which appear to be partly clustered, but they are not intermingled with the cluster of genes for 28S and 18S rRNA.

Thus, in the normal somatic cell, the genes for the rRNAs are redundant, that is, there are repeated sequences. These presumably occur in all the cells of an organism and at all times in its life history. For convenience, we shall refer to the genes coding for rRNA as the "ribosomal DNA" or rDNA.

However, we are concerned with an even more striking phenomenon, *specific gene amplification*. Oöcytes of several amphibians, an echiuroid worm, and the surf clam increase their synthesis of ribosomes by replicating specifically the genes for 28S and 18S early in oögenesis, subsequently using these extra genes as templates for massive rRNA synthesis.

An oöcyte of *Xenopus* has about 1000 times as many genes for 28S and 18S rRNA as does the nucleus of the somatic cell. These extra copies are made during a relatively short interval of oögenesis, between two to four weeks after metamorphosis. Moreover, once made, these extra copies are isolated in multiple nucleoli. Since the growing oöcyte persists in the first meiotic prophase for an extended period it is tetraploid. Thus one would expect to find four nucleoli. Instead, in *Xenopus* oöcytes, there are about 1000 (Fig. 8-1). Several techniques prove not only that these nucleoli contain DNA, but that the DNA contains sequences homologous to the 28S and 18S rRNAs; in other words, it is, at least in part, rDNA.

These multiple copies function only during oögenesis. They are not replicated during cleavage, and ultimately are somehow rendered nonfunctional and discarded.

Why do oöcytes need multiple copies of these genes? The most obvious explanation is that they are needed to support the extraordinarily high rate of rRNA synthesis. An immature oöcyte of *Xenopus* can synthesize rRNA at a rate comparable to an equal weight of liver tissue containing about 200,000 cells. It is interesting that the genes for 5S rRNA are not amplified. However, as we have noted, they are already highly redundant (20,000 copies per haploid complement, compared to 450 copies each for 28S and 18S). Clearly the 28S – 18S gene cluster is somehow coordinated with the 5S genes, for the latter, although present, do not function in the anucleolate mutant.

The presence of extrachromosomal nucleoli in amphibian oöcytes stimulated Miller and Beatty to isolate the genes coding for ribosomal RNA precursor molecules and to observe them in the electron microscope (Frontispiece). The isolated nucleolus contains a compact fibrous core, containing DNA, RNA, and proteins, and a granular cortex that

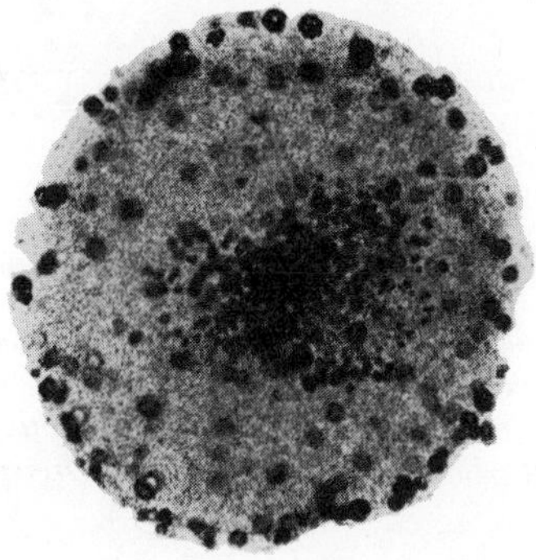

***Fig. 8-1*** *Photomicrograph of an isolated germinal vesicle of* X. laevis. *The germinal vesicle was dissected from a mature oöcyte in 0.01* M *$MgCl_2$ and 0.02* M *tris buffer, pH 7.4, and flooded with cresyl violet stain. Its diameter is about 400 μ. The deeply stained spots are some of the hundreds of nucleoli. (From D. D. Brown and I. B. Dawid, in* Science, *by courtesy of the authors and American Association for the Advancement of Science. Copyright © 1968 by the American Association for the Advancement of Science.)*

lacks DNA. When the fibrous core is isolated and "unwound," it is seen to contain a thin axial fiber, 100 to 300 Å in diameter, that is periodically coated with matrix material (Frontispiece and Fig. 8-2). From studies using enzymatic digestion, Miller and Beatty conclude that the core fiber is a double-helix DNA molecule coated with protein. Further studies with enzymes, coupled with electron microscope autoradiography, indicate that the matrix material is RNA, most likely a precursor of ribosomal RNA. Therefore it appears that each matrix-covered DNA segment codes for the pre-rRNA. The redundant structural arrangement of genes and intergene segments visually confirms the biochemical evidence for the organization of the genes coding for the rRNAs.

Not all of the species studied present as striking a picture. In the echiuroid worm, *Urechis*, and the surf clam, *Spisula*, the amplification, although real, is small in comparison with that in the amphibians, being only 5- to 10-fold. The oöcytes of these species have only a single nucleolus.

There can be no question that differentiation can result, in these examples, from differential gene content. Moreover, amplification of these genes is observed in *Drosophila*. However, at the moment, we do not know how general this phenomenon may be. A few additional examples of differential replication of parts of the genome have been described recently, but we know little about them. One of the more

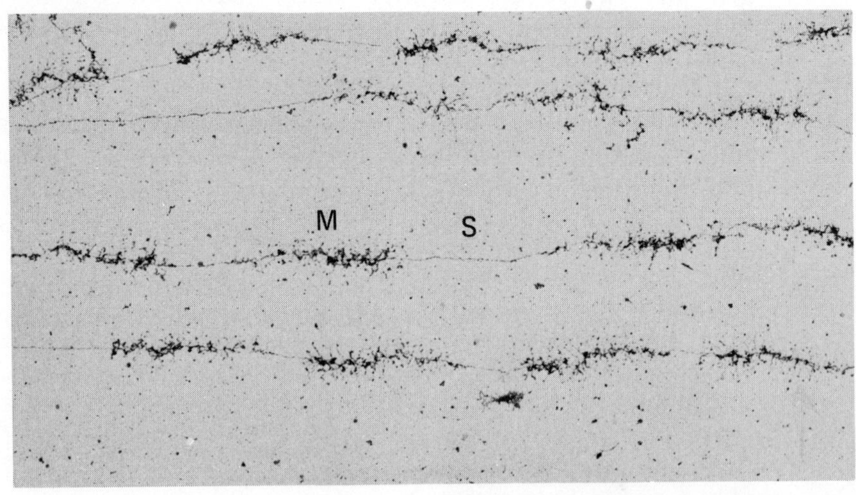


**Fig. 8-2** *Portion of nucleolar core isolated from* Triturus viridescens *oöcyte, showing matrix units* (M) *separated by matrix-free segments* (S) *of the core axis. Intermatrix segments of various lengths are present. Electron micrograph (× 10,000) by Drs. O. L. Miller, Jr. and Barbara R. Beatty, Biology Division, Oak Ridge National Laboratory.*

striking is the DNA puffing observed in sciarid polytene chromosomes. These DNA puffs are a characteristic feature of the chromosomes of both *Rhynchosciara angelae* and *Sciara coprophila*. They have been studied less intensively than the RNA puffs discussed earlier (pp. 161 and 168), but like RNA puffs, DNA puffs appear to be visible manifestations of gene activity, in this case amplification of localized parts of the genome.

## CONTROL AT TRANSCRIPTION

The control of cellular activities at the level of transcription appears to be universal. It is hardly necessary to expand on this point, for we have already treated the subject, under several headings, e.g., "When do genes begin to act?" (page 157). Most of the information we have available about early embryogenesis comes from studies on whole embryos. We know very little about differences between parts of embryos. What evidence we do have permits two conclusions; namely, that (1) the parts of the genome transcribed change during embryonic development, and (2) differentiative events involve at some point, a transcription step, which may immediately precede the differentiation or may occur some time in advance of it, on a fixed schedule. We shall consider further examples when we take up the synthesis and assembly of proteins and organelles (Chapter 9).

Since control at transcription is a common event, how is it accomplished? We do not know. It is commonly assumed that since chromosomal DNA exists as a highly ordered complex with proteins, and perhaps with RNA, that one or more of the proteins "masks" the DNA and inhibits transcription. It is further assumed that the positioning, quality, or quantity of the "mask" is somehow regulated by molecular input from the cytoplasm.

This hypothesis is based, in part, on the well known regulator–operator–operon hypothesis. *It is not at all certain, however, that the regulatory mechanisms in the cells of animals and plants function in the same way as those in bacteria.* However, since the concept is so often discussed, and since so much of our experimental design and terminology have sprung from it, we should consider it here in the context of differentiation. In their far-reaching hypothesis, Jacob and Monod postulated that there are at least two fundamental classes of genes, structural and controlling. *Structural* genes are those which code for mRNAs and hence specify the amino acid sequences of proteins and are those genes on which our discussion thus far has been centered. *Controlling* genes are those which regulate DNA function, and in turn fall into two classes, *regulators* and *operators*. A set of closely linked genes (Fig. 8-3) forms an *operon*, so called because these genes are

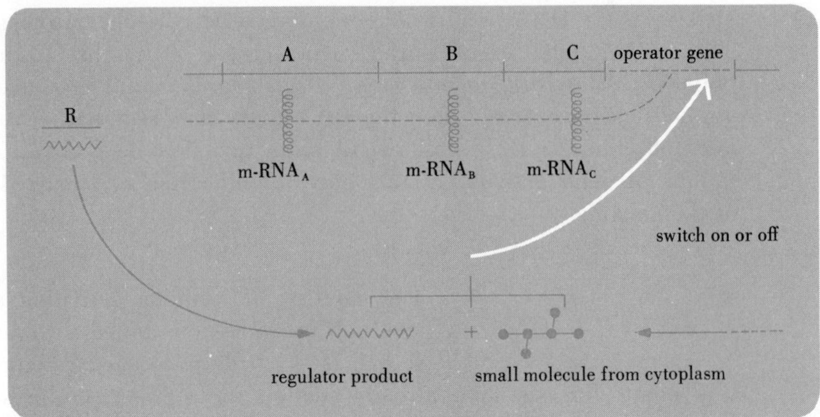


***Fig. 8-3*** *Cell regulatory mechanisms.* R: *regulator gene.* A, B, C: *linked genes determining enzymes of a synthetic pathway. $mRNA_A$ is messenger RNA synthesized by gene A, and so forth. Production and release of mRNA may be either permitted or inhibited by the operator gene which, in turn, is turned on or off by the reaction between the regulator product and a small molecule from the cytoplasm. (From D. Bonner and S. Mills,* Heredity, *2d ed., © 1964, by permission of Prentice-Hall, Inc., Englewood Cliffs, N.J.)*

under the immediate control of a common gene, their operator. When the operator is active, the structural genes under its control synthesize mRNA; when it is inactive, the structural genes are inactive. What determines whether the operator is active or inactive? It is inactive whenever it combines with a specific repressor, which is in turn the product of a regulator gene. It is the product of the regulator gene, which interacts with the "environment"—that is, with the specific metabolites in the cell, which are, in a sense, the effectors of enzyme induction and repression. The difference between induction and repression appears to reside in the way metabolites affect the operator–repressor complex. In enzyme induction, the inducing metabolite is believed to combine with the repressor to block its interaction with the operator. In enzyme repression, combination of the repressing metabolite and repressor must somehow strengthen interaction with the operator.

Many of the predictions of the model have now been verified for bacteria and viruses. Two repressors have been isolated and characterized as proteins (in the *lac* operon in *E. coli*, which controls the enzymes $\beta$-galactosidase, permease, and transacetylase; and in cells containing bacteriophage $\lambda$). These proteins have a high affinity for DNA, suggesting that the repressor blocks transcription by binding directly to DNA.

To what extent may this scheme be applied to embryos? Let us begin this discussion by briefly reviewing the organization of DNA in chromosomes. The complex of DNA with protein and RNA is termed *chromatin.* It has two states, "condensed" and "extended"—the terms meaning just what they say. A typical example of condensed chromatin is a chromosome at metaphase when it is discrete and small. A "puff" is an example of extended chromatin. Condensed chromatin is not active; that is, the DNA is not transcribed. The DNA in extended chromatin is transcribed. There is evidence that most of the DNA of somatic cells is not transcribed, and that the fraction that is varies from one cell type to another. Different cell types use different amounts and parts of the genome. In maturing red blood cells, narrowing their range of transcription to hemoglobin, the amount of condensed chromatin is high.

There is also biochemical evidence for organ-specificity of chromatin; that is, chromatin extracted from bone marrow and thymus, for example, contains readily transcribable DNA sequences specific for those organs.

This kind of evidence is taken to mean that most of the DNA is masked. What is the nature of the mask?

It appears reasonably certain that in mammalian chromatin, ionically linked proteins mask DNA. Some of the masking, possibly a major part, may be due to *histones.* Histones are logical candidates. These basic proteins (which are made in the cytoplasm) are found closely associated with DNA in chromosomes. Moreover, there is an impressive body of evidence that the removal of histones from chromosomes *in vitro* may derepress genes, leading to mRNA synthesis, and conversely that the addition of histones to DNA preparations will inhibit synthesis. There is also evidence that nonhistone, acidic proteins may also mask DNA. It is interesting that the repressor in the bacteriophage λ is an acidic protein. But how does one account for the specificity in the regulation of gene action? Several ideas have been advanced. According to one, the specificity is embodied in a special class of RNA molecules linked to histones. The base composition of the DNA to be masked would be recognized not by histone but by RNA. A second proposal suggests that DNA–histone interactions are modified according to the acetylation of histones. Acetyl groups are added to histones at times of extensive gene activation. A third scheme attempts to bring together ideas from bacterial (specific repressors) and animal (nonspecific masks or repressors) systems. Specific repressors would interact by reversible binding with one or a limited number of operators, and nonspecific histones or other nonspecific masks could interact with the DNA at any point.

As in the usual bacterial scheme, the regulatory action of a specific repressor would be reversible. Histone would turn a gene off only if it became complexed with the operator; only under these conditions would the initiation of RNA synthesis become sensitive to histone binding. If there are no inducers for histones, there would be only one time during the cell life cycle when a gene whose operator is complexed with histone could become free of histones. This would be during the period when both DNA and histone synthesis occur in the nucleus and cytoplasm, respectively. In other words, newly duplicated genes have the option of complexing with either a specific repressor or histones. A gene previously masked nonspecifically with histones may simply reassociate again with histones. However, if other factors have changed since the previous replication (change in the cytoplasm, and so forth), it may now associate with a specific repressor, opening up a new range of possibilities.

We must emphasize that all of these schemes, especially the third, are *highly speculative.* There is very little evidence of the existence of cytoplasmic inhibitors and stimulators. We will consider it only after we have taken up other levels of control.

## *THE PROCESSING LEVEL*

The immediate product of the genes specifying the 28S and 18S rRNAs is not identical with the final products. These genes are simultaneously transcribed, producing a single, larger precursor molecule, which sediments at 45S. In the nucleolus this molecule is "processed," or cleaved, into the two definitive smaller molecules. This much is definitely known, but the details are not understood. The process appears to be more than a simple splitting of the 45S molecule; some sequences are probably lost, and there may be changes in composition.

Moreover, these events are only the beginning of a larger story. The ribosomal RNAs are only a part of the ribosomes themselves which, in *Xenopus*, are composed of at least 30 kinds of proteins, organized into a particle consisting of two subunits. Little is known about the control of synthesis of these proteins except that in *Xenopus* it seems to be coordinated with the synthesis of the 28S and 18S rRNAs, for when rRNA synthesis is lacking in the anucleolate mutant, the synthesis of ribosomal proteins also fails. Thus one example of *processing* is the assembly of the ribosomes.

We shall consider only two further examples. First, among the products of transcription there are found some very large molecules of DNA-like RNA, much larger than the usual mRNA. Much of this material never leaves the nucleus. Its role is unknown. There are several possibilities: it may be functional within the nucleus; it may be a precursor to be cleaved into smaller mRNAs that will function in the cytoplasm; or it may be a nonfunctional by-product of mRNA synthesis.

Second, mechanisms must exist for "protecting" mRNA in the cytoplasm from the activity of cytoplasmic nucleases. Some mRNA's, for example, the "maternal" mRNAs, emerge into the cytoplasm well before they are translated, and there is no proof that they are incorporated into conventional polyribosomal complexes.

## *CONTROL AT TRANSLATION*

Translation-level control is as fundamental as control by transcription and by processing. Many examples could be cited. However, many of the more instructive illustrations must be drawn not from early embryogenesis, but from studies of tissue-specific proteins in the terminal stages of cell differentiation. These may be taken up more appropriately in the next chapter. Here we will confine ourselves to a consideration of maternal messages, in whose function translational controls play a part. As we have seen, the rate of protein synthesis in sea urchin eggs is low. However, "maternal" mRNA—that made during oögenesis—is stored in an inactive form. Following fertilization, before new transcription begins, the "maternal" mRNA is translated at a high rate. Even as new mRNAs are formed and begin to function, some maternal mRNAs apparently still are used. Gradually the pattern changes, until the genome of the zygote assumes full command.

## *THE SEARCH FOR CYTOPLASMIC INHIBITORS AND STIMULATORS*

It remains a common assumption that there must exist in the cytoplasm substances that interact with the genome or its products at these four levels, thus regulating cellular activities. The entire history of the study of nucleocytoplasmic interactions argues for their existence. We have discussed one recent example (page 179), the cytoplasmic control of the rRNA cistrons in *Xenopus*. This experiment argues for an inhibitor (or repressor), although it is not known whether it might act directly or indirectly on the genome.

Early attempts to isolate and characterize this inhibitor from *Xenopus* blastulae suggest that it is dialyzable and heat stable, indicating that it is a relatively small molecule. However, it is not yet clear whether it inhibits at the transcription or processing levels.

There is also preliminary evidence of stimulating factors in the cytoplasm. Nuclei taken from adult *Xenopus* brain, in which DNA synthesis has stopped, and injected into *mature*, unfertilized eggs, begin DNA synthesis. The cytoplasm in the mature unfertilized egg is poised to initiate DNA synthesis and somehow stimulates DNA synthesis in nuclei that do not normally make DNA. When brain nuclei are injected into *immature* oöcytes, they are not stimulated to divide. Thus during maturation of the oöcyte, factors responsible for the initiation of DNA synthesis emerge, and they are capable of acting on nuclei from adult cells.

## *SOMATIC CELL HYBRIDIZATION*

One of the more novel ways of approaching the interactions of nucleus and cytoplasm and, hopefully, ultimately identifying the factors involved, is the *fusion of animal cells* to form *heterokaryons* and *somatic cell hybrids*. A *heterokaryon* results from the fusion of two or more cells and consists of a single cytoplasmic mass containing two or more different nuclei. There are no further implications, that is, to be a heterokaryon, the mass does not have to propagate itself. A true *hybrid cell strain*, in contrast, is a propagating cell line containing the chromosomes from different parent cells within a single nucleus. In other words, it involves the *mating* of somatic cells. These recent innovations have already begun to prove their usefulness.

Heterokaryons are commonly formed by mixing two cell types with an inactivated virus, the so-called Sendai virus, named for one of the cities in Japan where it was studied. The exact manner in which the noninfective virus operates in promoting cell fusion is unknown; in some way, by modifying the cell membranes, cells are caused to fuse. The following example is especially instructive. Red blood cells of chickens were fused with human cells. The latter were either of two types—cultured cell lines known as HA and HeLa. The HA cells have the capacity to produce *interferon*, a specific product whose role in the animal is to inactivate viruses. We are interested in it here only as a "marker," indicative of the onset of gene function. HeLa cells do not produce interferon, nor do chicken erythrocytes. In fact, in the latter the nuclei are considered to be inactive (although they are not extruded as are the nuclei of mammalian erythrocytes). The heterokaryon of HeLa cells and erythrocytes produces no interferon. However, interferon is produced in the HA–erythrocyte combination. But what kind? Interferons are known to be species-specific. Both chicken and human interferons are made. At some level the erythrocyte has been reactivated to make a product it probably never made in the course of its life history. A long silent part of the genome appears to have been derepressed.

Propagating hybrid strains have usually originated "spontaneously" when the cells of two established lines are mixed in culture. The underlying cause of such fusions, mating, and subsequent establishment of propagating lines is not known. Evidence of interaction between the genomes in hybrids is given in the following experiment. Hybrids were made between cells of a pigmented Syrian hamster melanoma line (a melanoma is a highly malignant black tumor) and an unpigmented mouse cell line. The hybrid cells were isolated and maintained in culture for up to 100 cell generations. The hybrid cells remain *unpigmented* and lack one of the key enzymes involved in pigment production. Somehow the functions of the hamster melanoma cells are suppressed by factors in the mouse cells.

One of the limitations of the techniques of cell hybridization—the low rate of spontaneous occurrence—may soon be overcome. It seems possible to greatly increase the rate of true hybridization by using virus-induced fusion as the first step, coupling it with media that favor selective growth of hybrids. Such "virus-assisted hybridization" may prove to be of great value in another way. Most of the work thus far has involved hybridization of established cell lines grown in culture. These cell lines are not "normal" diploid cells. Cell lines that have become established—that is, capable of propagation *in vitro* indefinitely—often have undergone modifications, including changes in chromosome number (see page 153). Early indications are that "virus assistance" may make possible the hybridization of normal embryonic cells.

## FURTHER READING

Birnstiel, M., J. Speirs, I. Purdom, and K. Jones, "Properties and Composition of the Isolated Ribosomal DNA Satellite of *Xenopus laevis*," *Nature*, vol. 219 (1968), p. 454.

Bonner, J., M. E. Dahmus, D. Fambrough, R. C. Huang, K. Marushige, and D. Y. H. Tuan, "The Biology of Isolated Chromatin," *Science*, vol. 159 (1968), p. 47.

Britten, R. J., and E. H. Davidson, "Gene Regulation for Higher Cells: A Theory," *Science*, vol. 165 (1969), p. 349.

Brown, D. D., and I. B. Dawid, "Specific Gene Amplification in Oöcytes," *Science*, vol. 160 (1968), p. 272.

Ebert, J. D., "Levels of Control: A Useful Frame of Perception?" in *Current Topics in Developmental Biology*, vol. 3, A. Monroy and A. A. Moscona, eds. New York: Academic Press, 1968, p. xv.

Ephrussi, B., and M. C. Weiss, "Regulation of the Cell Cycle in Mammalian Cells: Inferences and Speculations Based on Observations of Interspecific Somatic Hybrids," in *Control Mechanisms in Developmental Processes*, M. Locke, ed. New York: Academic Press, 1968, p. 136.

Gall, J. G., "Differential Synthesis of the Genes for Ribosomal RNA during

Amphibian Oögenesis," *Proceedings of the National Academy of Sciences*, vol. 60 (1968), p. 553.

Maden, B. E. H., "Ribosome Formation in Animal Cells," *Nature*, vol. 219, (1968), p. 685.

Ritossa, F. M., "Unstable Redundancy of Genes for Ribosomal RNA," *Proceedings of the National Academy of Sciences*, vol. 60 (1968), p. 509.

Zubay, G., *Papers in Biochemical Genetics*. New York: Holt, Rinehart and Winston, 1968.

*chapter* **9**

# Beyond the Ribosome

In the preceding chapters our discussion has revolved chiefly about the nucleus and its relations with the cytoplasm, especially the ribosomes. It is fitting that the stress should be placed here, for this is the stage on which today's drama is being played. We may predict, with some degree of assurance, that in a few years the emphasis will have shifted to questions that we are only beginning to perceive today.

Our task is difficult, for it is not always easy to come to grips with these problems and to be able to state them in precise ways. In this chapter we shall begin to inquire into those factors that determine not only the shape of a cell but also the organization and distribution of the organelles within it.

## THE SYNTHESIS AND ASSEMBLY OF PROTEINS

### Synthesis

Protein synthesis is a sequential process; we have been looking at the beginning. Is anything to be gained by looking at the end products? There is merit in making an inventory of proteins and describing changes in them during development, not only because we want to learn all we can about differentiation, but also because we must constantly be on the lookout for systems in which the entire range of reactions may be studied.

But even changes in patterns of protein synthesis are difficult to recognize. Just a few years ago we were concerned only with questions such as the following: Are specific new proteins synthesized in cells before the cells can be recognized by morphological criteria as erythrocytes or feather cells or muscle cells? Or does the synthesis of specialized products begin only after morphologic criteria are distinguishable? Today we recognize that questions asked in this way are not truly meaningful. Obviously an erythrocyte is only an erythrocyte when it contains hemoglobin, and muscle cells, by definition, contain contractile proteins. With what precision can we say that a protein is being synthesized for the first time? The answer depends on the sensitivity of the techniques used—both biochemical and morphological. Now questions can be framed more precisely: At what point in the development of a prospective muscle cell (myoblast) is the segment of the genome directing the synthesis of the contractile protein myosin activated to produce the mRNA coding for it? Once produced, is the message short- or long-lived? If the latter, when does it come into play, and what factors regulate its translation?

Moreover, we must push our analysis back further in development. We referred to the myoblast. Obviously this cell not only has been determined, or "firmly biased" toward the synthesis of the contractile proteins, but it also has morphologic characteristics permitting us to recognize it. These characteristics must reflect the presence and interaction of molecular species that we have not begun to identify. Thus the embryologist studying protein synthesis must pay attention not only to the factors regulating the ultimate expression of genes for the already defined proteins, the hemoglobins, myosins and the like, but also to the ontogeny of the (thus far) obscure molecules that first mark a cell as a reticulocyte or a myoblast. We shall consider only a few examples of defined proteins that are currently being studied.

**Lactate dehydrogenases** For the embryologist, one of the more "useful" proteins is the enzyme lactate dehydrogenase (LDH), an oxidoreductase that appears to be ubiquitous in tissues of vertebrates. It catalyzes the interconversion of pyruvate and lactate and, simultaneously, of nicotinamide-adenine dinucleotide (NAD) and dihydronicotinamide-adenine dinucleotide ($NADH_2$). As Markert and Moller found, LDH exists in five distinct molecular forms, or *isozymes.* We now know, in fact, that a number of enzymes exist in several different isozymic forms, as resolved by electrophoresis in starch gels (Fig. 9-1). Nearly all tissues contain measurable amounts of all five LDH isozymes, designed simply LDH, 1 through 5 (Fig. 9-2).

The adult pattern of LDH isozymes in any tissue emerges gradually during embryonic development. In the mouse, all embryonic tissues first exhibit mainly LDH-5. During embryogenesis, enzyme activity

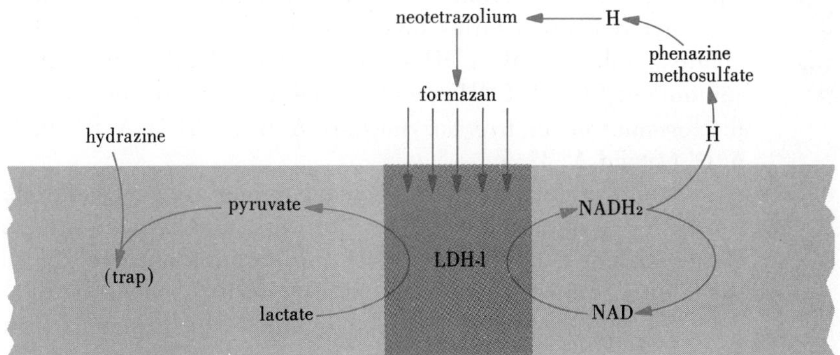


**Fig. 9-1** *Diagram of the reactions involved in visualizing the isozymes of LDH after resolution by starch gel electrophoresis. The method essentially provides a colorimetric test for the production of $NADH_2$. (After C. L. Markert, in* Cytodifferentiation and Macromolecular Synthesis, *by courtesy of Academic Press.)*

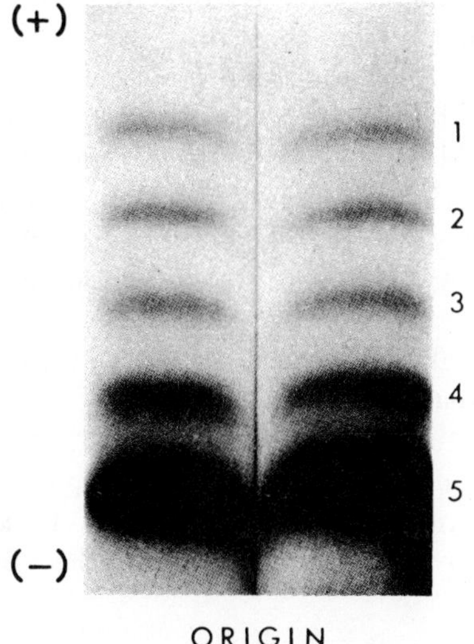


**Fig. 9-2** *Diagram of zymograms of two independent preparations of mouse LDH from skeletal muscle. Note that all five isozymes are present. Electrophoresis occurred at pH 7.0 at room temperature for about six hours at a voltage gradient of 6 V/cm. (After C. L. Markert, in* Cytodifferentiation and Macromolecular Synthesis, *by courtesy of Academic Press.)*

progressively shifts toward LDH-1. However, there is much variation: in skeletal muscle LDH-5 predominates in both embryo and adult, whereas in the heart a pronounced shift is found. In embryonic heart, LDH-5 predominates; in adult heart, LDH-1 and LDH-2 predominate. Each tissue matures at its own rate.

In the chick embryo, LDH-5 is not the principal isozyme; instead LDH-1 appears first. The pattern does not change in the heart throughout development, but in skeletal muscle the shift is from LDH-1 to LDH-5.

The LDH molecule may be split into four polypeptide chains of equal size. These subunits are inactive enzymatically, but two electrophoretically distinct forms may be distinguished. LDH-1 contains four identical subunits (B), LDH-5 contains four identical subunits (A). The subunits of LDH-2, LDH-3, and LDH-4 are mixtures of A and B. Thus, the formulas of the five isozymes are $A^0B^4$ (LDH-1), $A^1B^3$ (2), $A^2B^2$ (3), $A^3B^1$ (4) and $A^4B^0$ (5).

It appears that the A and B polypeptides are gene-controlled; thus we are led to inquire how the levels of function of these two genes are regulated, not only to account for the patterns observed in adult tissues, but also for the changing patterns during development.

One interesting idea, which is still being explored, took shape from the observation that LDH-5 tended to predominate in tissues that are exposed to anaerobic conditions, whereas LDH-1 was found in tissues in aerobic, well-oxygenated environments. It appears, then, that the control of LDH synthesis may be related to oxygen tension in the cells.

Evidence for metabolic differences among embryonic cells abounds. For example, the metabolic pathways operating in the early development of the brain and heart differ markedly.

Early chick embryos have been cultivated in a medium containing traces of a metabolic inhibitor, *antimycin A*, a substance produced by a species of the mold *Streptomyces*. Concentrations of this inhibitor as low as three tenths of a microgram per embryo, applied via the endoderm, inhibit the development of the cells destined to form heart almost completely, while leaving the developing brain and spinal cord intact (Fig. 9-3).

At the same or slightly higher concentrations, in older embryos, somites form but do not persist. However, hemoglobin is formed in the presence of antimycin A.

Antimycin A inhibits oxidative metabolism at a point in the electron transport chain contained in the mitochondria. Extracts of mitochondria contain a protein capable of preventing the effects of antimycin A.

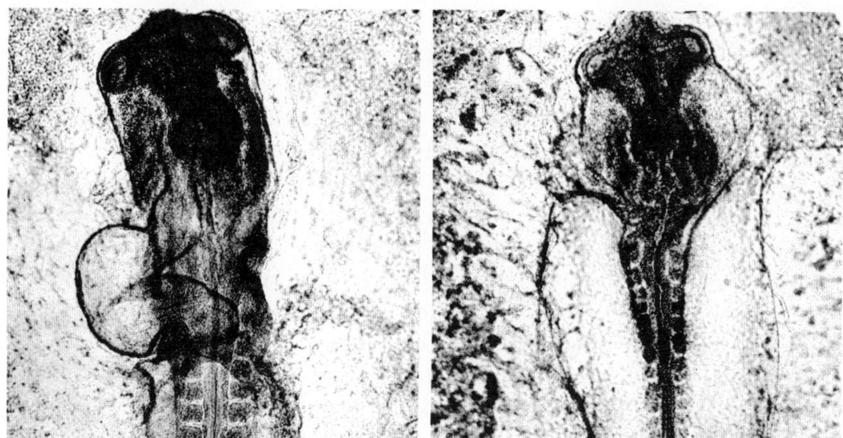

***Fig. 9-3*** (Left) *Anterior end of a chick embryo cultivated on an albumen saline-agar medium for 24 hours after explantation at primitive streak stage.* (Right) *Embryo cultivated for a similar period in a similar medium to which 0.03 microgram of antimycin A was added.*

As we have said earlier, different cells may have alternate pathways for performing the same task; often each of two cell types may be capable of using both pathways, but normally use only one. We may conjecture then that the early chick embryo has two pathways of oxidative metabolism, one sensitive and one insensitive to antimycin A. Under the usual conditions of explantation, the heart-forming regions and, to a lesser extent, the somites, are operating via a pathway sensitive to antimycin A, whereas the remainder of the embryo operates via predominantly insensitive pathways. However, if the embryo is grown under high levels of oxygen, it adjusts to the former pathway.

***Hemoglobin*** Hemoglobin has also been studied extensively; or we should say, "the hemoglobins," for there are a number of different hemoglobins known in every species studied thus far, recognizable by virtue of differences in electrical charges. In man, who has been studied more fully than other animals, the different hemoglobins reflect single gene differences (see *Genetics*, in this series). Moreover, studies of amino acid composition and sequence of the hemoglobins demonstrate that many of the abnormal hemoglobins differ from the normal by only a single amino acid at the same point in a limited fragment of the molecule.

Moreover, as we have related earlier (page 167), the machinery for hemoglobin synthesis, the polyribosomes of reticulocytes, is being actively investigated. Of special concern to the student of development, however, is the discovery that in all of the species that have been studied carefully, the hemoglobins of the embryo and adult differ. We cannot generalize about the number of embryonic hemoglobins, but the fact of their existence is clear. In man, fetal hemoglobin (HbF) is the predominant hemoglobin of the fetus. Even at birth, it comprises 60 to 80 percent of the total hemoglobin. Most HbF has disappeared by the fourth month after birth, but adults carry traces of it, usually less than one percent. The fetal protein differs from that of the adult in one pair of its polypeptide chains, but it can be distinguished from the adult fraction by its resistance to alkaline denaturation.

When is hemoglobin first synthesized? Obviously, man is not the experimental material of choice. More work on this question has been done on the chick embryo than on that of any other species. Hemoglobin is a tetramer, composed of four polypeptide chains and nonprotein heme groups containing iron. In the chick embryo, it is not possible to say when the "very first" molecules are synthesized, but synthesis of embryonic hemoglobins has clearly commenced by the six- to seven-somite stage. It is also clear that transcription of the genes controlling embryonic hemoglobins has occurred substantially earlier. Actinomycin D does not inhibit hemoglobin synthesis in early somite stage embryos. In fact, inhibition is not observed if embryos are exposed after the late head-fold stage. When younger embryos are exposed, hemoglobin synthesis is prevented. These observations, which are representative of a larger number, certainly suggest that the "hemoglobin message" (mRNA) has been produced by the head-fold stage but that it is inactive until later, indicating control of synthesis at the translational level. If so, how is it regulated? At the present time an unequivocal answer is not possible. It was thought at one time that the study of the synthesis of the heme group held the key. It was proposed that the availability of the heme precursor delta amino levulinic acid (DAL), resulting from the activity of the enzyme DAL synthetase, is the central regulator of hemoglobin production. Heme is clearly necessary for hemoglobin synthesis, but is it the controlling element?

Apparently not, for when three- to six-somite embryos are cultured on a rich whole-egg medium, added DAL does not result in enhanced hemoglobin synthesis as compared with controls. DAL does enhance hemoglobin formation in embryos cultured on minimal media. Although DAL may be rate limiting under conditions of deprivation, it does not appear to be *the* controlling factor in the normal initiation of hemoglobin synthesis.

***Crystallins and keratins*** The *crystallins* are the principal proteins of the ocular lens. We know that the lens is formed as a consequence of an inductive interaction between the optic vesicle and the overlying ectoderm.

The *keratins* are the principal products of epidermal cells; keratins are the materials from which hair and, in birds, feathers and their homologues, scales, are made. Normally these definitive structures result from an inductive interaction involving the epidermis and its underlying dermis. Rawles and others have studied these interactions, using the technique of reciprocal exchange of epidermis and dermis from prospective scaled and feathered regions. It is clear that both the capacity of the dermis to call forth a response in epidermis and the capacity of the epidermis to respond vary with the time of development as well as location on the body (Fig. 9-4).

The cells of the developing lens and feather have several properties in common with reticulocytes. Each of these cell types makes large quantities of a small number of proteins: crystallins, keratins, and hemoglobins, respectively. Moreover, the cells convert virtually their entire substance into their specific products. We have learned

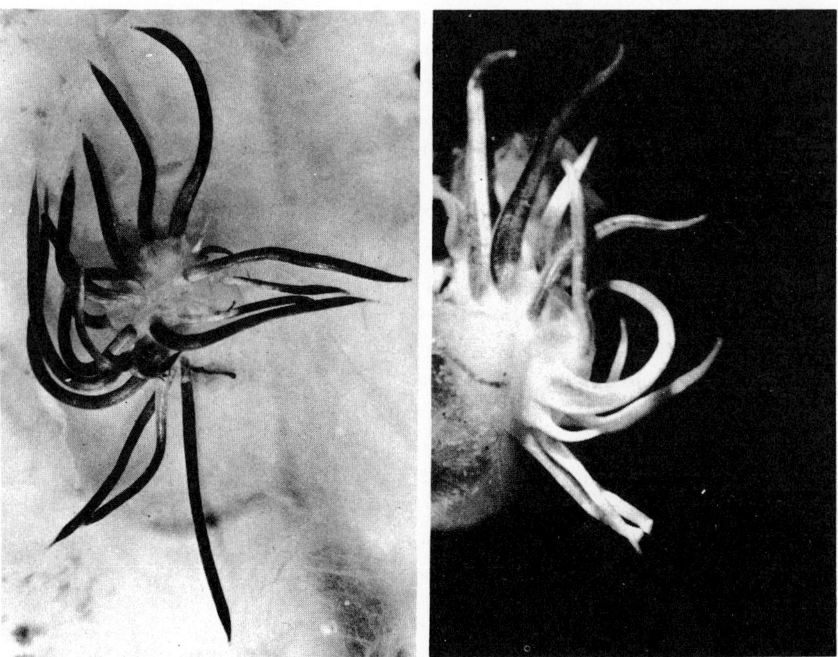

**Fig. 9-4** *Grafts showing normal down feathers produced by the recombination of epidermis and dermis and transplantation to the chorioallantoic membrane.* (Left) *8-day chick embryo back epidermis combined with 11-day foot dermis;* (Right) *10-day epidermis plus 6-day back dermis. (From M. E. Rawles, in* Journal of Embryology and Experimental Morphology, *by courtesy of the author and the Company of Biologists.)*

that in reticulocytes hemoglobin synthesis continues after the nucleus disappears, through the intervention of a stable messenger. Lens and feather cells also utilize long-lived mRNAs, as shown by experiments in which the synthesis of their specific proteins continues in polyribosomes prepared from them after the synthesis of DNA-dependent RNA has been prevented by actinomycin D.

However, in focusing attention on the activities of prospective feather and lens cells, we have ignored a large question. How does interaction with dermis stimulate an epidermal cell to produce not just keratins, but *specific* keratins of beak, feather, or scale? How does interaction with optic vesicle stimulate an ectodermal cell to form crystallins? Where in the responding cell's inner controls does the "signal" from dermis or optic vesicle impinge? Not all of the facts are in, but the ideas we do have will be taken up in Chapter 10.

## Assembly

### Self-assembly at the molecular level

We begin by noting that up till now we have avoided several key questions. For example, you may have wondered why we did not take up the fate of a protein after it was produced. True, cells are characterized by specific differences in macromolecules, but once a protein rolls off the polyribosomal assembly line, then what? If it is a hormone, ACTH perhaps, or a digestive enzyme, say trypsin, it is ultimately secreted. But suppose it is a structural protein, such as collagen, actin, or myosin? How are these molecules ordered into threads and the threads woven into fabrics?

One of the cardinal principles of biological organization is that of *self-assembly*. Once peptides are formed, they have the ability to aggregate spontaneously to form larger units. We cannot hope to treat the chemical forces involved in this brief book. It is essential, however, that we illustrate the problems of macromolecular organization.

We begin by taking up an exception to the "rule" that LDH exists in five isozymic forms. Certain fish do not produce five isozymes, but only two or three. In the "two-isozyme fish," the isozymic forms present are the homopolymers, $A^4B^0$ and $A^0B^4$. To appreciate what is to follow we need to add one additional experimental observation. The enzyme may be dissociated chemically *in vitro* into its tetramers. The tetramers or subunits can then be recombined. The recombination ordinarily can be extended to mixtures of isozymes from widely differing species.

We have seen that in the fish we are discussing, the A and B subunits apparently do not interact *in vivo*, since the only isozymes found are the homopolymers. Moreover, the A and B subunits of this fish do not interact in the test tube either. However, what is remarkable is the fact that when they are combined *in vitro* with mammalian subunits, functional hybrid molecules are produced! Something about the subunit structures prevents combination of heterologous subunits from the same fish, but not from mammals.

There are other examples. We might have elected to consider the spontaneous aggregation of two $\alpha$ and $\beta$ chains to form the tetrameric hemoglobin molecule; or the aggregation of the molecules of the enzyme pyruvic decarboxylase with six molecules of dihydrolipoyl dehydrogenase and 24 of a transacetylase, each with its own coenzyme, to form a "supermolecule" of molecular weight 5 million, organized to accomplish the metabolism of pyruvic acid; or of actin, myosin, and other contractile proteins to form the muscle filament.

However, possibly the most interesting model for studies on morphogenesis is the organization of *collagen*. This structural protein is widely distributed throughout the animal kingdom, and in vertebrates it may account for up to 30 percent of the total protein. Many details of its structure are understood, and the protein can now be readily isolated, purified, and identified, especially by virtue of its high content of the amino acids hydroxyproline and hydroxylysine.

The elementary molecule of collagen, called *tropocollagen*, which is synthesized on polyribosomes, is a long rod, composed of three helical polypeptide chains; in the body these rods are aggregated into fibrillar arrays (Fig. 9-5).

Once synthesized, tropocollagen moves from the cell to the intercellular space. The manner in which this movement is accomplished is virtually unknown. However, some information about the fate of tropocollagen outside the cell is available. The most recently formed tropocollagen is soluble in cold salt solutions under physiologic conditions. Progressively, however, the tropocollagen molecules are organized into larger aggregates, collagen fibrils, with a characteristic repeating axial structure (Fig. 9-6) having a native repeating period of 640 Å. As aggregation occurs, the protein becomes more and more insoluble. The large bulk of collagen in an old animal is insoluble except under very unusual conditions.

In considering aggregation of the basic units, we must bear two facts in mind. First, as these identical molecules come together, order emerges. This self-ordering of macromolecules appears to be a major force in morphogenesis. Second, the ordering requires a compatible environment. Although it has not yet been established that changes in the extracellular microenvironment determine the aggregation patterns, it is clear that varying conditions do affect them. In the presence of serum acid glycoprotein, extracted collagen fibrils reconstitute with a period of 3000 Å rather than 640 Å. With adenosine triphosphate (ATP), 3000-Å fibrils are again reconstituted, but with a different fine structure. Different salt concentrations produce different periodicities. The developmental implications are intriguing.

Collagen fibrils occur in bone and cornea and in the dermis of the skin, and in each they are assembled and arranged in a geometric pattern in a mucopolysaccharide matrix. In the dermis of fishes, for example, the pattern is highly ordered; in land animals, it is randomly interlacing. In amphibians, the larvae show a pattern like that of fishes; but during reorganization at metamorphosis the pattern changes to resemble the form characteristic of land animals.

***Morphogenesis of bacteriophage T4*** Assembly mechanisms may be studied effectively in microorganisms. One of the more striking ex-

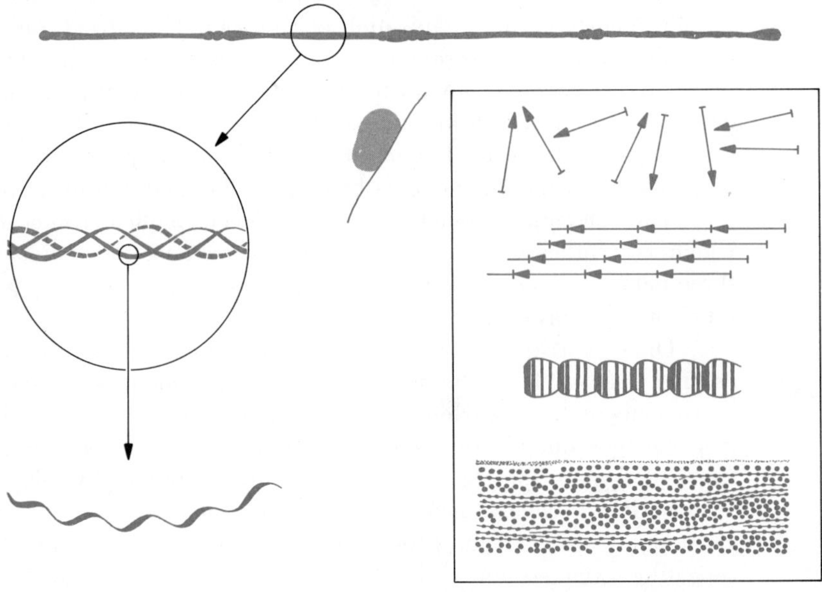

*Fig. 9-5 Collagen fine structure. The elongated tropocollagen molecule shown at top has an asymmetric fine structure roughly subdividing the molecule in quarters. It is compared with a 70S ribosome on a membrane sketched to scale below it. The enlargement of the section of the molecule to the left illustrates the three polypeptide chains, one of which, the dashed line, is different in amino acid composition from the other two. A further enlargement of a section of one of the chains illustrates the helical configuration of each of the polypeptides. In the box at right is a diagram of the manner in which tropocollagen units are postulated to aggregate, overlapping each other in a staggered array by about one quarter of their length, thereby giving rise to a collagen fibril with a characteristic repeating period. The fine structure within each of these periods would be a reflection of the fine details of asymmetry of the tropocollagen units arrayed in register. The lower portion of this block illustrates the manner in which collagen fibrils are found organized in plywoodlike sheets in a variety of tissues. (From J. Gross, in* Cytodifferentiation and Macromolecular Synthesis, *by courtesy of Academic Press.)*

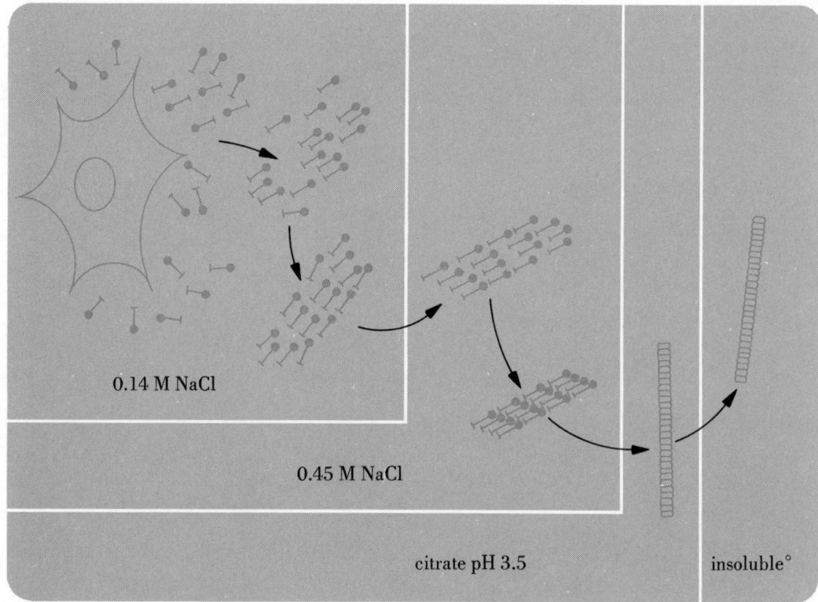


*Fig. 9-6 Diagram of hypothesis explaining the significance of the different extractable collagen fractions. Rodlike units represent tropocollagen molecules. Cold physiological saline extracts the most recently formed collagen; hypertonic salt solution extracts the same material plus older collagen in a more ordered state of aggregation; acid citrate buffer extracts all of the above plus some of the older collagen in the typical fibrillar form. The insoluble fibrils are older and their degree of cross-linking has prevented solubilization. (From J. Gross, in* Cytodifferentiation and Macromolecular Synthesis, *by courtesy of Academic Press.)*

amples is found in the studies of Edgar and Wood on the morphogenesis of bacteriophage T4.

The structure of this bacteriophage is diagrammed in Fig. 9-7. Note that in addition to the diagram representing the morphology of the phage as a whole, the parts are figured separately. There is good reason for drawing the figure just this way, for the parts clearly are made as separate units and subsequently assembled in stepwise fashion. Moreover, several of the steps have been demonstrated to be controlled by individual genes.

Over 40 genes are known to be involved in T4 morphogenesis. In Fig. 9-8 we see a map of the circular genome. Let us consider the effects of mutations at a few loci. Gene 9 mutants produce inactive phage particles with contracted sheaths. Mutations of genes 11 and 12 produce fragile particles which dissociate to free heads and free tails.

The manner in which these mutants are exploited to study morphogenesis is illustrated by the following example. A multiple mutant

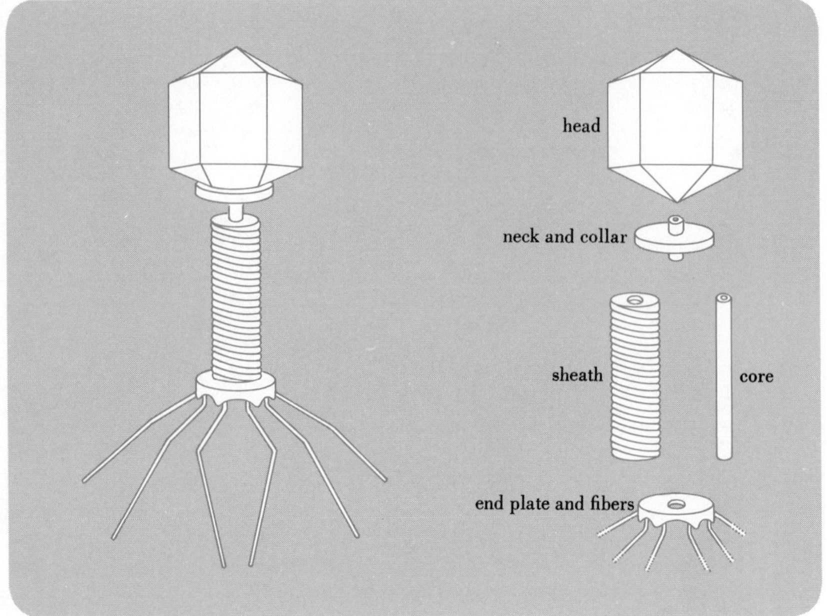


*Fig. 9-7 Parts and structure of the T4 bacteriophage.*

(X4E) was developed, defective in tail fiber genes, 34, 35, 37, and 38. When a culture of *E. coli* was inoculated with X4E, phage particles lacking tail fibers were synthesized. In contrast, another gene, 23, controls the synthesis of the major structural component of the head membrane. A mutant, defective in this gene *am*B17 produces tail fibers, but no heads or complete particles.

When purified tail-fiberless particles are incubated with an extract prepared from "headless" mutant *am*B17, complete, active phage are produced.

The results of many such *complementation* experiments tell us that at least some of the major steps in the morphogenesis of phage T4 involve the gene-controlled assembly of completed units. Such studies suggest that other morphogenetic pathways may be similarly open for attack.

## THE FORMATION OF ORGANELLES

It is a large jump from the ordering of tropocollagen molecules into collagen fibrils to the ordering of the fibril into the species- or stage-specific geometric pattern in the skin. The organization

***Fig. 9-8*** *Defective phenotypes of conditional lethal mutants of T4D under restrictive conditions. Characterized genes are represented by shaded areas illustrating relative locations and, if known, approximate map lengths. The enclosed symbols indicate defective phenotypes as follows: DNA neg., no DNA synthesis; DNA arrest, DNA synthesis arrested after a short time; DNA delay, DNA synthesis commences after some delay; mat. def., maturation defective, DNA synthesis is normal but late functions are not expressed; a hexagon indicates that free heads are produced; an inverted T indicates that free tails are produced; tail fiber, fiberless particles produced; gene 9 mutants produce inactive particles with contracted sheaths; gene 11 and 12 mutants produce fragile particles which dissociate to free heads and free tails. (From R. S. Edgar and W. B. Wood,* Proceedings of the National Academy of Sciences, *vol. 55, 1966, by courtesy of the authors and the Academy.)*

of these plywoodlike lattices is baffling. The pattern is orthogonal, but related somehow to the contours of the body surface. As Edds wrote, it is still impossible to conjure up the shape of a tadpole from any known properties of collagen molecules or fibrils.

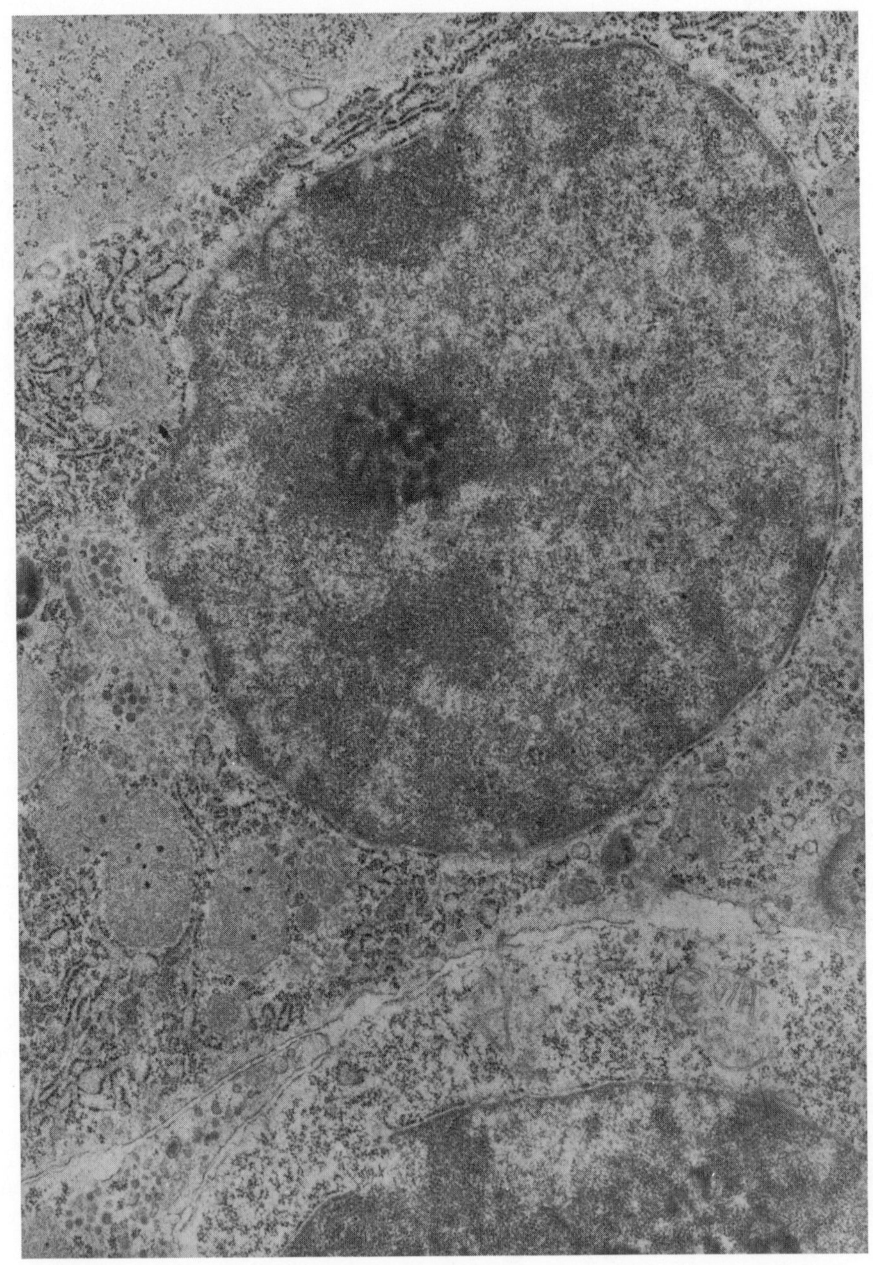

***Fig. 9-9*** *Electron micrograph of a section of liver from a 12-mm rat embryo. This cell shows the beginnings of differentiation. A few profiles of endoplasmic reticulum are evident, but many ribosomes are unattached.* (Upper left) *Part of an erythroblast.* (Lower right) *A less differentiated hepatoblast.* × *12,000 (approximately). (By courtesy of K. R. Porter.)*

Or, to put it another way, compare Figs. 9-9 and 9-10, which are electron micrographs of the embryonic liver cell or hepatoblast and the adult liver cell. In the hepatoblast we see the beginning of differentiation. A few profiles of endoplasmic reticulum are evident, but many ribosomes are unattached. In the adult liver, the endoplasmic reticulum is highly organized and there are few free ribosomes. As we look at the

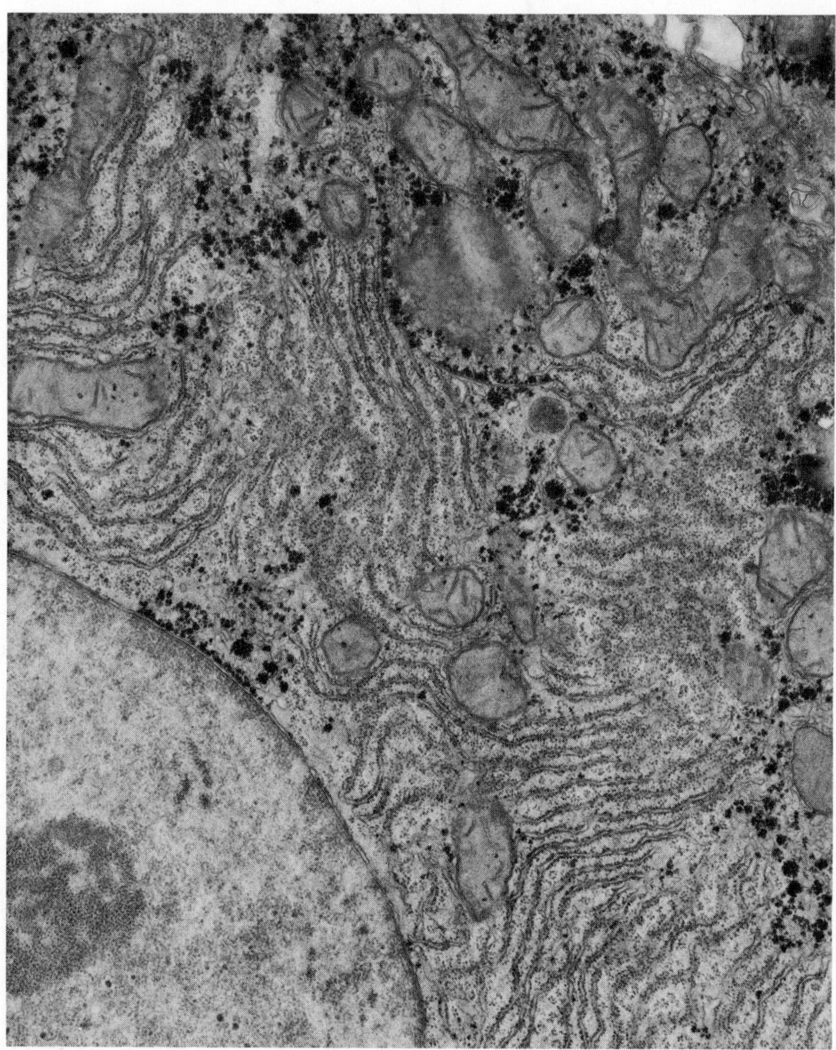

**Fig. 9-10** *Electron micrograph of section of adult rat liver, including part of a bile canaliculus* (upper right). *Note part of nucleus* (lower left) *and numerous profiles of endoplasmic reticulum.* × *15,000 (approximately). (By courtesy of K. R. Porter.)*

fine structure of the cell—the nucleus and nuclear envelope, endoplasmic reticulum and ribosomes, mitochondria and lysosomes—we are bound to ask, how do they originate? How are they generated and so intricately organized?

We have already touched upon the principal steps in the elaboration of ribosomes (page 186). Although studies of the other organelles are not precisely comparable, the steps in their construction have been described, largely by the technique of electron microscopy. Figure 9-11 depicts the stages in the development of a *chloroplast.* It is highly diagrammatic—deliberately so, for the new principle we wish to emphasize emerges from another approach.

We know that both chloroplasts and mitochondria have a regular lamellar fine structure (Fig. 9-11). Plastids are semiautonomous units with their own hereditary apparatus and protein-synthesizing machinery. Chloroplasts of *Acetabularia*, *Euglena*, and several higher plants are known to contain DNA. Studies by density-gradient ultracentrifugation show that the DNAs of nuclei and chloroplasts differ. The evidence that DNA is made in plastids, rather than derived from nuclear DNA, is impressive. It is also clear that plastids contain RNA. In addition,

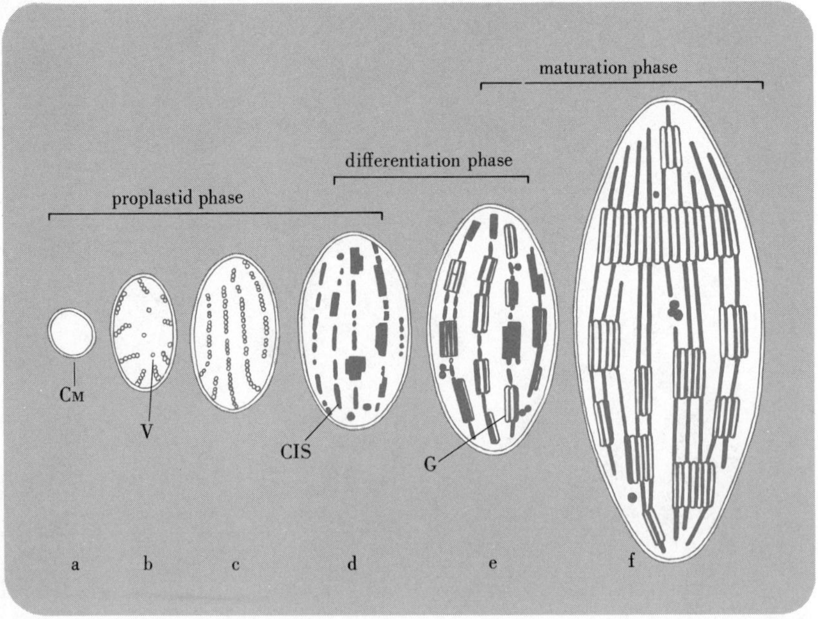


**Fig. 9-11** *Stages of development of a chloroplast in the light.* CM: *double chloroplast membrane;* V: *vesicle;* CIS: *flattened cisternum;* G: *granum. (After S. Granick, in* Cytodifferentiation and Macromolecular Synthesis, *by courtesy of Academic Press.)*

there is evidence that the DNA of plastids serves as a template for DNA-dependent RNA synthesis. Ribosomes have been isolated from chloroplasts of several plants.

The relations between the nuclear genome and the plastid are not completely understood. Nuclear genes are known to control plastid development. Some barley mutants, for example, fail to make complete chloroplasts. Normal plastid development will occur, however, if amino acids (aspartic acid in one mutant, leucine in another) are administered to the seedling.

Another example is a mutant of maize (yellow stripe 1) in which the ability of the roots to absorb iron is impaired and some chloroplasts fail to mature. When iron is "forced" into the plant by repeated spraying, normal development is restored.

It is not known whether other kinds of relations exist between the nuclear genome and the plastid. What is known is that the plastid contains all the machinery required to code and synthesize at least some of its components. Possibly the most striking fact is that this machinery is set into motion by light. Plastids will develop to the proplastid stage (Fig. 9-11C) in the dark. To go further, light is required. The usual speculation about the role of light is that it somehow activates the plastid genome, leading to mRNA synthesis. There is no evidence to permit us to evaluate the hypothesis. There may be a long chain of events from the absorption of light by *phytochrome* to the initiation of development. Phytochrome is a protein-pigment complex, universal in higher plants. Its action is still unknown. It is believed that the molecule changes shape in going from one form to another as a result of light absorption, and that this change of shape is related to a crucial catalytic role.

*Mitochondria* from all sources examined also contain double-stranded high molecular weight DNA. Its density is characteristically different from that of nuclear DNA. In animal cells it is also commonly observed to be circular (Fig. 9-12). Thus far there is little evidence for circularity in mitochondrial DNA in microorganisms and plants. It should also be added that it is difficult to see what advantage results from circular, versus linear, DNA, unless in affording greater resistance to nucleases, circularity allows DNAs to exist in sites where otherwise they would be destroyed.

In *Neurospora* the evidence is strong that mitochondria replicate by division. *Neurospora* mitochondrial DNA appears to replicate by the same mechanisms as those observed in nuclear DNA.

In crosses between two species of *Neurospora*, *crassa* and *sitophila*, each containing mitochondrial DNA with its own characteristic density, the inheritance of the two mitochondrial DNAs shows a clear cytoplasmic—not a nuclear—pattern. The evidence for both *Neurospora*

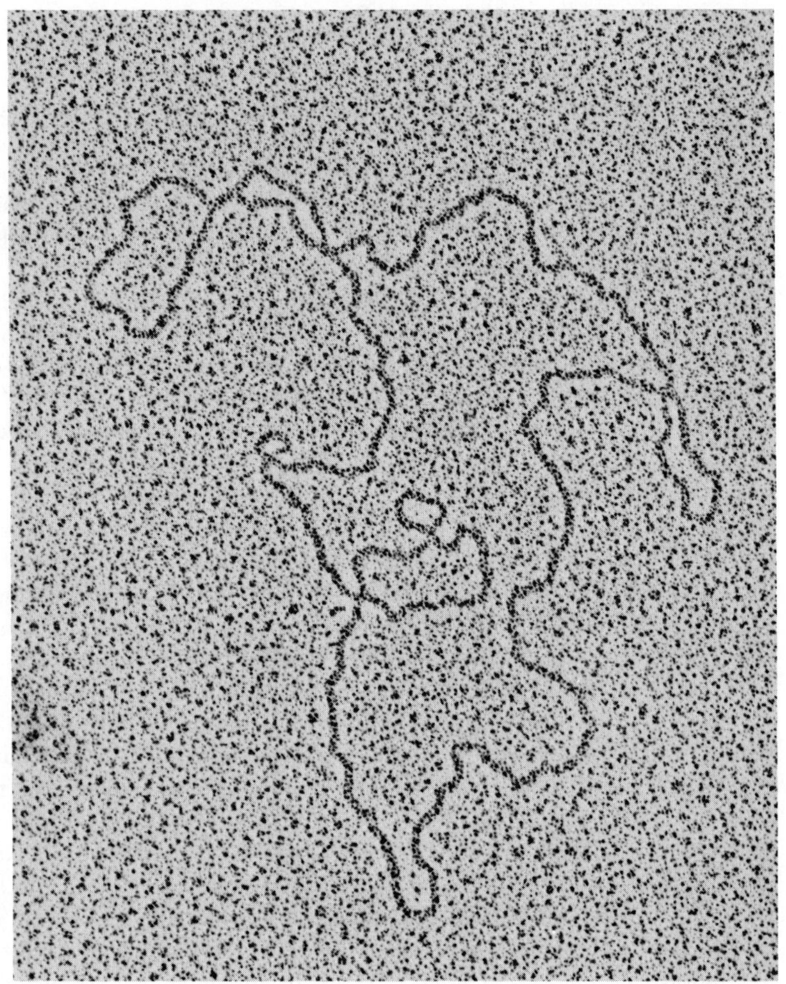

*Fig. 9-12 Electron micrograph of DNA from oöcyte mitochondria of* X. laevis. *× 85,000. (By courtesy of I. B. Dawid and D. B. Wolstenholme.)*

and yeasts shows that mitochondria do contain genetic information. It is believed that mitochondrial DNA replicates independently of nuclear DNA, and that it controls some of the constituents of mitochondria, probably including some of the structural components.

In *Neurospora* the morphological mutant *abnormal* shows how nuclear and mitochondrially directed components might interact. This mutant not only shows gross abnormalities in morphology, but also alterations in patterns of the respiratory pigments, the cytochromes. Now the synthesis of the cytochromes appears to be under nuclear control (at least this has been established in yeasts). However, in *Neurospora* the pattern of expression of these genes is somehow under the control of mitochondrial genes. One can envision two possibilities: products of mitochondrial DNA might control the expression of nuclear genes for cytochrome synthesis, or they might control the formation of mitochondrial membranes, thereby affecting the maintenance of the respiratory enzymes located on them.

We know also that the eggs of several animal species, including sea urchins and the amphibians, *Rana pipiens* and *Xenopus laevis*, contain far more DNA than the amount we would expect to find in the nucleus. What is the role of this "extra" DNA? One idea, advanced several years ago, held that such "preformed" DNA might serve as raw material for the new chromosomes formed during cleavage. However, recent evidence suggests that the chromosomal DNA is formed from the usual low molecular weight precursors. The bulk of the cytoplasmic DNA has been shown to be mitochondrial; it exists in the circular form (Fig. 9-12). The fate of mitochondrial DNA during development is not known, but it seems highly probable that it remains an integral part of the mitochondria and is distributed along with these particles into the cells of the embryo. Mitochondrial DNA is expected to be replicated whenever the multiplication of mitochondria is initiated during development.

A specific role in differentiation cannot at present be ascribed to cytoplasmic DNA. Because of its probable function in the reproduction of mitochondria, this DNA is very likely to be of importance in the growth of the embryo; if the thesis regarding mitochondrial DNA as a carrier of genetic information for part of these particles' structure is correct, it follows that no multiplication of mitochondria could proceed without it. There can be no doubt that mitochondrial multiplication is necessary for development. Analogous statements are, of course, also true for many other cell constituents, and it is impossible to derive from such a fact indications for a specific control function in embryogenesis for any substance. We must therefore refrain from invoking such specific functions for cytoplasmic DNA unless some evidence that would suggest it can be adduced.

One final point should be made. If it is correct that mitochondria are inherently capable of protein synthesis, as the foregoing discussion implies, they should contain ribosomal particles. Are the components of mitochondrial ribosomes the same as the other cytoplasmic ribosomes of which we have been speaking? They appear not to be. Evidence is now available for yeast, *Neurospora*, and *Xenopus*: two distinct mitochondrial ribosomal RNAs have been characterized. They are very different from the other cytoplasmic rRNAs and have base compositions reflecting that of the mitochondrial DNA. These observations further strengthen our belief that mitochondria possess an endogeneous and distinct protein-synthesizing apparatus.

## THE CELL CORTEX

The cortex of a cell is the outer part of the cytoplasm, including the cell membrane. This definition is imprecise to be sure; in fact, in eggs the cortex is morphologically ill-defined. In protozoans with a well-defined cortex, such as *Paramecium* and *Stentor*, as Sonneborn and Tartar and others have shown, the cortex plays an important role in controlling the specific pattern of differentiation of structural features, including the cilia and the mouth. Grafting experiments reveal that in *Stentor*, one part of the cortex, the oral primordium, is chiefly involved in this control.

In the eggs of both invertebrates and vertebrates, there is evidence that the cortex is physiologically active. We might recall that the early formative movements in amphibian eggs begin at the cell surface. Do cortical factors play a role in initiating them? Some years ago the hypothesis was advanced that the interaction of a cortical factor located in the gray crescent with another cytoplasmic factor was essential for gastrulation. Recently the hypothesis has been tested.

Cortex from the gray crescent region of uncleaved fertilized eggs of *Xenopus* was grafted to the opposite side of another egg (Fig. 9-13). The embryo now acquired two gray crescents, underwent two sets of movements, and produced two central nervous systems with associated structures.

When this part of the cortex is removed, by excising it from fertilized eggs, although cleavage continues, gastrulation does not occur.

The regions of the cortex grafted in these experiments are thicker than the plasma membrane, being about 0.5 to 3 $\mu$ in thickness, containing a hyaline layer with a few mitochondria and pigment layers below the plasma membrane. In embryos of other species—mollusks,

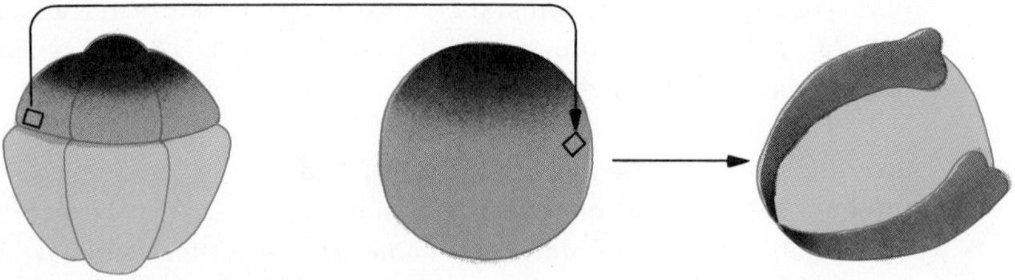

*Fig. 9-13 Diagram illustrating grafts of gray crescent cortex placed in an uncleaved fertilized egg. This egg will acquire two gray crescents, leading to the production of a secondary embryo, which is twinned with the normal one. Graft may come from any stage between one and eight cells. (After A. S. G. Curtis in* Endeavour, *22, by courtesy of the author.)*

insects, echinoderms—the cortex consists of little more than the membrane itself.

How could the critical structural information be contained in this thin layer?

## POLARITY AND PATTERNS

In Chapters 2 and 5 we discussed, briefly, the factors underlying polarity. We concluded that the factors controlling polarity must be sought in the expression of the "maternal" genome during oögenesis, in the relations between the maturing oöcyte and its environment.

There is no more pertinent example than the direction of coiling of the shell of the snail *Limnea*. The direction is determined at the time of the first cleavage division of the egg. It depends solely upon the genotype of the mother. In Fig. 9-14 we see that in crosses between snails

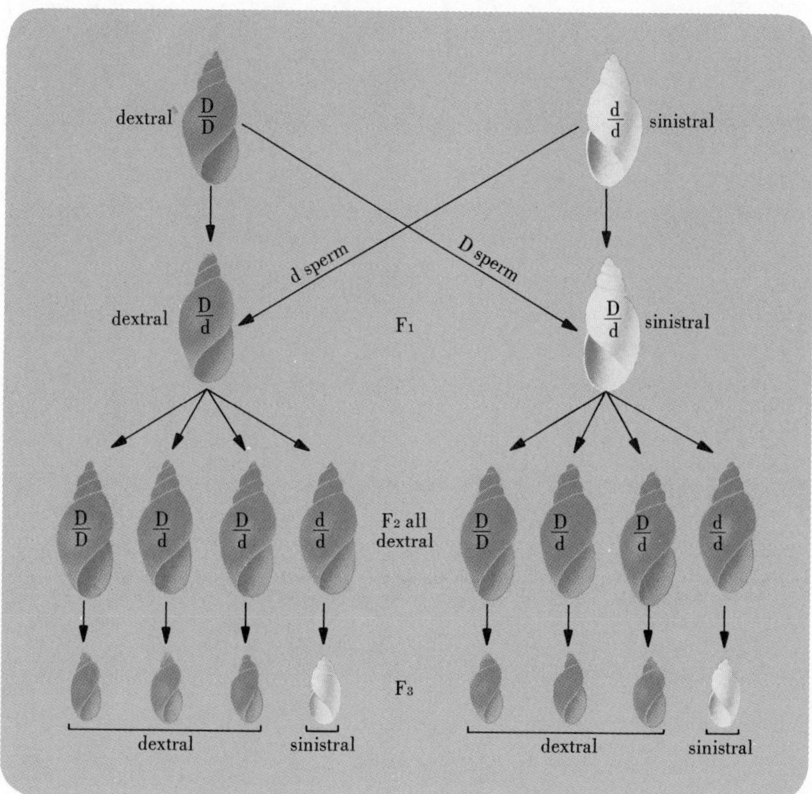


***Fig. 9-14*** *The inheritance of coiling direction in the shells of the snail* Limnaea. *(After Sinnott, et al.,* Principles of Genetics, *5th ed., McGraw-Hill Book Company, Inc., 1958.)*

with left-handed (sinistral) coiling and right-handed (dextral) coiling, the mother's genotype determines the phenotype of the offspring.

However, even this elegant example, like those previously discussed, does not give us better understanding as to the manner in which the genotype is expressed at the first cleavage: the forces immediately directing the positioning of the cell's organelles and the orientation of the cleavage spindle. We are so accustomed to thinking of polarity as a

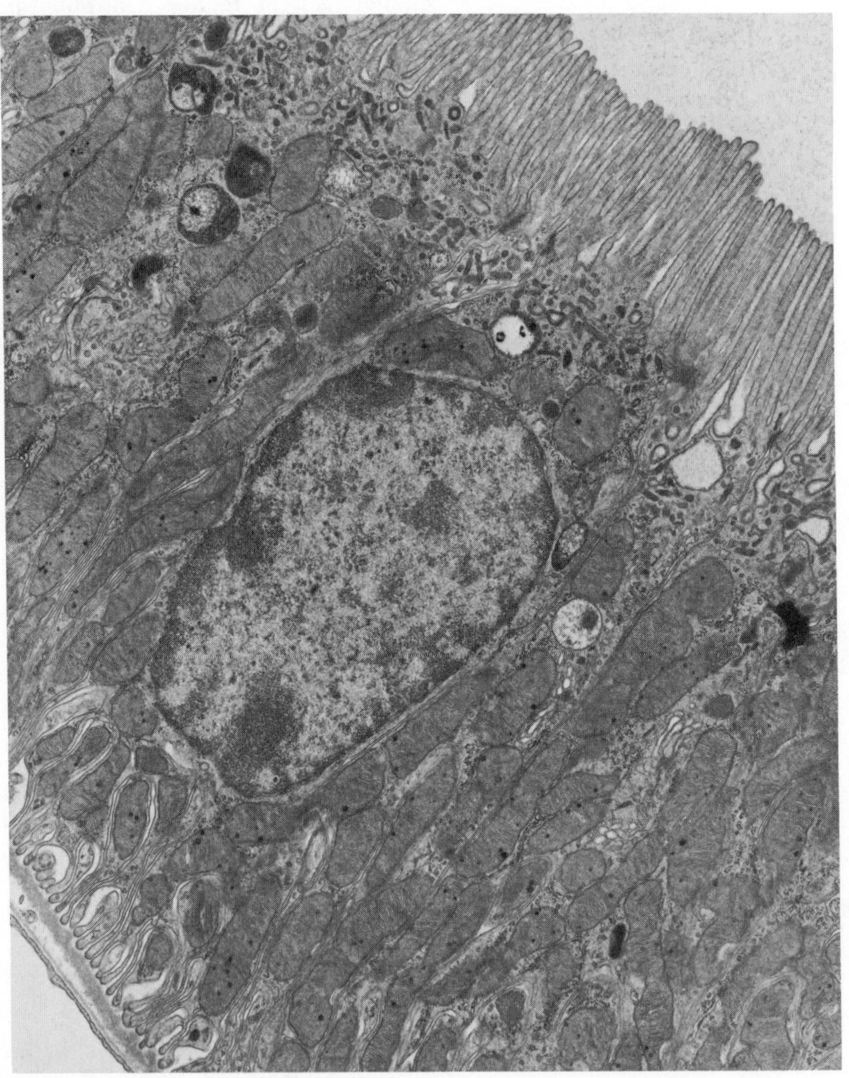

***Fig. 9-15*** *Electron micrograph of cells in the proximal tubule of mouse kidney.* × *9000 (approximately). (By courtesy of K. R. Porter.)*

property of organisms that we are inclined to forget that cells are polarized, their organization reflecting their activity. Compare, for example, the polarities of the kidney and pancreatic cells in Figs. 9-15 and 9-16.

But what controls the shape of a cell, and the distribution of organelles? Does a cell have a "cytoskeleton"? If so, how is it generated?

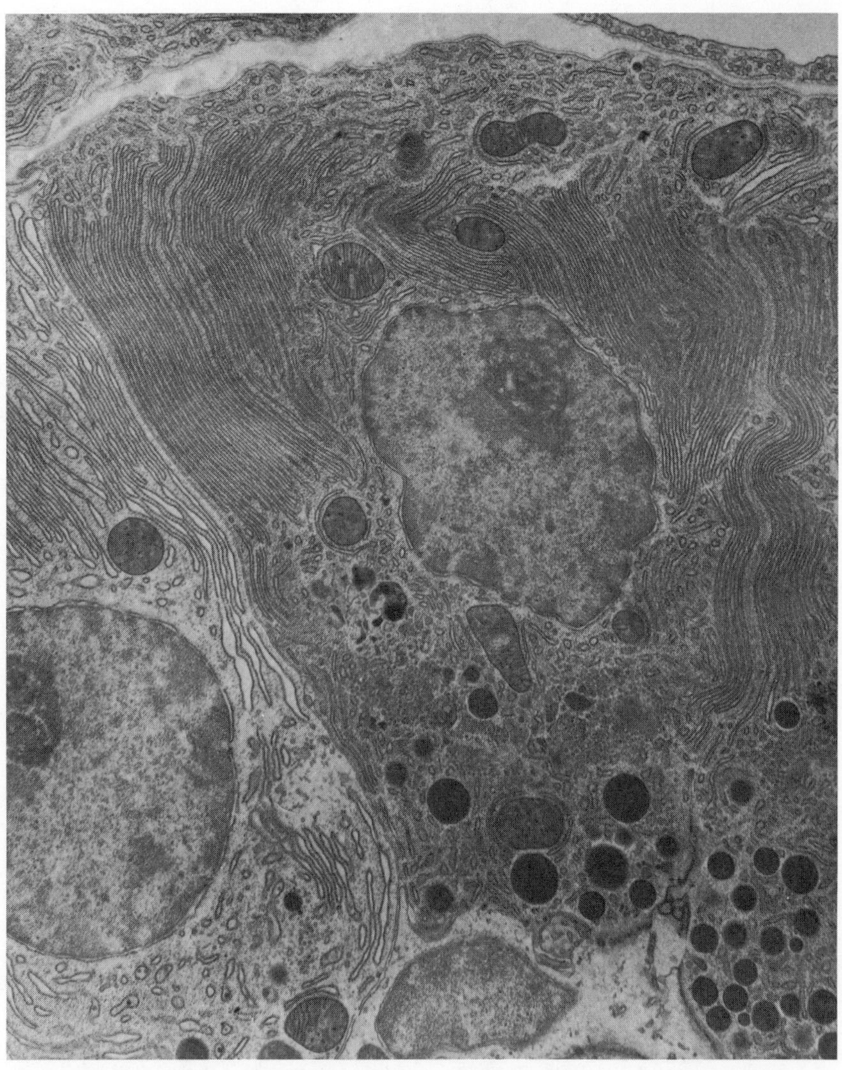

***Fig. 9-16*** *Electron micrograph of a pancreatic acinar cell as it occurs in the hibernating bat. The fine structure is more highly ordered than in the nonhibernating animal.* × *6000 (approximately). (Part of plate 7 in K. R. Porter and M. Bonneville,* An Introduction to the Fine Structure of Cells and Tissues, *1963. Reprinted with permission of the publisher, Lee and Febiger.)*

What are its properties? The latter question looms large when we consider that a "cytoskeleton" must have sufficient rigidity to give shape to the cell, yet be sufficiently elastic to permit movement of the cell and the extension, and movement or streaming within the cell.

It appears that the shape of the cell is influenced by factors within the cell itself. A candidate (or class of candidates) for the role of cytoskeleton is now emerging in the demonstration of the presence of "*microtubules*" in many cells (Fig. 9-17).

These long cylindrical structures, which range from about 200 Å to 300 Å in diameter, have walls composed of globular subunits arranged as filaments (50 Å in diameter) running parallel to the long axis of the "tubule." It is not resolved whether the microtubule is in fact a true tubule—that is, whether its center is hollow. The distribution of microtubules suggests a second role, related to their putative role as a cytoskeleton: participation in cell movement. We speculate on their role in cell shape because we observe them so often in elongate processes,

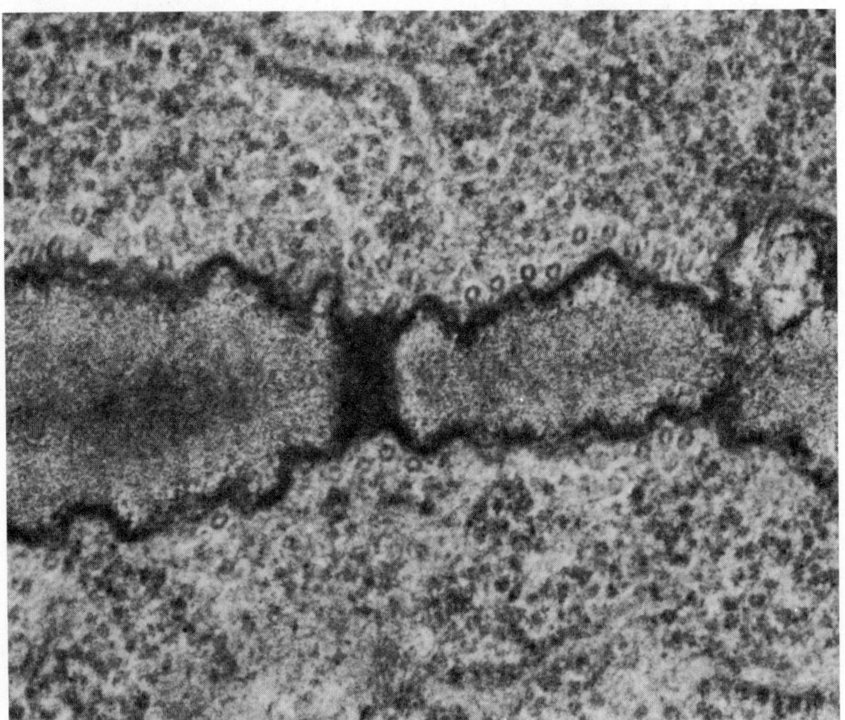

**Fig. 9-17** *Electron micrograph of part of lateral wall of a cell root tip of the plant* Phleum *as it appears in longitudinal section, showing microtubules which appear, end on, as circles with prominent limiting lines. (From M. C. Ledbetter and K. R. Porter, in* Journal of Cell Biology, *19, by courtesy of the authors and The Rockefeller University Press.)*

such as axons, and in flattened disks, such as blood platelets. We speculate about their role in movement because they occur in mitotic spindles, cilia, and flagella, including sperm tails. In plant cells they appear to be somehow related to cytoplasmic streaming. Populations of microtubules are also observed in the cortices of cells in the lens placode during induction by the optic vesicle.

It is suggested that either the microtubules themselves or the matrices between them contain (or are) adenosine triphosphatase, the enzyme intimately involved in contraction.

Thus these microtubules have at least some of the properties needed by a "cytoskeleton."

In the next chapter we shall take up the interactions of cells, leading to organogenesis and the elaboration of complex patterns. As we do so, we should bear in mind that as the cells we have just been discussing are taking shape, they are doing so within a larger whole, contributing to a kidney or pancreas or limb, or to a colored patch on a butterfly's wing, or a feather or a hair.

## FURTHER READING

Bogorad, L., "Control Mechanisms in Plastid Development," in *Control Mechanisms in Developmental Processes*, M. Locke, ed. New York: Academic Press, 1967, p. 1.

Curtis, A. S. G., "The Cell Cortex," *Endeavor*, vol. 22 (1963), p. 134.

Dawid, I. B., and D. R. Wolstenholme, "Ultracentrifuge and Electron Microscope Studies on the Structure of Mitochondrial DNA," *Journal of Molecular Biology*, vol. 28 (1967), p. 233.

deDuve, C., "The Lysosome," *Scientific American*, May 1963, p. 64.

Edds, M. V., Jr., "Animal Morphogenesis," in *This is Life*, W. Johnson and W. Steere, eds. New York: Holt, Rinehart and Winston, 1962, p. 271.

Edgar, R. S., and W. B. Wood, "Morphogenesis of Bacteriophage T4 in Extracts of Mutant-Infected Cells," *Proceedings of National Academy of Sciences*, vol. 55 (1966), p. 498.

Gross, J., C. M. Lapiere, and M. L. Tanzer, "Organization and Disorganization of Extracellular Substances: The Collagen System," in *Cytodifferentiation and Macromolecular Synthesis*, M. Locke, ed. New York: Academic Press, 1963, p. 175.

Markert, C. L., "Epigenetic Control of Specific Protein Synthesis in Differentiating Cells," in *Cytodifferentiation and Macromolecular Synthesis*, M. Locke, ed. New York: Academic Press, 1963, p. 65.

Novikoff, A. B., and E. Holtzman, *Cells and Organelles*. New York: Holt, Rinehart and Winston, 1970.

Papaconstantinou, J., "Molecular Aspects of Lens Cell Differentiation," *Science*, vol. 156 (1967), p. 338.

Rawles, M. E., "Tissue Interactions in Scale and Feather Development as Studied in Dermal–Epidermal Recombinations," *Journal of Embryology and Experimental Morphology*, vol. 11 (1963), p. 765.

Tatum, E. L., and D. J. L. Luck, "Nuclear and Cytoplasmic Control of Morphology in *Neurospora*," in *Control Mechanisms in Developmental Processes*, M. Locke, ed. New York: Academic Press, 1967, p. 32.

Ursprung, H., "The Formation of Patterns in Development." in *Major Problems in Developmental Biology*, M. Locke, ed. New York: Academic Press, 1966, p. 177.

Wilt, F. H., "The Control of Embryonic Hemoglobin Synthesis," in *Advances in Morphogenesis*, vol. 6, M. Abercrombie and J. Brachet, eds. New York: Academic Press, 1967, p. 89.

# chapter 10

# Mechanisms of Cell and Tissue Interactions in Animals

In discussing the organization of the collagen fibril as an example of the self-assembly of macromolecules, we observed that this ordering required a compatible environment. Conditions in the extracellular environment undoubtedly somehow determine the pattern of aggregation. Similar self-ordering processes must play a role in the organization of the many components that constitute a cell, but we have only begun to search for ways of exploring the molecular basis of the organization, shape, and polarity of individual cells.

A form of self-assembly is also apparent at a higher level of organization, in the aggregation of cells to form tissues. When an organism containing several tissue-types is experimentally dissociated into a mixture of individual cells, these cells can often reassemble into a pattern resembling that which they had in the original organ. The stability of cells in the embryo reflects both the properties of the cells themselves, and the stability of their environment. Cells impinge upon and influence each other within that environment. We are just beginning to perceive the nature of those influences. Some may be diffusible, humoral influences; others may require local specializations of the membrane at the site of cell to cell contact. Electrical communications between cells appear to be correlated with some of these specializations, and thus it is suggested that *junctional connections* may be involved in the exchange of signals between cells.

## THE HISTORY OF A MUSCLE CELL

In Chapter 4 we considered the life history of a neuron. Its biography was incomplete, however. Although we were able to recount the beginning and the end, and here and there a glimpse of intermediate events, we could say next to nothing about the factors that shaped the cell's history.

During the earliest stages of development, changes may occur synchronously in all of the cells of the embryo, irrespective of their eventual differentiation. At successively later stages, however, cells become progressively more divergent in their properties. During these phases we must focus our attention upon specific embryonic cell types. In what Abercrombie has called "an astonishing stride forward in the history of biology," over 60 years ago Ross G. Harrison unleashed a new technique of tremendous power, that of tissue culture, establishing that cells could be grown outside the body (and proving, as we have already described, that peripheral nerve fibers originate from nerve-cell bodies.) Only recently have culture techniques been sufficiently refined to permit the derivation of *clones*, populations of cells derived by division from a single isolated cell. However, most "established" strains of cells isolated originally from animal tissues and maintained continuously in culture are unsuitable for studies of cellular differentiation since it is exceedingly rare that such cell strains bear even the most tenuous resemblance to the major cell type of the tissue of origin. Such cell populations, during the course of their cultivation, lose the cell-specific properties that characterize the cells of the original tissue. In order to study cell-specific properties, and the manner in which they are acquired, it is necessary to apply cloning techniques to newly isolated embryonic cells. This goal was first achieved by Konigsberg in studies of the growth and differentiation of embryonic skeletal muscle cells.

Early in the development of tissue culture as a research tool, Margaret and Warren Lewis demonstrated that fragments of embryonic chick skeletal muscle, embedded in clotted plasma, not only grow but form striated muscle. It is now clear that such cells develop equally well when grown by newer methods of cell culture, in which the tissue is first enzymatically dissociated into its component cells. Cell suspensions prepared from the leg muscle of 12-day-old chick embryos attach to the bottom of the culture chamber, grow, and form a continuous sheet of cells. However, as the cell layer approaches confluency, large numbers of long fibers appear, which within a few days begin to contract spontaneously. The contractions tell us that functional muscle has differentiated, and morphologic and biochemical evidence is provided by the presence in the elongated cells of the cross-striated pattern and the contractile proteins typical of muscle. Thus embryonic muscle cells grown in cultures of randomly distributed individual cells are still capable of differentiating into structurally and functionally recognizable units.

Can a single cell give rise to a colony? By physically isolating a single myoblast (Fig. 10-1) and culturing it in a small glass cylinder, under conditions which exclude contact with any other cell, it was proved that the single cell can produce a colony of differentiated muscle.

During the first four days of culture, such cells divide every 12 to 18 hours, producing small colonies of roughly 50 cells (Fig. 10-1). The first indication of further differentiation is observed on the fifth or sixth day of culture, when cells fuse to form multinuclear "myotubes." At successively later stages the multinuclear myotubes increase in length and number until by the end of the second week they form a colony of interlaced fiberlike cells (Fig. 10-2). These colonies, which measure several millimeters in diameter, are apparent to the naked eye when appropriately stained. Moreover, many of the myofibrils contain cross-striations; the level of differentiation attained in such cultures is illustrated in Fig. 10-3. Not all of the single cells fuse, however; even at the end of the second week some remain, and continue to divide. However, the abrupt appearance of multinucleated myotubes is paralled by an equally abrupt drop in the overall rate of proliferation. The nuclei in the developing myotubes no longer divide. The fusion of myoblasts to form a multinucleated myotube exemplifies, albeit in a specialized form, the surface interactions of cells. It is now well-established, by several techniques, that skeletal muscle is a *syncytium*, many nuclei lying in a common cytoplasm, and that it is formed by the fusion of myoblasts. Until recently all of the evidence was derived from the study of cultured cells. Now, however, crucial evidence has been presented that this process does occur in the intact animal, through the

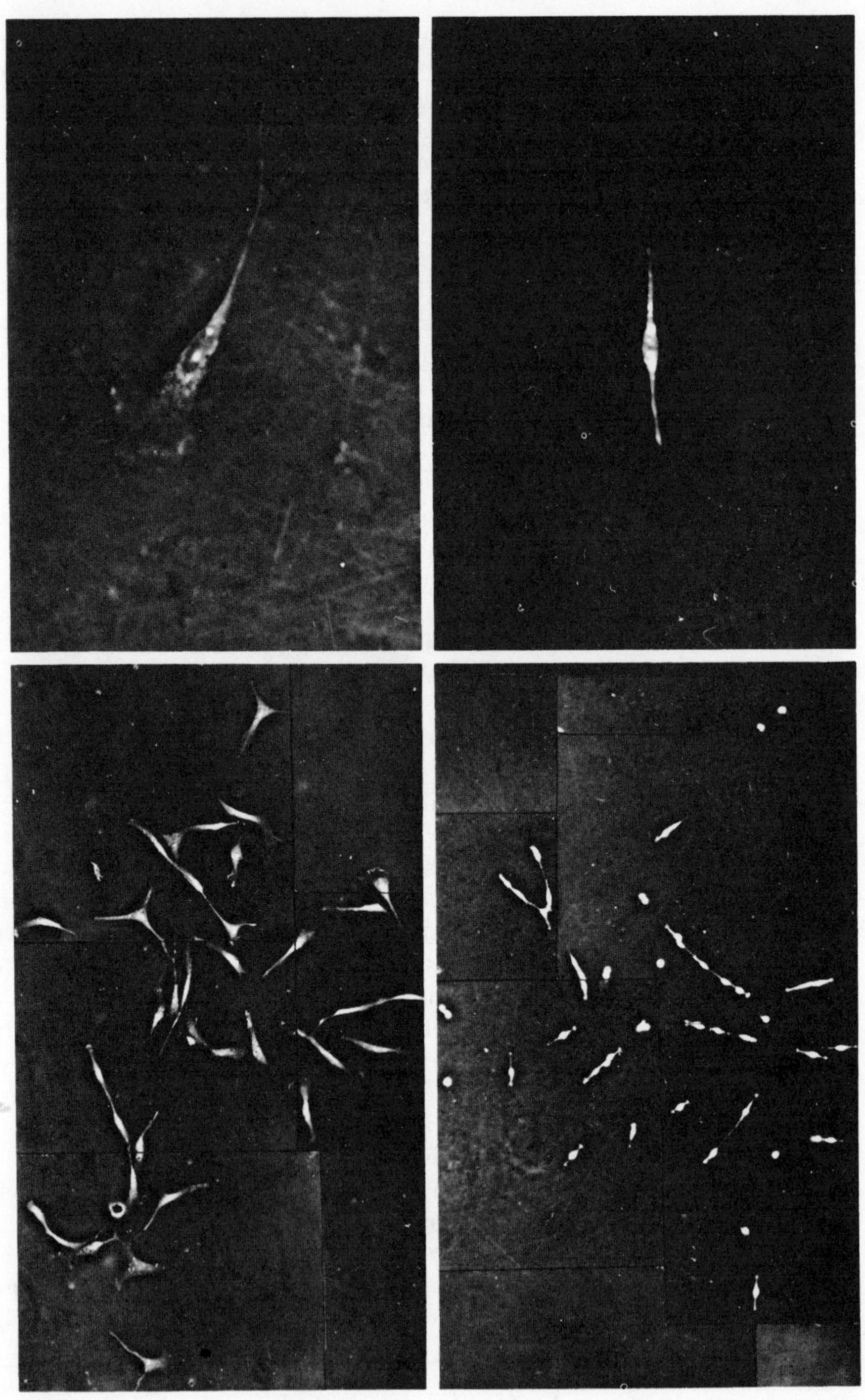

◄ **Fig. 10-1** (Top) *Single myoblast* (right) *and fibroblastic cell* (left). *As soon as single cells attach and flatten on the bottom of the petri plate, the myoblast can be distinguished by its spindle shape.* (Bottom) *Two living colonies photographed on fourth day of culture.* (Right) *embryonic muscle;* (left) *fibroblastic cells. The latter are more extensively attached and appear larger. (By courtesy of I. R. Konigsberg.)*

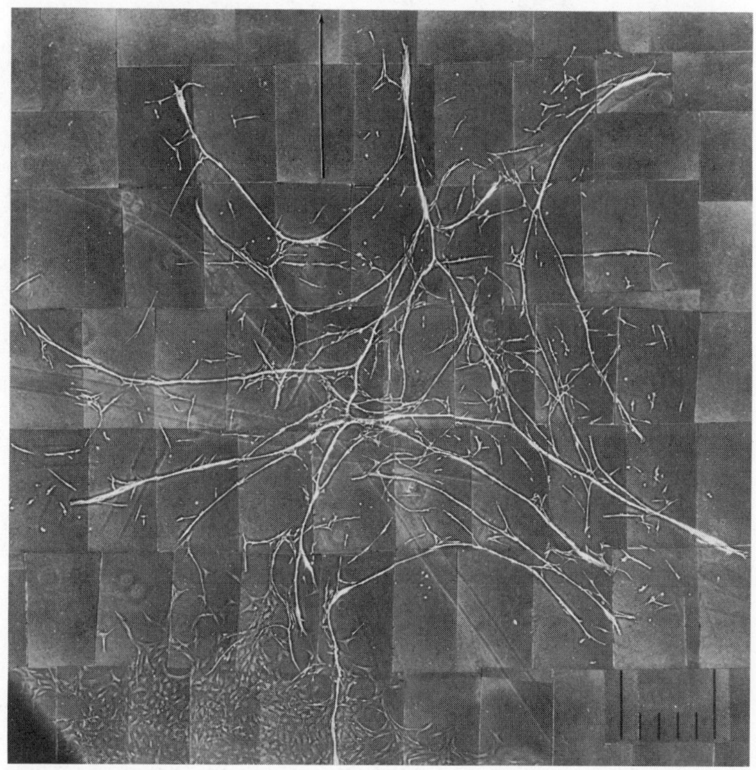

**Fig. 10-2** *A muscle colony on 13th day in culture. Cells of the leading edge of an invading colony of fibroblastic cells can be seen at the bottom. Each division of the scale equals 0.1 mm. Fig. 10-3 is a higher magnification of a part of the center of this muscle colony. (By courtesy of I. R. Konigsberg.)*

use of *allophenic mice* (page 149). You will recall that the allophenic mouse is a mosaic produced by the aggregation *in vitro* of cleaving eggs from two distinct strains, the composite blastocysts thus resulting being transferred to the uterus of a foster mother for further development. Here then is an ideal situation in which to test the hypothesis that the normal development of muscle involves fusion just as it does in culture. Earlier we discussed the fact that many enzymes are made up of subunits (page 193). We can now put that information to immediate use.

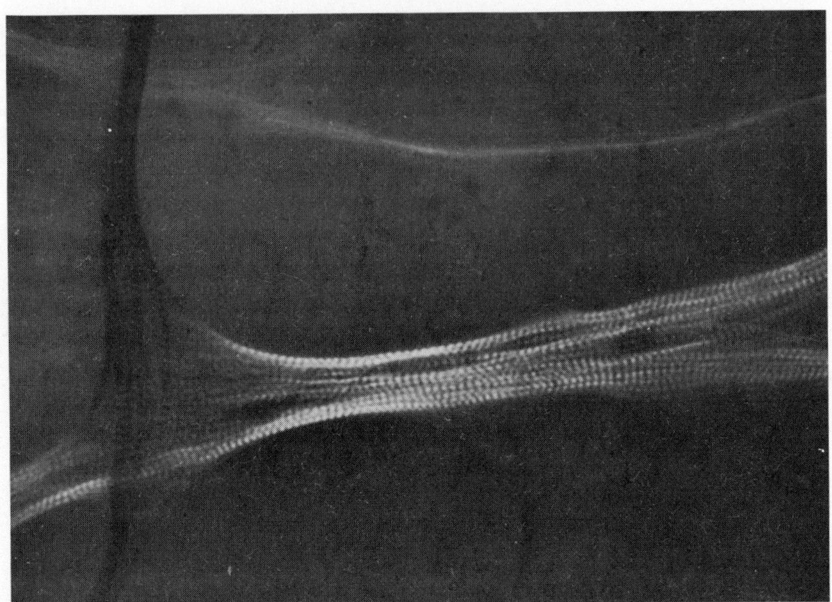

***Fig. 10-3*** *Polarizing optics demonstrate cross-striated myofibrils in chick embryo leg muscle growing* in vitro. *(By courtesy of I. R. Konigsberg.)*

The two strains of mice used in the experiment contain clearly different isozymes of the specific enzyme, isocitrate dehydrogenase, known to be made up of subunits. If muscle fibers are made by fusion, then in the mosaic animal myoblasts of both "parental" strains may contribute to the muscle fiber. Thus a myotube may contain two kinds of nuclei in a common cytoplasm. Since it is the cytoplasm in which the polymerization of enzyme subunits occurs, we should expect to find three kinds of enzyme molecules in a myotube formed by fusion of myoblasts of different origin: enzymes of each original type and "hybrid" enzyme. That is exactly what is found. *Skeletal* muscle from allophenic mice contained appreciable quantities of hybrid enzyme (Fig. 10-4). The hybrid enzyme was not found in tissues which are not syncytial—for example, *cardiac* muscle, liver, kidney, lung, and spleen.

In concluding this section, we return to one of the questions that lies at the very heart of developmental biology: the nature of cellular interactions. Our example, muscle formation, illustrates two of the major problems which confront us. We have just considered one: the interactions of myoblasts among themselves; the second is illustrated by the interactions of myoblasts with other cells, notably fibroblasts.

Muscle tissue contains two major cell types, the muscle cell itself and the fibroblast, which lays down the connective tissue framework of

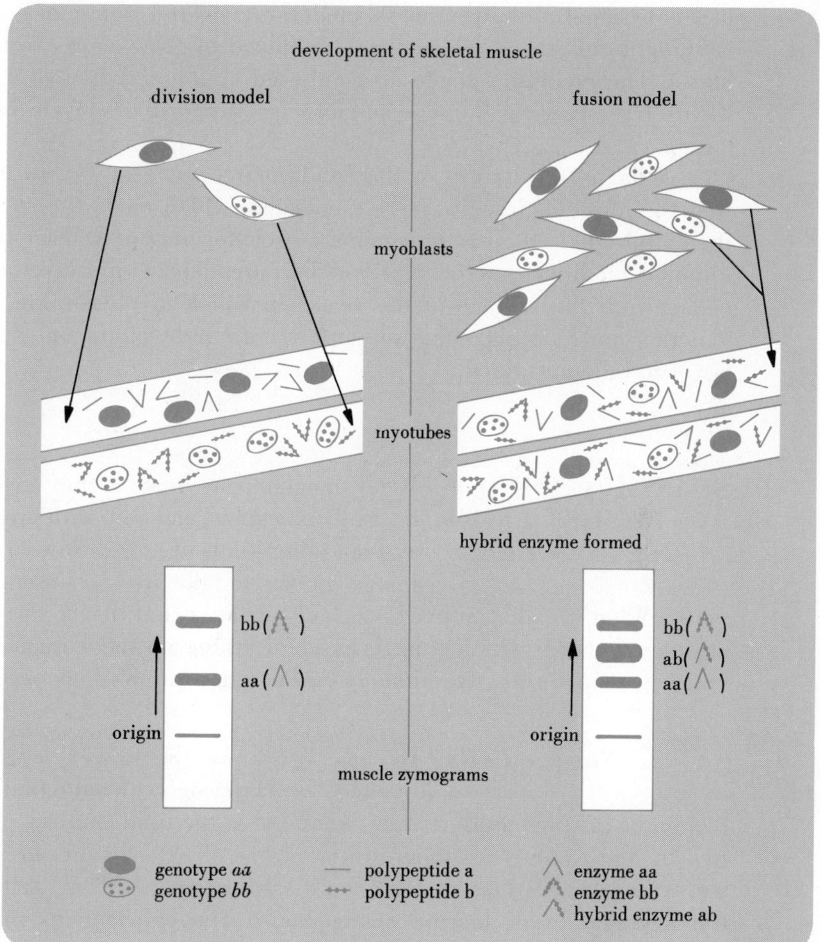


*Fig. 10-4 Diagram of expected isozyme results in* allophenic *mice, on the "division"* (left) *versus "fusion"* (right) *models of skeletal muscle development. When homozygous cells of different NADP-isocitrate dehydrogenase genotypes coexist, heterokaryons would result in the event of myoblast fusion, and hybrid enzyme could be formed. Enzyme molecules are represented as dimers formed in the cytoplasm from polypeptide subunits. (From B. Mintz and W. W. Baker,* Proceedings of National Academy of Sciences, *58, 1967, by courtesy of the authors and the Academy.)*

the organ. Mass cultures prepared by disaggregating embryonic muscle contain both cell types. Each type can be grown in the absence of the other. It has long been known that fibroblastic cells can be grown clonally, and we have just noted that clones of muscle cells have been developed, suggesting they do not require an interaction with fibroblasts. However, to clone muscle cells successfully, a "conditioned" medium had to be used initially—that is, medium recovered from mass cultures containing muscle fibroblasts was reutilized to grow clones from myoblasts. The medium appears to be altered by the activities of the cells themselves, making it more suitable for supporting muscle differentiation.

What is the nature of the conditioning process? It now appears that the fibroblastic cells secrete collagen, which coats the surface of the culture dish, thereby enhancing muscle formation. At least, the conditioned medium may be replaced by spreading a thin layer of pure collagen on the surface of the vessel in which myoblasts are cloned. Whether collagen plays this role in the normal development of muscle is yet to be determined.

## RECONSTRUCTION OF TISSUES FROM DISSOCIATED CELLS

What mechanisms insure that myoblasts normally recognize, and fuse with myoblasts, and not other kinds of cells? How do migrating cells come to take up their ultimate positions? What are the properties of cell surfaces that insure the orderly arrangements of cells within tissues? In taking up these questions we begin by considering experiments on the aggregation of sponge cells.

### Sponge Cells

By now, we know the far-reaching consequences of Harrison's introduction of the technique of tissue culture. Less heralded at the time and less provocative in immediately opening up new vistas of experiments to be done were the observations of Harrison's contemporary, H. V. Wilson, on the aggregation of marine sponge cells. His experiments, although simple in conception and execution, revealed unsuspected properties of cells, and produced some of the germinal ideas of an approach that is only now flourishing. It would be difficult to trace all the antecedents of the field. Rather than try to interweave the history of work on sponge cells with that on the aggregation of embryonic tissue cells, we will begin by analyzing the work on sponge cells—our discussion encompassing the pioneering work of Wilson and Galtsoff as well as recent contributions of Moscona, Humphreys, and others.

If a sponge, say the bright red species *Microciona prolifera* (referred to as M), is pressed through fine cloth into fresh seawater, the water is seen to contain individual cells and cell clusters. Allowed to stand, the cells migrate actively and reaggregate, reconstructing a sponge, complete with canals, spicules, and spongin.

A similar pattern is observed with another sponge, the purple-brown *Haliclona oculata* (H). Suppose the cells of the two species are intermingled. How will they reaggregate? As Wilson first observed, reaggregation is species-specific; from a mixture of H and M cells, separate H and M sponges are formed.

Such mechanically dissociated cells reaggregate effectively at 24°C; when the temperature is lowered to 5°C, they will still reaggregate, although more slowly.

What factors are involved in the reconstruction of a sponge? Do living sponge cells make a species-specific product, which acts at the cell surface?

Attempts to demonstrate the existence of such a product hinge on two technical advances: (1) A more consistent and reproducible method of dissociation is required. Therefore, sponge tissue is dissociated in calcium–magnesium-free seawater, producing a suspension of single cells. (2) It is necessary to distinguish between migratory and adhesive properties of the cells–in other words, to bring cells together in a consistent manner so that the size, number, and rate of formation of aggregates will be directly related to the adhesive properties of the cells. This aim is accomplished by having the cells reaggregate in a gyratory shaker, under constant rotation, the cells being swept into contact by the liquid.

Chemically dissociated cells, both H and M, behave as follows: (1) Kept in seawater free of divalent cations, they remain dissociated. (2) Returned to normal seawater at 24°C, they reconstitute a species-specific aggregate. Thus they behave like mechanically dissociated cells. (3) However, if returned to normal seawater at 5°C they do not reaggregate. Therefore, reaggregation of the cells requires both divalent cations and a temperature-dependent process that does not appear to operate at 5°C.

It was postulated that chemical dissociation removes a species-specific factor. In normal seawater at 24°C, cells would synthesize or secrete more of the factor; at 5°C they would not.

Can such a factor be isolated? If it is removed by chemical dissociation, the seawater in which tissues are dissociated should contain it. It does. A factor can be concentrated and assayed quantitatively, with some precision. Added to chemically dissociated cells in normal seawater at 5°C, it promotes their aggregation. It is species-specific. The factor obtained from H cells promotes only H aggregates; that from M cells promotes only M aggregates. When H factor is added to interspecific mixtures, only H cells aggregate; addition of M factor now brings about the adhesion of M cells.

These observations are important; we must take care, however, to insure that we distinguish between observations and interpretations. The evidence is clear that seawater in which sponges are dissociated contains a product that promotes species-specific aggregation when it is added to sponge cells kept at 5°C. The evidence is interpreted to mean that the product is synthesized or secreted at 24°C but not at 5°C. It would be premature to conclude that the product released into the water resides normally at the cell surface and plays a role in maintaining normal relations between sponge cells. It is the experimenters' goal to determine the chemical nature of these species-specific factors, their localization within, or on the surfaces of, cells, and their roles. Substantial progress has been made toward the first of these objectives. When purified preparations of the specific binding factors are examined in the electron microscope, they are found to consist of roughly spherical units 20 to 25 Å in diameter. Biochemical data suggest that these particles consist predominantly of glycoprotein—that is, a combination of carbohydrate and protein (the possibility has not been ruled out that lipid plays a role). An approximation (and it is only that) has been made of the minimal molecular weight; it is calculated to be on the order of 13,000. The relative roles of carbohydrate and protein, and how they are linked, and how calcium fits into the situation (except that it is essential for the function or stability of the ligants) are essential targets for future study.

### Embryonic Tissue Cells

It was nearly three decades before the ideas stemming from the initial observations on sponge cells began to have a significant impact on concepts of morphogenesis and nearly five decades before the approaches converged effectively. Beginning with his classic article "Tissue Affinity, a Means of Embryonic Morphogenesis," that ingenious student of amphibian development, Johannes Holtfreter, developed the concept of *selective affinities* between embryonic cells and tissues. He showed that when tissues of amphibian embryos are exposed to solutions of high pH they dissociate. Upon being returned to saline at a physiological pH, they reconstruct the tissue of origin (Fig. 10-5).

In an experiment performed by Townes and Holtfreter, for example, a piece of medullary plate and a piece of prospective epidermis are excised and disaggregated. When the free cells are intermingled under physiological conditions, they reaggregate at random and subsequently segregate in such a way that the surface of the explant becomes entirely epidermal, the cells of medullary plate taking up an internal position (Fig. 10-6). A more complex interaction is exemplified by a mixture of cells from medullary plate, neural fold, and endoderm. After 20 hours the surface of the aggregate consists of patches of ectodermal and endodermal cells. By the fifth day there is a single cap of epidermis. The underlying neural tissue is segregated, surrounded by differentiating mesenchyme. The final distribution shows clearly that in the process of sorting out, the different cell types exhibit a cell-specific

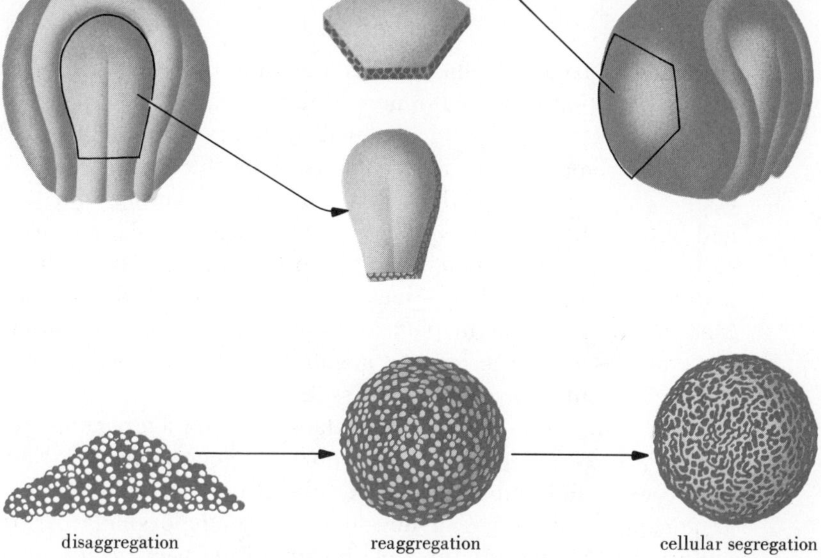


**Fig. 10-5** *A piece of medullary plate and a piece of prospective epidermis of an amphibian embryo are excised and disaggregated by means of alkali. The free cells are intermingled (epidermal cells indicated in color). Under readjusted conditions the cells reaggregate and subsequently segregate so that the surface of the explant becomes entirely epidermal. (After P. L. Townes and J. Holtfreter, in* Journal of Experimental Zoology, *128, by courtesy of the Wistar Institute.)*

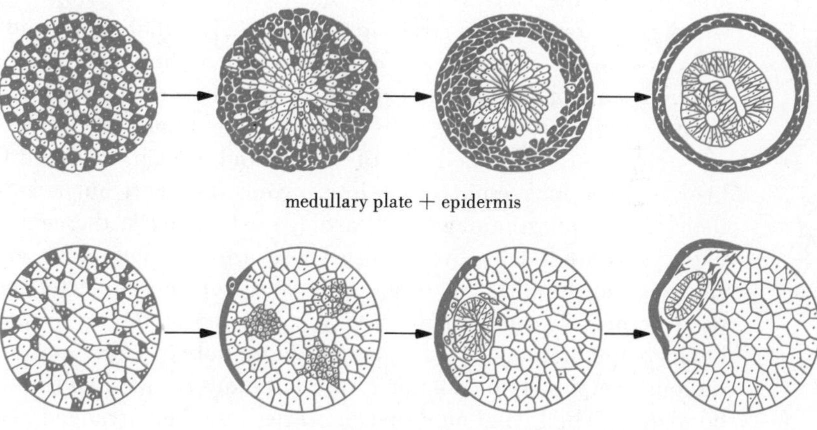


**Fig. 10-6** *Diagrammatic sections through successive stages of composite reaggregates.* (Upper) *Randomly arranged cells of epidermis (color) and medullary plate (white) move in opposite directions and reestablish homogeneous tissues.* (Lower) *The addition of neural fold cells to a mixture of medullary plate and endoderm cells produces epidermis and mesenchyme which prevent central allocation of the neural tissue and promote neurocoel formation. (After P. L. Townes and J. Holtfreter, in* Journal of Experimental Zoology, *128, by courtesy of the Wistar Institute.)*

tendency to arrange themselves in a definite tissue pattern, which corresponds to that observed in normal development.

Tissues of chick and mammalian embryos are dissociated readily by the removal of divalent cations or by the use of weak solutions of the enzyme trypsin, the method of choice varying with the tissue and aim of the experiment. In studies of tissue cells, as with sponge cells, the use of constant rotation permits one to study adhesion, as distinct from migration. Many experiments, in which Weiss and Moscona were among the pioneers, show that cells of chick or mammalian embryos also discriminate among one another, being able to sort themselves out and reconstruct tissues.

In contrast to sponge cells, however, the aggregation reactions among embryonic cells are not always species-specific. Embryonic mouse cells and embryonic chick cells, if derived from similar tissues at a comparable stage of development, produce similar aggregation patterns. Moreover, mixtures of similar cells of two species may, under some conditions, form a mosaic tissue. Although it has been reported that under some conditions chick and mouse embryonic myocardial cells exhibit species-specificity, under others heterospecific aggregates clearly are formed; moreover, the mixture of neural retina cells of chick and mouse produce not a chick neural retina and a mouse neural retina, but a single combined structure. Mixtures of chick and mouse cartilage cells behave similarly, but it is not altogether clear whether a *true chimeric* structure is formed, in which chick and mouse cells are mixed entirely randomly, or whether there are small groups of chick and mouse cells, trapped in the extracellular materials produced rapidly by cartilage cells under the conditions of cultivation.

However, when cells of a four-day limb precartilage and five-day liver of the chick embryo are intermixed, they sort out according to their functional groupings, the cartilage cells lying at the center of the mass, surrounded by liver cells. There are now numerous examples showing that embryonic tissue cells sort out according to functional tissue types. Tissue specificity is the rule. In Steinberg's study of 11 binary combinations of six types of chick embryo cells, an outer continuous component and an inner, discontinuous component could always be distinguished. He found that the series could be arranged in a "hierarchy," in which each tissue segregates internally to all those below it and externally to all those above it.

Mixtures of cells are capable of a surprisingly high order of normal morphogenesis. In experiments performed by Weiss and Taylor, cell suspensions of skin, liver, and kidney from 8- to 14-day chick embryos were mixed and deposited as random aggregates on the chorioallantoic membrane. Structures were obtained whose complexity rivaled that of the normal organs.

How are these events to be explained? Consider the following facts: (1) the initial aggregation appears to be nonspecific, cells forming randomly mixed aggregates. (2) There is then a reshuffling and shifting as the different cell types sort out and gradually group together (Fig. 10-7). (3) Each cell type retains its identity throughout the process, there being no transformation of one type into the other, as shown by studies in which one of the cell types in the mixture was radioactively labeled. For example, if labeled pigmented retina cells were mixed with cells of the mesonephros, the label appeared only in retinal parts of the aggregates. (4) The mobile and adhesive capacities of cells change with time. So far there is little evidence that adult tissue cells can behave

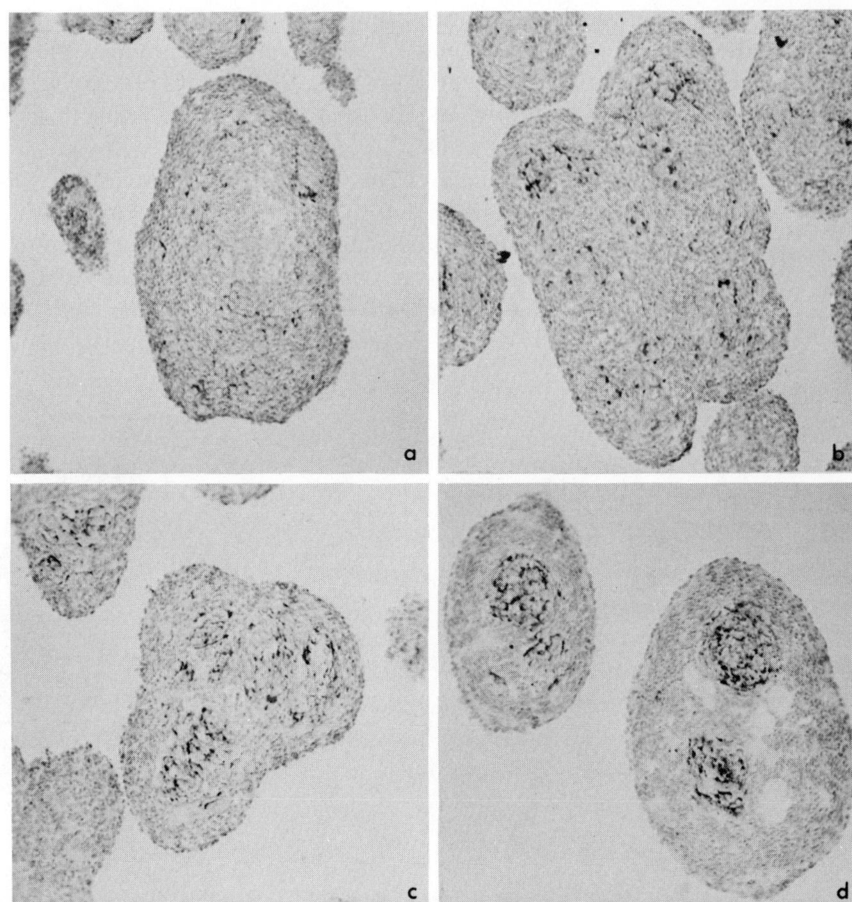


***Fig. 10-7*** *Time course of segregation of chick embryonic heart cells from chick embryonic retinal cells in gyratory shaker culture. Heart cells contain darkly stained glycogen granules.* (a) *After 17 hours;* (b) *after 24 hours;* (c) *after 31 hours;* (d) *after 66 hours.* × *130. (From M. S. Steinberg, in* Cellular Membranes in Development, *by courtesy of the author and Academic Press.)*

in this way, the capacity of embryonic cells seemingly declining with increasing age.

What kinds of mechanisms underlie the demonstrated order in the sorting-out process? Consider the experiment illustrated in Fig. 10-8. When liver cells and precartilage cells are mixed, liver cells take up the outer position. Starting with intermixed cells, liver cells gradually give up connections to other liver cells. If the experiment is started with a sphere of liver tissue, apposed to a sphere of precartilage tissue, the outcome is the same—that is, liver cells outside precartilage cells. However, the steps by which the final configuration is reached appear to differ significantly from those observed in the sorting out of intermixed cells. If precartilage cells are insinuating themselves into the center of the liver mass, at least some of them are adopting connections with unlike cells. If liver cells are spreading to surround precartilage, some of them are giving up connections to liver cells in order to establish connections with unlike cells.

Are there intercellular ligants in embryonic tissues? If they do exist, is their presence sufficient to account for the formation of characteristically organized multicellular patterns both *in vitro* and *in vivo*? There is evidence indicating that embryonic cells produce and release extracellular products, which serve as binding agents and promote orientation in histogenesis. In this approach emphasis is placed on

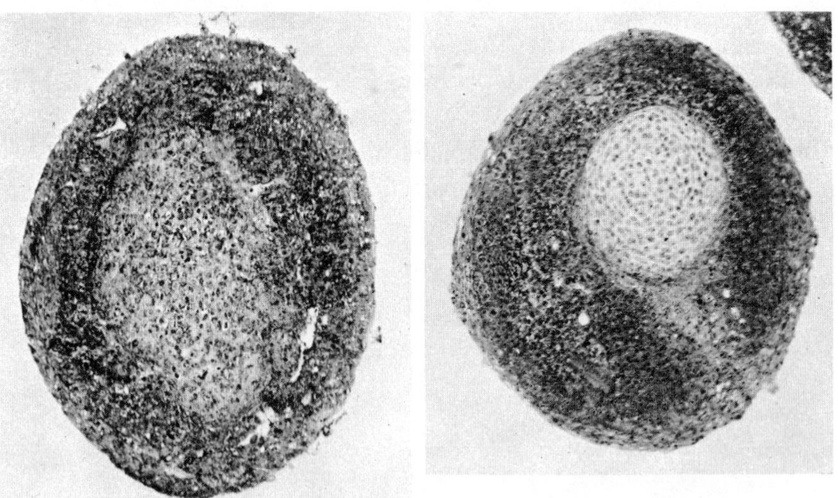

**Fig. 10-8** *Equilibria in self-ordering of tissues: 4-day chick embryo limb bud precartilage is surrounded by 5-day chick embryo liver.* (Left) *the result of sorting-out of intermixed cells;* (Right) *the result of spreading of fused intact tissue fragments. (From M. S. Steinberg, in* Cellular Membranes in Development, *by courtesy of the author and Academic Press.)*

the isolation of molecules that have the capacity to promote tissue-specific aggregation.

A substance prepared from the culture medium in which embryonic chick neural retina cells had been grown specifically enhances the aggregation of freshly dissociated retinal cells. The specificity of the preparation is shown by the fact that cells from other tissues do not respond to it. The binding activity is not removed from the preparation by liver cells, although it is by retinal cells. Moreover, antibodies made in rabbits against the substance do not cross-react with liver cells. These and other lines of evidence suggest that such materials may play an important part in the formation of tissues. Again, we need to know the nature of such factors and their origin and location within or on the surface of the cells. Only after these questions are answered can the larger question of the manner in which they act be examined.

Others would focus attention not on the products that may be produced and secreted by cells, but on the cell membrane itself. In this approach it is held that selectivity can be explained by differences in adhesive and cohesive properties characteristic of it. In short, this explanation requires differential adhesiveness. Can the characteristic behavior of dissociated and aggregating cells during sorting and tissue reconstruction *in vitro* be explained by quantitative differences in "strength of adhesion"?

We would prefer, however, to direct the spotlight not on conflicting approaches but on the central issue of the mechanisms underlying morphogenesis and on the manner in which they may be elucidated by studies of tissue reconstruction. Thus we would ask, in conclusion, where does the cell membrane end and a "product" begin? Is the distinction meaningful? Finally, we would correct any impression that might lead us to regard interacting cells as simple spheres, with smooth membranes. Interacting cells are enormously far removed from this description. They are constantly in motion and constantly changing shape—extending pseudopodia and filamentous extensions—and yet they are ultimately selective in their associations. Moreover, all of these properties are constantly subject to change during the course of differentiation.

## *DEVELOPMENTAL FAILURE IN AGGREGATELESS VARIANTS OF A CELLULAR SLIME MOLD*

DNA, mRNA, ribosome, . . . cell surface: what are the connections? How do we relate the information for synthesizing specific proteins to another level of organization, the interactions between cells? The gap is a large one and we are only beginning to inch across it. Let us consider first the genetic control of molecules at cell surfaces. In at least some cells, there is ample evidence that components of, or products at, their surfaces are gene-controlled. The blood group substances (A, B, Rh, and so on) fall in this category. We speak of them as the blood group antigens because when first discovered they were recognized by their capacity to evoke a specific antibody, the antibody being used as a diagnostic reagent.

An example more pertinent to the question of aggregation, however, is drawn from studies of the cellular slime mold, *Dictyostelium discoideum*. The life cycle of this species is shown in Fig. 10-9. Let us examine the cycle beginning with the spore stage. Spores germinate, producing amoeboid myxamoebae, which live in the soil (or, in the laboratory, on agar) feeding on bacteria and reproducing by binary fission. To initiate morphogenesis and differentiation, the investigator need only allow the supply of food to run out. Now the cells stop growing and enter the aggregation stage, streaming toward central collecting points.

Outlying cells are attracted toward aggregation centers by specific substances produced at the centers. Initially the amoebae come in separately, but they soon tend to flow together in streams. There are several unresolved questions: (1) What is the source of the agent (given the generic name *acrasin*, the cellular slime molds being the Acrasiales)? At one time it was thought that specific initiator cells could be recognized in the population of amoebae, but the identification is now considered doubtful. (2) What is the nature of the material? It is freely diffusible and water soluble; initially it appeared to be a highly

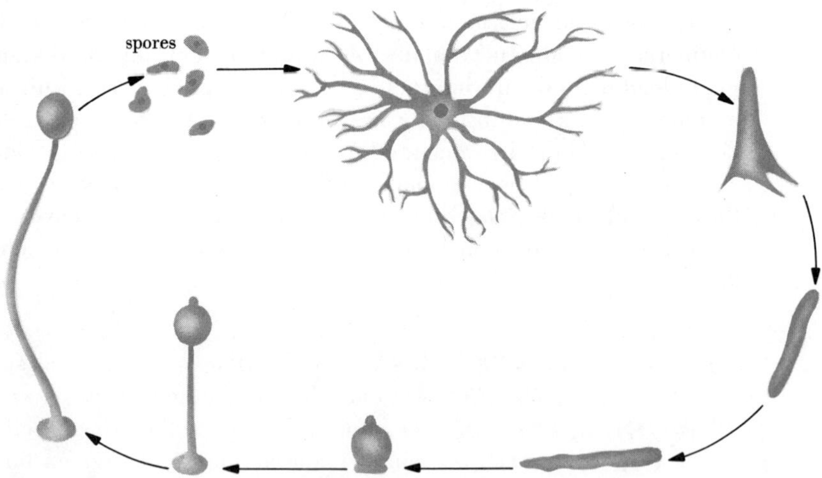


**Fig. 10-9** *Life cycle of* Dictyostelium discoideum. *The cycle begins and ends with a unicellular spore. (After J. T. Bonner,* The Cellular Slime Molds, *by courtesy of Princeton University Press. Copyright 1959 by Princeton University Press.)*

unstable molecule, but soon it was found that the instability resulted from the rapid enzymatic destruction of the agent by the cells. Not only the synthesis, but also the destruction, of acrasin must be regulated in order to achieve effective gradients. The chemical nature of acrasin is being investigated; although its action can be mimicked by steroids, including some of the sex hormones, the best available evidence suggests that cyclic 3′,5′-adenosine monophosphate (cyclic AMP) is an acrasin (Fig. 10-10). At least it is clear that cyclic AMP does attract amoebae. In another species, *Polysphondylium pallidum*, cyclic AMP is produced, but similar demonstrations have not yet succeeded in *Dictyostelium*, for the reason that these amoebae produce large quantities of an enzyme, phosphodiesterase, which destroys cyclic AMP.

When the aggregating streams merge, the rising cone-shaped center turns over on its side and becomes a pseudoplasmodium, a wormlike slug (Fig. 10-11) that glides along the agar, assuming different

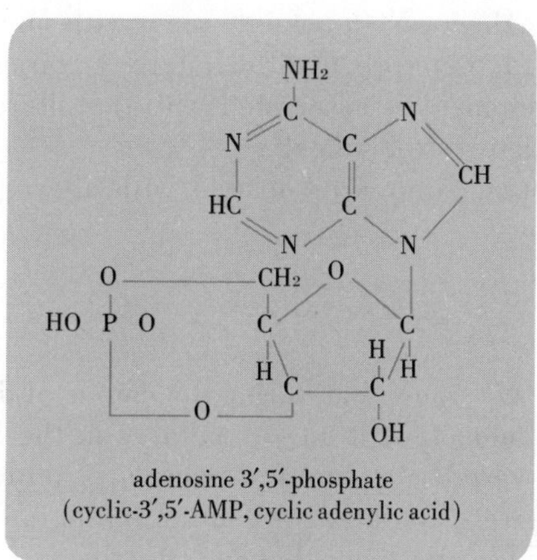

adenosine 3′,5′-phosphate
(cyclic-3′,5′-AMP, cyclic adenylic acid)

**Fig. 10-10** *Adenosine 3′,5′-phosphate (cyclic-3′,5′-AMP, cyclic adenylic acid).*

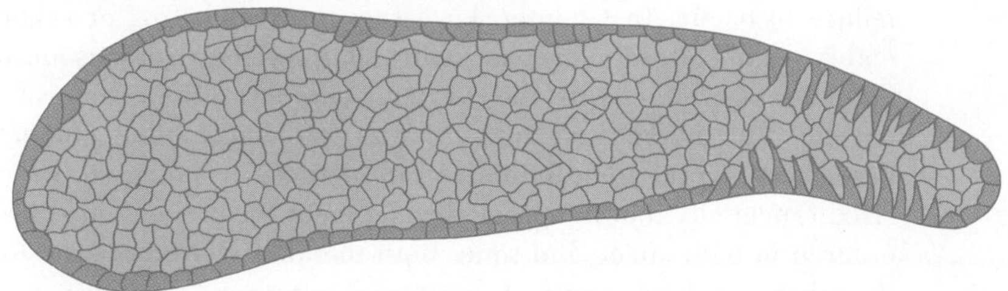

**Fig. 10-11** *A semidiagrammatic view of a section of a migrating pseudoplasmodium of* Dictyostelium discoideum *showing the typical transverse orientation of the cells in the narrow anterior portion. (From J. T. Bonner,* The Cellular Slime Molds, *by courtesy of Princeton University Press. Copyright 1959 by Princeton University Press.)*

shapes and having a tip that resembles the human tongue. At the end of migration, the tip of the cell mass stops moving but the posterior end continues to gather in, and the slug thus becomes round. The subsequent changes—resulting in the culmination stage—are shown dramatically in Fig. 10-12. Stalk and sorus, characteristic of the mature fruiting body, are formed.

As much as we would like to examine the process of cellular differentiation in *Dictyostelium*, we cannot allow ourselves the luxury of another diversion. Suffice it to say that all of the approaches we are discussing have been (or may be) applied to these aggregative organisms—and that many of them are discussed, knowledgeably and readably, in John Tyler Bonner's book, *The Cellular Slime Molds*.

We must return to our main point: the cell surfaces of the myxamoebae. By the use of ultraviolet irradiation, variants of *Dictyostelium* may be produced which lack the ability to aggregate. Can clues to the mechanisms of adhesion be found by studying them?

Antibodies made to normal, wild-type *Dictyostelium* react with antigens of the cell surface. However, they fail to react with some of the variants. It may be suggested, therefore, that the failure of aggregateless variants to complete development is associated with the alteration or elimination of surface antigens in the amoebae.

Are other developmental failures associated with aberrations at the cell surface?

## *INDUCTIVE TISSUE INTERACTIONS*

We know that during the course of development few, if any, structures in the body of vertebrates—and of many invertebrates, as far as the evidence goes—are elaborated without an initial interaction of their constituent tissues. We also know that one effective way of illustrating such an interaction is by indicating the consequences of its failure to occur. In Chapter 4 we saw that when frog or salamander embryos are caused to exogastrulate, so that the chordamesoderm does not come in contact with the overlying ectoderm, the ectoderm fails to develop into nervous system. In that and succeeding chapters, we described several such inductive interactions; in doing so, we emphasized repeatedly that in normal development the process must be highly ordered in both space and time. Both the ability of one tissue to affect the other and the capacity of the second to respond must be considered.

Often such interactions are spoken of as "contact mediated"; this expression may have two implications. The first possibility to be considered is that there is no exchange of material between the cells—the surface of one cell reacting with the surface of an adjacent cell,

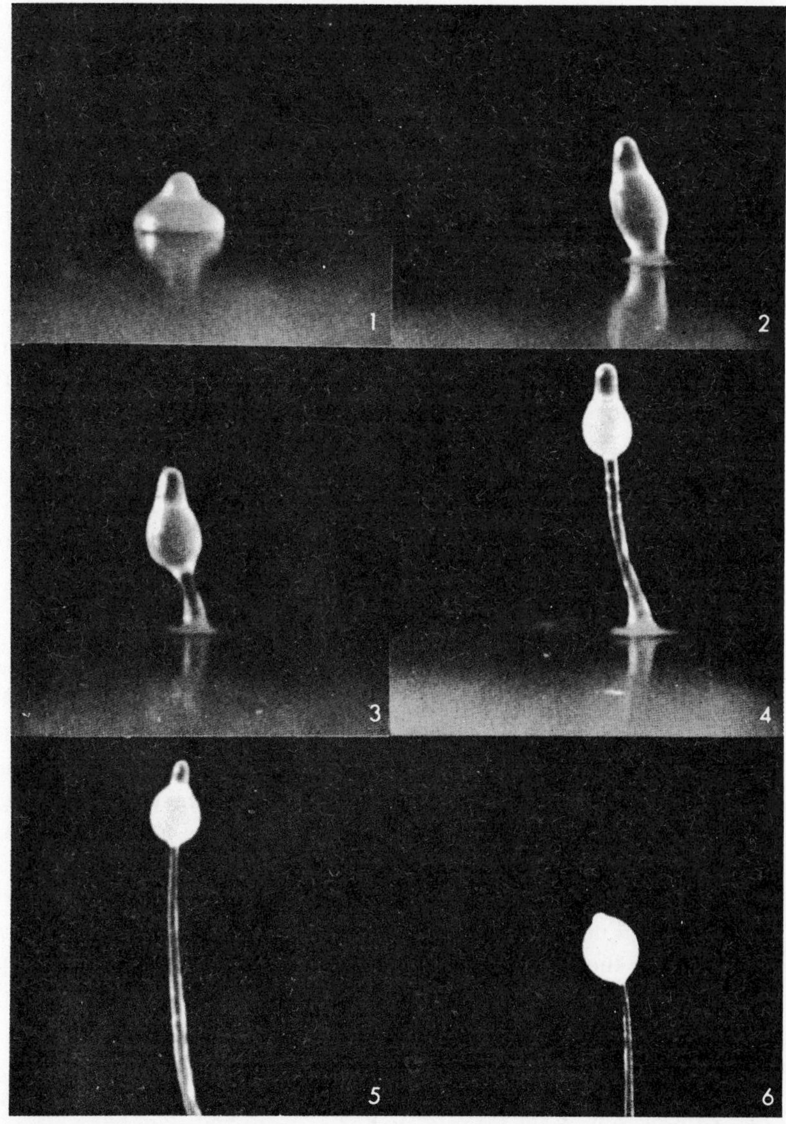


**Fig. 10-12** *Culmination of* Dictyostelium discoideum. *Each photograph represents a time interval of about 1 1/2 hours. (From J. T. Bonner,* The Cellular Slime Molds, *by courtesy of Princeton University Press. Copyright 1959 by Princeton University Press.)*

and the resulting surface changes being reflected in changes within the cell. We have already presented some of the ideas about the interactions of cell surfaces resulting in aggregation of similar cells. In such interactions one cell does not appear to influence, or change, the direction of differentiation of its neighbor. We see no convincing evidence that inductive interactions, which by definition result in an alteration of the developmental course of the interactants, are mediated through surface contacts without an exchange of material.

The second implication of the expression "contact mediated" seems more plausible—namely, the implication that the reacting tissues remain in intimate association in such a manner, and for periods long enough, that an exchange or interaction of products is permitted. An increasing familiarity with the membrane systems of cells should warn us that we are using the simple word "contact" to refer to what must be not one, but rather an array, of complex interactions: the juxtaposition of pores in apposed membranes with a closely timed transfer of particulates, macromolecules, or small molecules; the elaboration of processes or products from apposed cells which interact; or the uptake by one cell of small molecules from another. There may be as many mechanisms as there are combinations; there may be a few large classes of mechanisms; or there may be one mechanism common to all.

## *Epithelio-mesenchymal Interactions*

The differentiation of epithelial tissues—such as kidney, pancreas, salivary gland, skin, thyroid, and thymus—depends on interaction between their epithelial and mesenchymal layers. An epithelium is a continuous cell layer, in which the cells are typically held together by tight junctions. A mesenchymal layer is typically embryonic connective tissue, a network of branched fibroblasts in a fluid ground substance. Epithelium grown alone in culture does not differentiate. When epithelium and mesenchyme are grown in culture—separated by a thin, porous membrane filter, which permits the passage of molecules but not cells—the epithelium differentiates (Fig. 10-13). Here, then, is evidence of an inductive interaction,

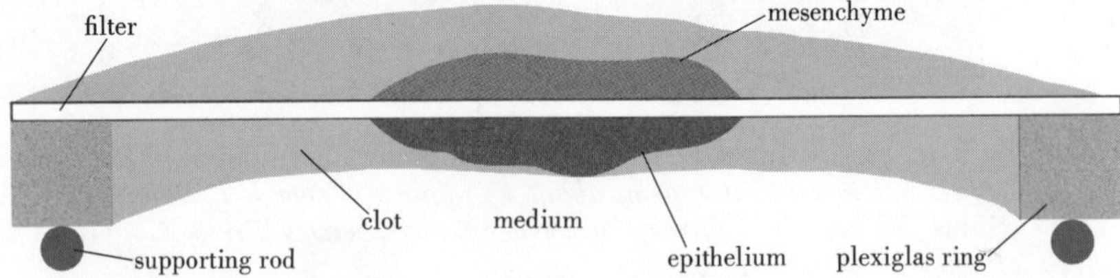


***Fig. 10-13*** *Filter assembly for preparing standard transfilter cultures of epithelium and mesenchyme. Intact rudiments may be cultured either "in the clot" in the position of the epithelium, or "on the platform" in the position of the mesenchyme. (After C. Grobstein, in* Science, *143, by courtesy of the author and the American Association for the Advancement of Science. Copyright © 1964 by the American Association for the Advancement of Science.)*

involving the passage of some substance or substances across the filter.

Let us consider one example: the mammalian exocrine pancreas. The pancreatic rudiment evaginates from the midgut of the nine-day mouse embryo having 21 to 23 somites. It is at this time that mesoderm cells appear over the endoderm. Three stages can be identified before the diverticulum forms. In the first, the pancreas cannot be formed *in vitro*. No cells can be demonstrated to have the capacity to form pancreas. To put it another way, no cells have been determined to form pancreas. In the second phase, endoderm plus its adjacent tissues (neural plate, somite, dorsal aorta) can form pancreas *in vitro*. Clearly determination has already occurred. Finally, the endoderm will form pancreas if it is combined with mesoderm not only from pancreas but from other tissues—for example, the salivary gland. When the diverticulum first evaginates, traces of the specific products characteristic of the pancreas—insulin and the enzymes, lipase, ribonuclease, amylase, and others—can be detected. Low levels of these proteins are maintained during the succeeding three days, during which the diverticulum undergoes rapid mitosis. At about 14 to 15 days, cell division stops in the central acini and specific synthesis is accelerated (Fig. 10-14).

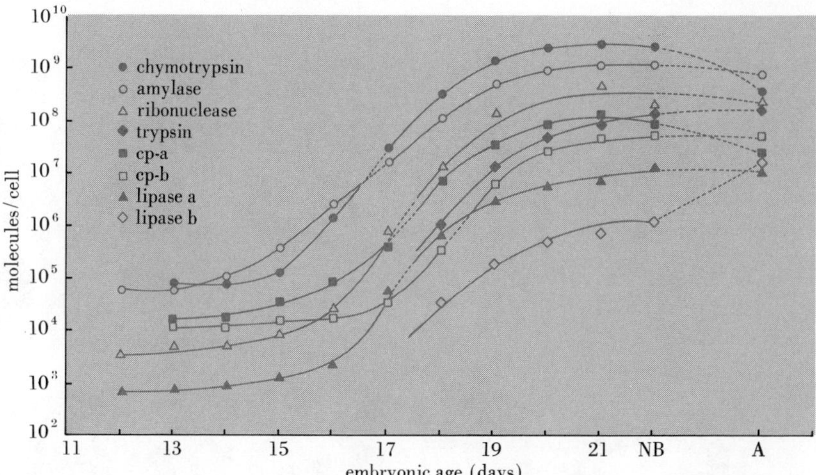


***Fig. 10-14*** *Developmental profiles of specific pancreatic exocrine proteins. Pancreatic rudiments of various ages were excised in Earle's balanced salts solution, washed and sonicated, after which DNA and protein content and enzyme activities were determined. Ordinate: molecules of each protein per pancreatic cell, calculated from the quantity of enzyme based on the rate constants of crystalline porcine or bovine preparations and from the number of cells based on the value 7.12 × $10^{-12}$ g DNA/cell. Abscissa: embryonic age in days; NB indicates newborn; A, adult; CP-A, carboxypeptidase A; CP-B, carboxypeptidase B. (After W. J. Rutter, et al.,* J. Cell. Physiol, *72, suppl. 1, by courtesy of the Wistar Institute Press.)*

What is the nature of these tissue interactions? In the case of pancreas, evidence obtained by Grobstein and his colleagues from such "transfilter" inductions (Fig. 10-13) shows that the effective material is noncellular and that it can cross a 20-$\mu$ interspace. In this experimental system at least 30 hours' interaction appear to be required. If the mesenchyme is removed before 30 hours, the epithelium does not differentiate. From 30 to 48 hours, little or no change can be detected in the epithelial cells by current methods, but shortly thereafter they begin to differentiate. What is occurring in the critical 18-hour period?

As we said earlier, pancreatic epithelium requires the presence of mesenchyme, as do many epithelia, but it does not require pancreatic mesenchyme. Any embryonic mesenchyme will do. Even chick embryo mesenchyme will induce embryonic mouse pancreatic epithelium to differentiate. The effective factor in the mesenchyme appears thus far to be large molecular material sensitive to trypsin (suggesting that it contains protein). The complexity of the problem is further emphasized by the fact that in some epithelio-mesenchymal interactions, such as in the salivary gland, a specific mesenchyme is required; in others, "nondescript" mesenchyme will suffice.

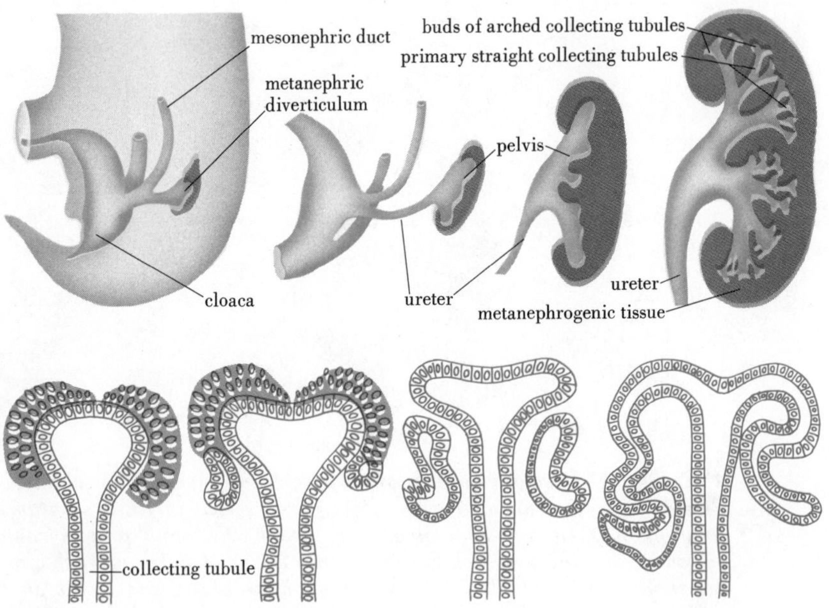


**Fig. 10-15** *Formation of kidney and kidney tubules.* (Upper row) *Morphogenesis of metanephric diverticulum in human embryos 4, 8, 11, and 20 mm in length.* (Lower) *Development of metanephric tubules (see text). (After L. B. Arey,* Developmental Anatomy, *6th ed., by courtesy of W. B. Saunders Company.)*

The submandibular salivary gland, especially, shows the complexities of these interactions. Condensed salivary gland mesoderm is required, specifically. It manufactures collagen that is polymerized at the epithelial surface. This collagen plays an essential role in maintaining the morphology of the epithelium, since treatment with collagenase results in a loss of lobulation. However, although collagen is necessary, it is not a *sufficient* cause. Collagen alone does not lead to morphogenesis. In fact, three factors seem to be required: collagen, a growth stimulator to support mitosis, and some unknown "specific" agent.

In the pancreas, the induction is one-way; that is, in this combination, mesenchyme influences epithelium, but epithelium does not appear to influence mesenchyme. In other combinations, each affects the other. The metanephric kidney rudiment of the 11-day mouse embryo consists of a ureteric bud and surrounding nephrogenic mesenchyme (Fig. 10-15). Together, both differentiate: the bud into the collecting system, the nephrogenic tissue into tubules. Separated in culture, they do not differentiate. Again, tissue specificity is not absolute: metanephric mesenchyme may be combined with ureteric bud, or salivary gland epithelium, or even the dorsal part of the spinal cord, and it will differentiate into tubules.

## *An Artificial System of Embryonic Induction*

If a piece of prospective epidermis of an early gastrula of a newt is cultured in a salt solution at physiological pH, it forms a cluster of ciliated epidermal cells. If the same kind of explant is cultured for three hours in a solution containing a protein derived from the bone marrow of a guinea pig and is then transferred to salt solution, it develops mesodermal and some endodermal tissues, including striated muscle and notochord. An agent that acts similarly may be extracted from chick embryos. In addition, another complex of factors that has been prepared from chick embryos induces forebrain and eye and other ectodermal structures. This second factor is believed to be a ribonucleoprotein.

Are these useful kinds of experiments? They are artificial in the sense that conversion of prospective epidermis into mesodermal tissues does not occur normally, and that guinea pig and chick molecules normally are not operating in amphibian cells. What do they tell us?

It will be helpful in framing an answer to this question to have the following additional information. In other amphibian species, such diverse agents as methylene blue, steroids, even abnormal pH, can evoke "neurulation" in prospective epidermis. Usually, formless masses —but occasionally more organized structures of the forebrain—are produced. For most species, however, the induction of mesodermal or endodermal structures requires that macromolecules be present.

These kinds of experiments can be useful, to a degree. At least we can make some interesting speculations about the action of such agents. What can substances like methylene blue or steroids be doing? We might speculate that normally in the formation of the neural tube interaction with mesoderm modifies a repressor, initiating a chain of reactions. It is tempting to postulate that nonspecific agents might trigger a derepression, setting at least a part of the normal chain of events in motion.

But the protein fraction of bone marrow and the protein or ribonucleoprotein fractions of chick embryo change the course of the reaction; prospective epidermis now makes mesodermal structures. Thus, these substances must act differently from methylene blue. There are two major possibilities. First, in those systems in which proteins are implicated, it is not too far-fetched to suggest that the active molecule is enough like one of the normal repressors to substitute for it. Thus, in prospective epidermis, an operator gene may be repressed, and the structural changes characterizing neurulation blocked. Since developmental processes are sequential and interrelated, this repression might permit the next step toward higher organization in another direction. And, as we have mentioned earlier, there is evidence that bacterial repressors are proteins.

As a second possibility, what if the active substance is a nucleic acid? This idea is an old favorite among embryologists, who find it hard to give up; we should add hastily that the evidence doesn't *require* that it be dropped! But despite a large body of circumstantial evidence favoring a role of RNA in some of these interactions, the evidence is far from convincing. Brachet's early observations had a significant impact; he brought to bear evidence that suggested the transfer of RNA between interacting tissues; even if the transfer were established, however, we would have to prove that the RNA played a directive role.

Note that we are speaking of RNA broadly, not of mRNA or ribosomes or polyribosomes, for two reasons. First, much of the evidence was obtained before these newer concepts evolved; and second, the interacting systems employed have presented technical difficulties that make it difficult to be more specific. There appears to be no indisputable example in which differentiation has been experimentally "directed" by specific RNA in embryonic cells.

### *The Changing Chemical Environment*

Although we have offered only a sample of it, abundant and compelling evidence exists to show how the changing chemical environment results in the differential expression of genes, with telling effect on the course of differentiation. Enough has been said to emphasize that the conditions for differentiation, achieved through the interactions we have referred to as inductive, may differ widely, and thus a search for a single, universal "inductive agent" is illusory. Molecules exchanged between cells probably affect mitosis, cell participation in morphogenesis, and synthesis and assembly of specific products. All three kinds of processes are essential for organogenesis. Thus far we have focused on associations in which large molecules appear to be involved. Lest this emphasis be misleading, we should present at least one example in which a cell with at least two alternative capacities is "signaled" by a small molecule to take one course to the exclusion of another. When epidermis of the chick embryo is explanted to a medium of blood plasma, it produces stratified epithelium that is characteristic of skin, the outer layers of which become keratinized. When vitamin A is added to the medium, the cells form a columnar epithelium, which produces cilia and secretes mucus. Only a short treatment with vitamin A is necessary; cells exposed briefly and returned to normal medium proceed to make mucus.

## *CONCLUSIONS*

Each differentiating tissue cell has had its own inner controls. Yet in its development the cell is part of a larger whole; during its differentiation the cell must respond to control factors extrinsic to it. Thus, one of the large tasks of developmental biology is to identify those external controls and understand the ways in which they impinge on the cell's inner controls.

In examining the external controls in tissue interactions, we have found first that they involve intimate association between cells. This association does not require "contact" in the mechanical sense of immediate juxtaposition and touching of surfaces; it does require that the cells communicate in a common microenvironment. This communication may take many forms. We have seen that it would be unwise to exclude small molecules as agents, for under the proper conditions they could operate as substrates in differentiation triggered by the induction of an enzyme, or in feedback inhibition, or in repressing an operator. Large molecules contain more "possibilities" for accounting for changing specificities. But large or small, the agents act over short distances and, like acrasin, are probably inactivated rapidly. Differentiation requires mechanisms not only of synthesis but of degradation. Throughout this chapter we have stressed the sequential nature of differentiation. All of the schemes proposed rely heavily on that premise. Once a given reaction in the sequence is initiated experimentally, the normal train of events follows unbroken.

It is true that bacterial ribosomes will make hemoglobin if they are directed by mRNA from reticulocytes. But the "heterologous inductors" such as the protein from guinea pig bone marrow, influence not just the synthesis of a single protein but a long chain of events, not the least of which are changes in cell morphology. Earlier we saw (Fig. 4-16) that one of the first manifestations of induction of lens in ectoderm by underlying optic vesicle is the elongation of ectodermal cells. Byers and Porter have now shown that subsequent to lens induction in the chick, many microtubules, oriented parallel to the direction of elongation, appear in the prospective lens cells. Thus we have further evidence of an oriented reorganization of cell components.

## FURTHER READING

Bonner, J. T., *The Cellular Slime Molds,* 2d ed. Princeton, N.J.: Princeton University Press, 1967.

Byers, R., and K. R. Porter, "Oriented Microtubes in Elongating Cells of the Developing Lens Rudiments after Induction," *Proceedings of the National Academy of Sciences,* vol. 52 (1964), p. 1091.

Coleman, J. R., and A. W. Coleman, "Muscle Differentiation and Macromolecular Synthesis," *Journal of Cell Physiology,* vol. 72, suppl. 1 (1968), p. 19.

DeHaan, R. L., and J. D. Ebert, "Morphogenesis," *Annual Review of Physiology,* vol. 26 (1964), p. 15.

Ebert, J. D., "Morphogenetic Movements and Congenital Defects," in *First Inter-American Conference on Congenital Defects,* M. Fishbein, ed. Philadelphia: Lippincott, 1963, p. 61.

Fell, H. B., and E. Mellanby, "Metaplasia Produced in Cultures of Chick Ectoderm by High Vitamin A," *Journal of Physiology (London),* vol. 119 (1963), p. 470.

Fleischmajer, R., ed. *Epithelial-Mesenchymal Interactions.* Baltimore: Williams & Wilkins, 1968.

Gregg, J. H., and C. W. Trygstad, "Surface Antigen Defects Contributing to Developmental Failure in Aggregateless Mutants of the Slime Mold, *Dictyostelium discoideum,*" *Experimental Cell Research,* vol. 15 (1959), p. 358.

Grobstein, C., "Cytodifferentiation and Its Controls," *Science,* vol. 143, (1964), p. 643.

Holtfreter, J., "Tissue Affinity, a Means of Embryonic Morphogenesis" (1939), reprinted in *Foundations of Experimental Embryology,* B. H. Willier and J. M. Oppenheimer, eds. Englewood Cliffs, N.J.: Prentice-Hall, 1964, p. 186.

Humphreys, T., "Chemical Dissolution and *in vitro* Reconstruction of Sponge Cell Adhesions. I. Isolation and Functional Demonstration of the Components Involved," *Developmental Biology,* vol. 8 (1963), p. 27.

Konijn, T. M., D. S. Barkley, Y. Y. Chang, and J. T. Bonner, "Cyclic AMP: A Naturally Occurring Acrasin in the Cellular Slime Molds," *American Naturalist,* vol. 102 (1968), p. 225.

Konigsberg, I. R., "Clonal analysis of myogenesis," *Science,* vol. 140 (1963), p. 1273.

Lash, J. W., "Chondrogenesis: Genotypic and Phenotypic Expression," *Journal of Cellular Physiology*, vol. 72, suppl. 1 (1968), p. 35.

Mintz, B., and W. W. Baker, "Normal Mammalian Muscle Differentiation and Gene Control of Isocitrate Dehydrogenase Synthesis," *Proceedings of the National Academy of Sciences*, vol. 58 (1967), p. 592.

Moscona, A. A., "Aggregation of Sponge Cells: Cell-Linking Macromolecules and Their Role in the Formation of Multicellular Systems, *In Vitro*, vol. 3, (1968), p. 13.

Rutter, W. J., J. D. Kemp, W. C. Bradshaw, W. R. Clark, R. A. Ronzio, and T. G. Sanders, "Regulation of Specific Protein Synthesis in Cytodifferentiation," *Journal of Cell Physiology*, vol. 72, suppl. 1 (1968), p. 1.

Steinberg, M. S., "The Problem of Adhesive Selectivity in Cellular Interactions," in *Cellular Membranes in Development*, M. Locke, ed. New York: Academic Press, 1964, p. 321.

Torrey, T. W., "Morphogenesis of the Vertebrate Kidney," in *Organogenesis*, R. L. DeHaan and H. Ursprung, eds. New York: Holt, Rinehart and Winston, 1965, p. 559.

chapter **11**

# Remodeling Processes and Their Controls

In Chapters 9 and 10 we have emphasized the problems of the assembly, not only of structural elements such as fibrous proteins and organelles, like chloroplasts and mitochondria, but of tissues themselves. However the problem is even more complex, for many structural units must be built not just once, but repeatedly, in the same constant basic pattern. For example, some cells can repeatedly re-form typical patterns of cilia after their loss or resorption. Moreover, to permit growth and change, structures must be "dismantled" periodically, and then reorganized. The most familiar example is the mitotic apparatus, which must be disassembled and assembled regularly.

Although remodeling and reshaping of

tissues must occur throughout life, it is during the period of development that they occur most strikingly. These processes differ widely in the nature of the changes observed, and in the factors that initiate and control them. In a number of systems, the morphogenetic events are hormonally controlled. Thus in this chapter we shall first consider cell death as a morphogenetic mechanism, after which we shall bring together a few examples illustrating the major principles of hormone action in relation to differentiation and morphogenesis.

## CELL DEATH IN MORPHOGENESIS AND SENESCENCE

### In Limb Development

One of the ways in which tissues and organs take shape is through differential cell death. Cell death is, in fact, a commonplace mechanism, as illustrated in the sculpturing of the limb. In an elegant morphologic and experimental study Saunders showed that the shaping of the chick wing was accomplished in part by the occurrence of localized zones of cell death. In particular, the shaping of the upper arm and forearm, especially the prospective elbow, and the elimination of tissues between the digits are effected in this manner (Figs. 11-1 and 11-2).

What is the utility of cell death? Is it mechanical? What is the fate of the breakdown products? We don't know. What factors control the onset of morphogenetic death? Its location in space? What "sets the death clock"? If a group of cells destined to die in the normal course of events at stage 24 in the wing of a chick are removed at stage 17 and grafted to the region of the somites, they die on schedule. If they are grafted to the dorsal side of the limb bud, they survive However, if they are grafted later, at stage 22, they die in either site. These experiments suggest that although the "death clock" is set by stage 17, it can be turned off, by the imposition of external controls in an appropriate environment, up to stage 22. Thereafter it proceeds inexorably.

The necrotic zone of the wing bud ("determined" to die at stage 24) dies on schedule when cultured *in vitro*. However, death may be prevented by combining the necrotic zone with wing or leg mesoderm. In fact, limb mesoderm prevents death even when the necrotic zone and the mesoderm are cultured by transfilter techniques (page 236). The filters preclude cell contact, but allow some "factors" to pass. Death does not occur in explants from embryos up to stage 20 cultivated with limb mesoderm of three- to four-day embryos. After stage 22, death is not prevented. Results are variable at stages 20–21.

How is the death sentence executed? In a number of systems (see below), cells prepare for their own demise by producing or activat-

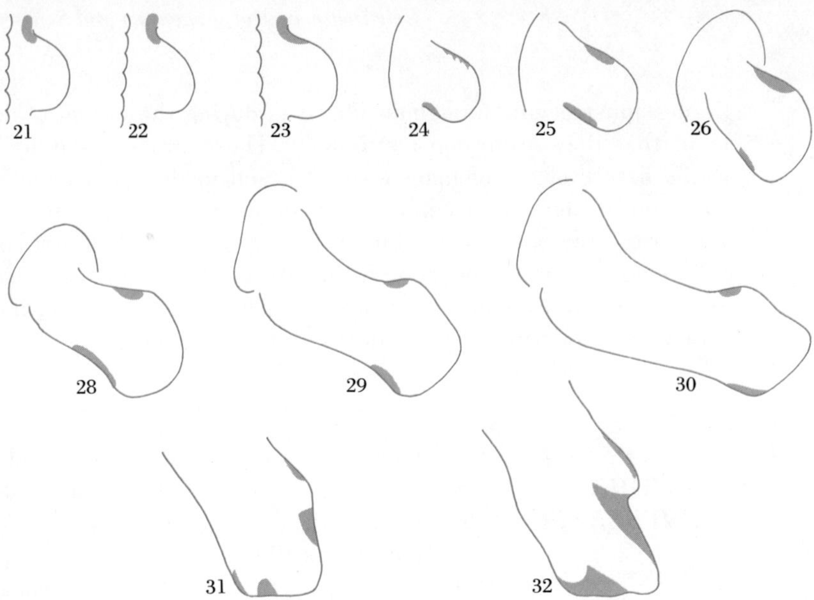


**Fig. 11-1** *Sketches of stages in development of the chick embryo wing bud showing regions of massive cell death and necrosis in the superficial mesoderm. (After J. W. Saunders, Jr., M. T. Gasseling, and L. C. Saunders, in* Developmental Biology, *5, by courtesy of Academic Press.)*

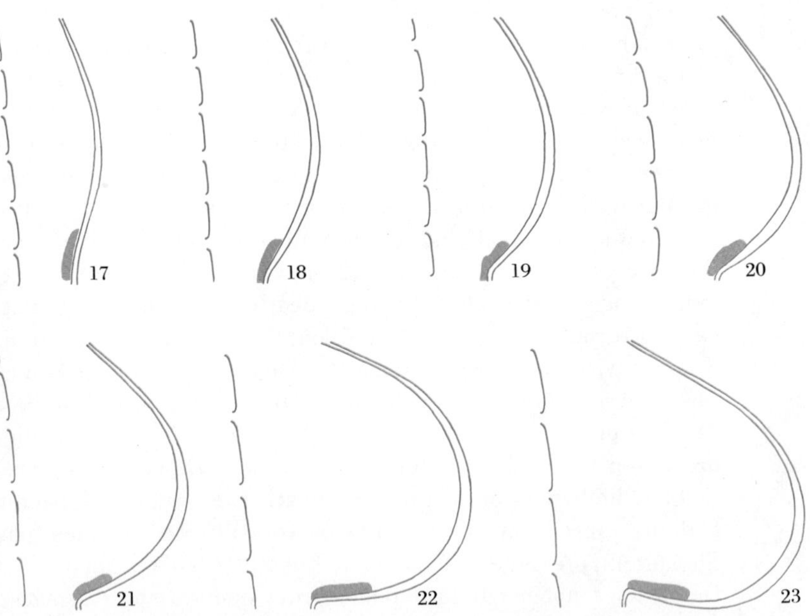


**Fig. 11-2** *Stippling shows the approximate distribution of prospective degenerating cells of the posterior necrotic zone in the developing chick wing bud. (After J. W. Saunders, Jr., M. T. Gasseling, and L. C. Saunders, in* Developmental Biology, *5, by courtesy of Academic Press.)*

ing a battery of hydrolytic enzymes, to be used in breaking down their "remains." These enzymes are usually contained in distinct organelles, the *lysosomes.* However, in the limb, lysosomal enzyme activity does not seem to play an important role. The initial changes have not been recognized with certainty. Eventually, once general decay is advanced, cells and cell debris are engulfed by phagocytes.

### *In the Developing Gonad and Accessory Structures*

Massive cell death occurs in the oviduct of the chick embryo. In this system we do find a synthesis of lysosomal enzymes in preparation for death. By way of background we should note that although the sex of the vertebrate embryo is determined genetically at fertilization, virtually all of the embryonic structures necessary for *either* sex are laid down morphologically and are present for a time during early embryogenesis as discrete primordia. This "indifferent" phase is observed in every individual, regardless of its genetic sex. It is this bisexuality of the early embryo that accounts for the relative ease with which embryos undergo a reversal of sex, as in naturally occurring developmental anomalies (hermaphroditism) or as a result of experiment.

The gonad itself is bisexually organized, containing two distinct components, a central *medulla* and a *cortex* surrounding it. In the male it is the medulla, and in the female it is the cortex (Fig. 11-3), that gradually predominates. The differentiation of the gonad does not in-

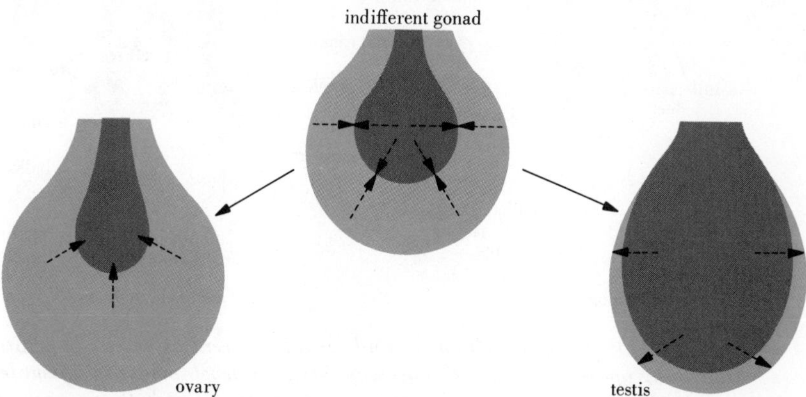


*Fig. 11-3 Diagrammatic representation of the male and female components of the sexually undifferentiated amphibian gonad and their roles in sex differentiation: the medulla is dark color; the cortex is lighter color. The broken arrows indicate the mutually antagonistic or inhibitory actions exerted between the two components in the course of sexual differentiation. (From R. K. Burns (after Witschi), in* Survey of Biological Progress, *1, by courtesy of Academic Press.)*

volve the transformation of one component into the other, but rather the gradual predominance of one and regression of the other.

Likewise the accessory sex structures, including the sex ducts and organs, are laid down and develop in similar or identical fashion up to a point, each with the capacity to develop into either sex (Fig. 11-4). Depending on the species, this primitive condition may persist, with the ducts of the recessive sex functioning in other roles, or regressing. In *amniote* embryos (reptiles, birds, mammals), the ducts are typically transient structures. One of the pairs of primitive sex ducts, the Müllerian ducts, provide an especially interesting example. These are the forerunners of the oviducts in the female. In the male the Müllerian ducts regress. In the female chicken the right Müllerian duct regresses, leaving only the left duct which persists to form the definitive oviduct.

Prior to the breakdown of these structures, they show a considerable synthesis of lysosomal enzymes—for example, acid phosphatase, cathepsins, and ribonuclease, among others. Acid ribonuclease

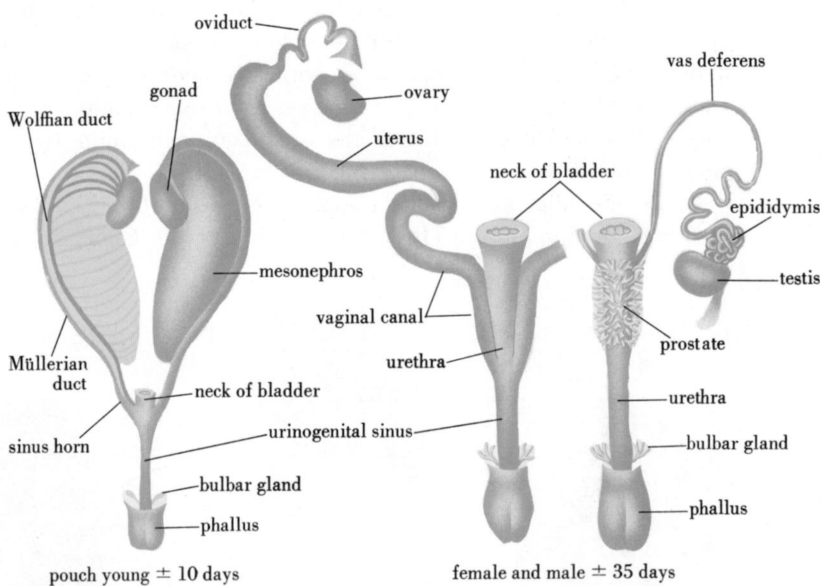


*Fig. 11-4 Early development and sexual differentiation in the genital tracts of young opossums.* (Left) *The bisexual stage of development in a female embryo about 10 days of age, showing the paired gonaducts of both sexes, the sexually indifferent stage of the urinogenital sinus, and the undifferentiated genital tubercle or phallus.* (Right) *Male and female at about 35 days, when sexual differentiation is far advanced, showing the structures that develop from the primitive sex ducts, and dimorphic development of the sinus region. The phallus shows chiefly a difference in size, without marked morphologic divergence. (From R. K. Burns, in* Survey of Biological Processes, *1, by courtesy of Academic Press.)*

increases in both ducts in the male, but only in the right duct of the female.

The precise mechanisms that trigger these syntheses, and the subsequent degeneration, are not known. Since 1935 it has been known that injection of estrogens (female sex hormones) into genetic male chick embryos before urinogenital differentiation results in feminization, with retention of Müllerian ducts. Conversely, injection of androgens (male hormones) into genetic females results in breakdown of the left, as well as the right, duct (masculinization). These facts certainly suggest that the sex hormones of the embryo control these events in some way. Müllerian ducts have been cultured *in vitro*, and the effects of exogenous hormones assayed. However, the results are too conflicting to permit a general conclusion; possibly, androgens do trigger autolysis of Müllerian ducts in males; whether androgens also cause normal breakdown of the right duct in the female is unclear. When and in what form do the sex hormones appear in development? Do hormones activate DNA-dependent RNA synthesis of lysosomal enzymes? Or do they inhibit some key metabolic process in the cells, leading secondarily to release of lysosomal enzymes?

These questions of mechanisms of hormone action have not been resolved. We can, however, consider the question of the embryonic origin of sex hormones. The modern era of the study of sex differentiation was initiated by studies of Lillie and Keller and Tandler on "one of Nature's experiments," the *freemartin* (Fig. 11-5). In cattle twins, the vascular connections of their fetal membranes usually permit the

**Fig. 11-5** *Twin calves removed from the uterus, showing chorionic fusion and anastomosis between major blood vessels.* (Left) *Male twin;* (right) *freemartin. (After F. R. Lillie, in* Journal of Experimental Zoology, *23, by courtesy of the Wistar Institute.*

intermingling of blood of the twins prior to, or about the time of, structural sex differentiation of the gonad. If the twins are of opposite sex, and vascular connections are established, the female partner is modified in the male direction, and is called a freemartin. It is inferred that the gonads of the male twin produce and release male hormones at an early stage; thus, the female twin would be modified in the male direction.

This concept stimulated the application of the technique of *parabiosis*—that is, duplication in the laboratory of Nature's experiment. When amphibian embryos of opposite sex are joined (Fig. 11-6), the male is usually dominant, as in the freemartin. However, in the laboratory one can "juggle" shapes and sizes, and by combining small, slower growing males with larger, rapidly growing females, the female can be made to predominate.

Although chick embryos can be parabiosed for other purposes, another technique is more effective in studies of sex differentiation. Fragments of embryonic gonad of one sex are implanted into the coelom of a host embryo of 50 hours. If the recipient is of the opposite sex, and if the embryonic gonad contains sex hormones, the host's organ

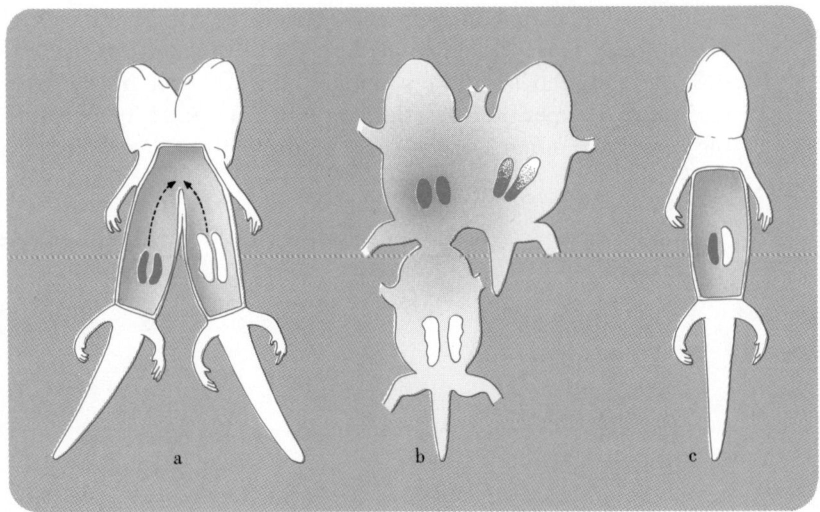


**Fig. 11-6** *Diagram illustrating different modes of grafting in amphibians in order to bring about vascular continuity between individuals, and association of gonads of different sex.* (a) *Homoplastic twins in salamanders. The body cavities are largely separated, and vascular communications between the gonads are remote.* (b) *Anuran twins, showing side-to-side or head-to-tail union; reversal changes appear only under the first condition, when the gonads are in close proximity. (After E. Witschi.)* (c) *Orthotopic transplantation of the gonad primordium resulting in two gonads of opposite sex resident in a single individual (Humphrey's method). (After R. K. Burns, in* Sex and Internal Secretions, *by courtesy of The Williams & Wilkins Company.)*

should be modified in the direction of the donor's sex. Positive effects are produced by embryonic gonads from donors 6 to 11 days old.

In mammals, the testis of a 15- to 16-day-old rat grafted upon the atrophied seminal vesicle of a castrated adult produces an intense activation of the seminal epithelium, indicating hormone action.

In all of these examples, the graft reproduces the action of known purified hormones obtained from the adult gland. However, we should add that this similarity in action does not prove that the hormones normally influencing the course of sex differentiation in the embryo are identical to those of the adult. In general they act in the same way, although there are variations in the behavior of the hormones in different species. For example, although it is possible to "transform" a developing genetic male opossum into a female by administering female sex hormones to the very young animals in the pouch, transformation in the opposite direction has not yet been accomplished. Ultimately, as micromethods are perfected, it should be possible to determine whether the early embryonic sex hormones and those elaborated later are identical.

The gonadotropic hormones secreted by the anterior pituitary gland control the maturation, and regulate the action, of the gonads. When does the pituitary-gonad "axis" become established?

The anterior lobe of the pituitary does not appear to be concerned in the primary differentiation of sex—that is, in the formation of the gonads themselves. If young embryos are hypophysectomized—in other words, if the anterior lobe is removed—the gonads develop normally, being recognizable as ovaries or testes. Later, however, development of the accessory structures, which depend on the sex hormones, may be retarded, and there is an indication of diminished secretory activities of the gonads through lack of gonadotropic stimulation. In general, the evidence in amphibians, birds, and mammals supports the view that gonadotropic activity begins shortly after the gonads differentiate, during the differentiation of sex ducts and accessory structures. In chick embryos, for example, hypophysectomy before the onset of sex differentiation may be accomplished in two ways: by partial decapitation, in which the forebrain area (origin of the anterior pituitary) is removed surgically after 33 to 38 hours of incubation (Fig. 11-7); and by irradiation. Histologic differentiation of the gonads proceeds normally, although the interstitial cells of the testis and the cortex of the left ovary are deficient.

Both the gonad and anterior pituitary have hormonal activity beginning at least during the later stages of embryonic life. The interlocking of functional activity appears to occur at about the time the gonads approach sexual maturity. The sex hormones appear to be formed and released first. Does their release trigger the release of previously accumulated gonadotropins?

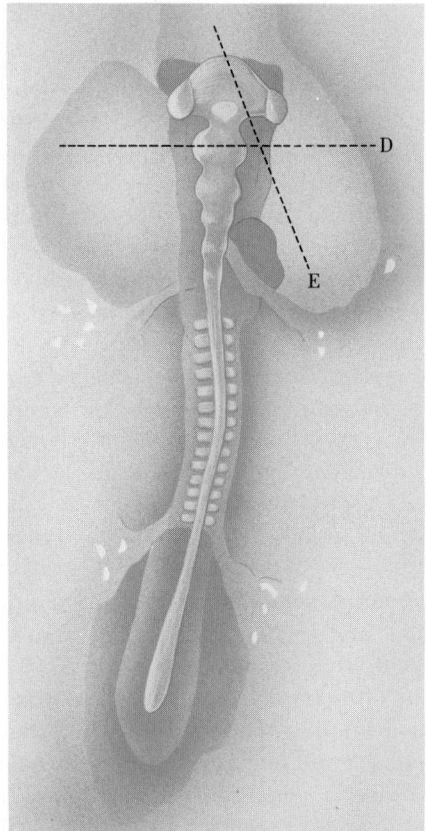


**Fig. 11-7** (Left) *Chick embryo at stage 11 + (38–40 hours), showing levels at which cuts are made for partial decapitation* (D) *and removal of an eye cup* (E). (Below) *Partially decapitated and normal chick embryos at 17 days. (By courtesy of Carnegie Institution of Washington.)*

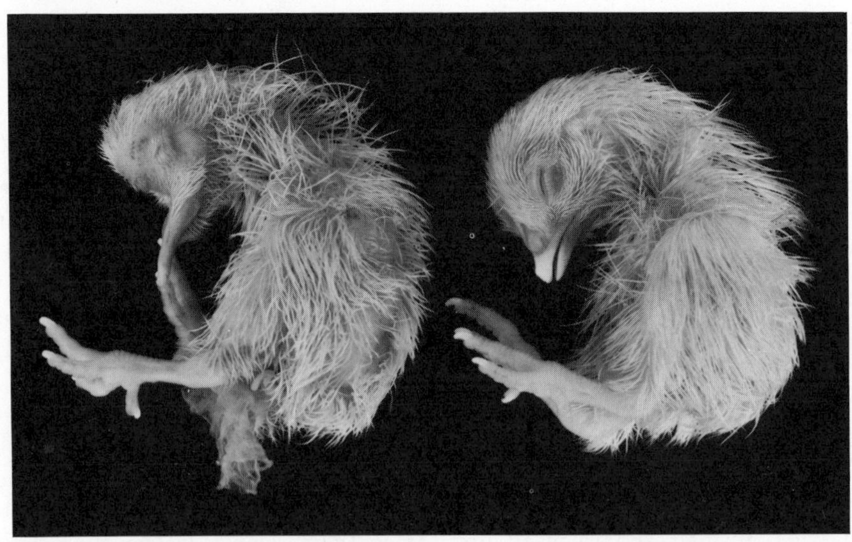

**In Amphibian Metamorphosis** By *metamorphosis* we mean a radical transformation in form taking place during a relatively brief period in development, as from a larva to the adult. Metamorphosis occurs in many phyla, in invertebrates, from sponges through echinoderms, and in several of the chordates as high as amphibia. We shall consider only two examples: amphibians and insects.

In the amphibians we shall limit ourselves to the principal features of metamorphosis in anurans—frogs and toads. Although we have been concentrating on *regressive* changes, we must bear in mind that *progressive* changes also occur in this change from an aquatic to a terrestrial form of life.

Regressive changes embrace the reduction or complete elimination of organs or tissues required in larval life but unnecessary in adults. Progressive changes include the development and functional maturation of new organs. We find, too, some structures that must be modified during metamorphosis.

The most dramatic regressive changes include the resorption of the tadpole tail (Fig. 11-8), the gills, and horny teeth. Among the progressive changes are the development of limbs, the middle ear, and the tongue. Modifications occur in the skin and intestine. The climactic events of metamorphosis in amphibians are under the control of the thyroid hormone. This fact is so well known that we will not elaborate on it. When the pituitary gland reaches a given level of differentiation it produces the thyroid-stimulating (thyrotropic) hormone, which in turn stimulates the thyroid gland to secrete the thyroid hormone. (In the laboratory we commonly use thyroxin or triiodothyronine, components or modified components of the thyroid hormone.) In nature it is the release of thyroid hormone that initiates metamorphosis, which may also be readily initiated in the laboratory by thyroxin.

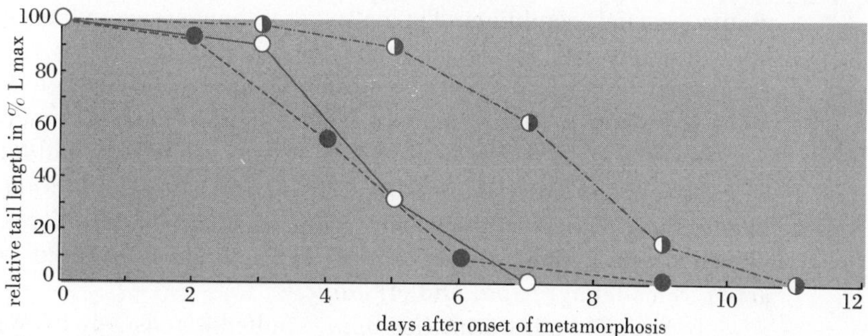


***Fig. 11-8*** *Reduction of tail length during spontaneous metamorphosis. Curves refer to three tadpoles undergoing metamorphosis at 20°C. (After R. Weber, in* Experientia, *13, by courtesy of Birkhäuser-Verlag.)*

The most remarkable aspect of a remarkable process is this: The *same hormone affects* virtually *every cell* in the tadpole's body, but *each type of cell responds differently*. Thyroid hormone initiates and accelerates tail resorption while promoting limb growth. Although the liver does not increase in size, a new set of proteins is induced, including new urea cycle and respiratory enzymes. Serum albumin synthesis is also affected. Adult rather than embryonic hemoglobins are made. The adult form of the visual pigment (visual purple, or rhodopsin) of the retina is synthesized. Moreover, the threshold of hormone required varies from site to site.

What is known of the mechanisms involved? There is no reason to believe that the constructive processes, once under way, differ in any fundamental way from those in any other developing system. On the side of regression, it has been known that tadpole tissues contain an enzyme, collagenase, which will break down collagen. It is also known that these resorbing tissues contain high quantities of cathepsins, which are enzymes of protein breakdown, nucleases, and other lysosomal enzymes (Fig. 11-9).

There are numerous suggestions that the hormone acts at the level of RNA synthesis in each responding cell, either directly or through some intermediary. But the nature of the "receptor" is obscure.

### In Insect Metamorphosis

Two types of metamorphosis can be distinguished among the various insect orders: *incomplete* and *complete*. In the former, although the sexually immature infant or *nymph* resembles the adult, there are significant differences (only adults have functional wings, for example). The grasshopper exemplifies the extreme of incomplete metamorphosis, the young hatching from the egg as nearly perfect adults, except for their smaller size, lack of wings, and immature sexual condition. Typically, grasshoppers copulate in early fall; after fertilization, development begins, and an embryo is formed. They then undergo a "rest" or *diapause* until spring, when development is resumed, with nymphs later hatching in early summer. Metamorphosis is said to be complete in insects such as Lepidoptera, in which the immature and mature forms are radically different. Commonly there are four stages: (1) *egg*; (2) *larva* (caterpillars, grubs, maggots), the feeding forms; (3) the outwardly quiescent (but inwardly active remodeling) *pupa*; and (4) *adult*.

In the wild silkworm, *Cecropia*, studied extensively by Williams, Schneiderman, and others, the caterpillar hatches from its egg and periodically molts, molting being the distinguishing feature of insect growth. The insect epidermis is bound to an outer layer of cuticle and thus cannot grow (unless it is detached from the cuticle). Periodically the epidermal cells of the larvae detach from their cuticle, grow by cell division or cell enlargement (or both), and secrete a new and extensible cuticle. While this new cuticle is being secreted, the epidermal cells produce and release enzymes that digest the inner layers which

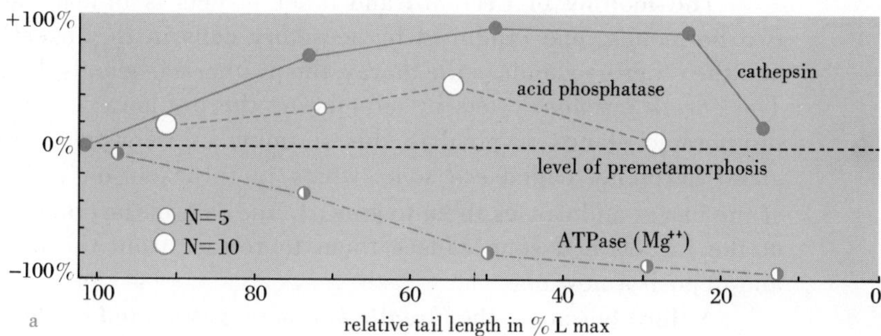


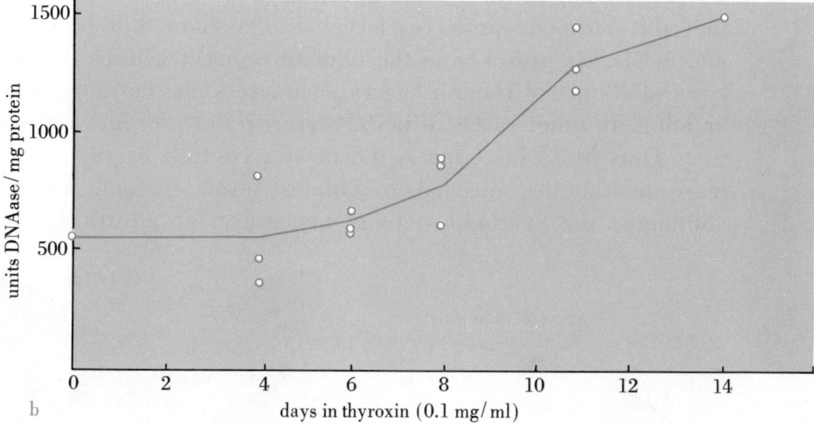


**Fig. 11-9** *Levels of enzyme activity in anuran tadpole tail tissue. Note increase in lysosomal enzymes.* (a) *During spontaneous metamorphosis. Ordinate: Relative increase or decrease in total activity per tail related to the average levels of activity in premetamorphic tadpoles. Abscissa: Relative tail length expressed in percentage of the tail length measured at the onset of metamorphosis. Each symbol represents the mean of a group of tadpoles and refers to a definite interval of metamorphosis. Areas of circles indicate number of individuals. (After R. Weber, in 13 Mosbacher Colloquium,* Induktion und Morphogenese, *Berlin, Gottingen, Heidelberg; Springer-Verlag, 1963.)* (b) *DNAase II activity per milligram of protein in isolated tadpole tails after exposure of tadpoles to thyroxin for various periods. Thyroxin was added to the medium in a final concentration of 0.1 μg/ml. The medium was changed every 2 to 3 days. No tail resorption was apparent prior to 10 days in thyroxin, but by 11 days tails exhibited some degree of resorption. (After J. R. Coleman, in* Biochim. Biophys. Acta, *68 (1963) 142 (Fig. 2) by courtesy of the author and Elsevier Publishing Company.)*

they then absorb. Next, epidermal cells "waterproof" the new cuticle and the old cuticle is shed. In the *Cecropia* silkworm there are four larval molts after hatching, a larval–pupal molt and a pupal–adult molt (Fig. 11-10). The stages between molts are known as *instars*.

The molting of *Cecropia* and other insects is brought about by two hormones, one produced by secretory cells in the insect's brain and the other by glands in the thorax, the prothoracic glands (Fig. 11-11). The "*brain hormone*" acts by stimulating the prothoracic glands. The prothoracic glands respond to this stimulus by releasing prothoracic gland hormone, called *ecdysone*, which in turn acts on various cells of the insect and causes them to grow. In the case of the epidermal cells of the insect, ecdysone causes them to retract from the old cuticle and deposit a new one.

A third hormone, the *juvenile hormone*, is secreted by the *corpora allata*, endocrine glands located near the brain. Sir Vincent Wigglesworth's studies in the South American bloodsucker *Rhodnius* showed that this hormone promotes larval development, but prevents metamorphosis. Its presence in the immature insect insures that when the larva molts it will retain its larval characters and not differentiate into an adult. In other words, it permits growth but prevents maturation.

Thus brain hormone is the master control: by regulating its release, molting is controlled. In simplest terms, in some way ecdysone stimulates the synthetic activity necessary for growth and molting.

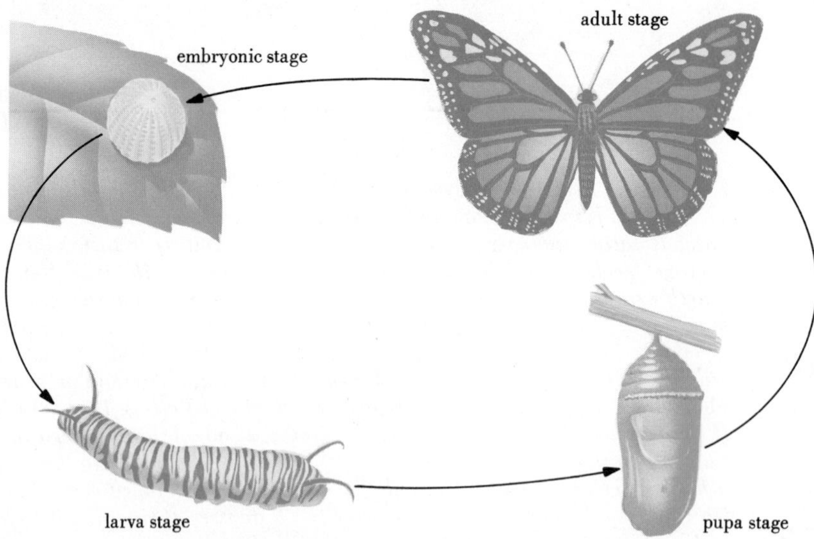


*Fig. 11-10 Principal stages in the life cycle of a moth or butterfly that undergoes complete metamorphosis.*

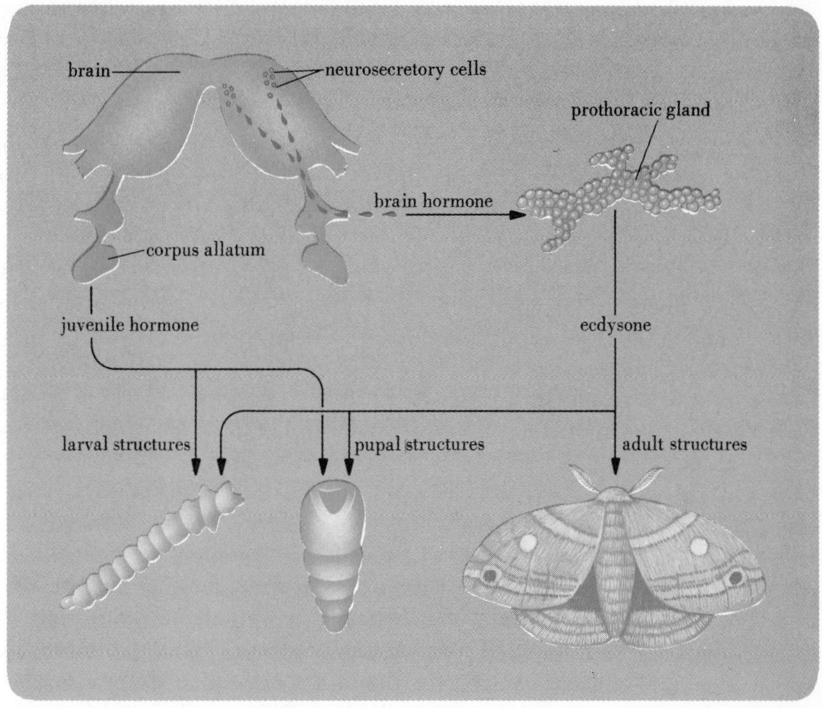


***Fig. 11-11*** *Diagram showing interactions of juvenile hormone, brain hormone, and molting hormone (ecdysone) in* Cecropia *silkworm. If much juvenile hormone is present when the insect molts, it will molt into another larval stage. If a low concentration of juvenile hormone is present, the larva will molt into a pupa. If no juvenile hormone is present, the pupa will molt into an adult. (Modified from H. A. Schneiderman and L. I. Gilbert,* Science, *vol. 143, 1964. Copyright © 1964 by the American Association for the Advancement of Science.)*

Juvenile hormone influences the kind of synthetic activities that occur. In high concentration it suppresses all synthetic activities not characteristic of the larva. In low concentration, synthetic activities characteristic of the pupa are permitted. When juvenile hormone is absent, ecdysone, acting alone, elicits syntheses characteristic of the adult.

Ecdysone, which acts on all insects—and has also been isolated from crustaceans—is a mixture of two closely related steroid hormones, alpha and beta ecdysone. As with the hormones of vertebrates, a number of experiments have led to the hypothesis that it ultimately regulates gene activity, but exactly how it acts is not yet known.

The nature of the evidence concerning ecdysone action is the following: When injected into *Chironomus* larvae, ecdysone induces the chromosomal RNA puffing pattern somewhat characteristic of the pupa. Intense local RNA synthesis is stimulated. In *Sciara* larvae, the hormone also stimulates DNA puffing. However, it is not clear whether the hormone acts directly on the genes, or whether it acts indirectly by causing changes in the nuclear membrane or in the cytoplasm.

It has also been suggested that some plant hormones may act directly on the genome and thus regulate the quantity of gene products. Gibberellic acid, the hormone that promotes stem elongation in dwarf peas, results in increased RNA synthesis in treated plants. Nuclei isolated from dwarf pea plants and incubated in precursors of RNA in the presence of gibberellic acid also show increased RNA synthesis; this RNA differs in composition from that synthesized in control nuclei, suggesting that gibberellic acid is acting selectively on parts of the genome. However, we must conclude that we still do not know the primary site of action of any hormone or its mode of action in molecular terms.

***In Senescence*** The study of *aging* properly lies within the scope of developmental biology. Not only is it difficult to state just when "development" ends and "aging" begins—the ill-defined term *maturity* usually being the border between the two—but it is actually difficult to distinguish development and aging in the most fundamental sense. Evidence is slowly accumulating that aging is more than just an accumulation of physiological "insults," the random deterioration of parts. We often read the charge that manufacturers are programming obsolescence in their products: it seems entirely possible that senescence is "built into" the genome. We have already seen that many cell populations undergo programmed senescence in normal development.

We have already remarked (page 189) that established cell lines apparently are capable of infinite proliferation in culture. However, they are usually grossly abnormal in *karyotypes*, the karyotype being the characteristic chromosome complex of an animal or plant. It is often suggested that the stability of these cell lines is somehow the result of their abnormal chromosomes.

How do normal diploid cells in primary cell strains behave in culture? The evidence is conflicting. It is frequently said that diploid cells have a limited lifetime in culture; that there is a period of active proliferation, followed by a decline to a state from which cells cannot be subcultured. Human cells are usually cited as examples. It is often said that the number of generations undergone by various strains of diploid human cells is about $50 \pm 10$, and that the only "immortal" cell strains are those that have undergone a change in chromosome number or composition. However, when one examines the literature carefully, he finds that the number of generations undergone by diploid human cells in culture varies widely with the medium used. Moreover several strains of diploid animal cells (derived from animals with shorter life-spans than man) are capable of subculturing for far longer periods.

At present, then, cell culture techniques have not provided compelling evidence for a built-in, finite life-span for normal cell strains.

However, in the intact animal, the deterioration of some functions with age clearly results from a decline in the number of cells that are capable of division. Especially noteworthy is the observation that old animals have a diminished capacity to make antibodies, and that their failure may be attributed to a decline in number of cells capable of proliferating in response to stimuli by foreign antigens.

## *FURTHER READING*

Burns, R. K., "Experimental Reversal of Sex in the Gonads of the Opossum, *Didelphys virginiana*," *Proceedings of the National Academy of Sciences*, vol. 41 (1955), p. 669.

———, "Role of Hormones in the Differentiation of Sex," in *Sex and Internal Secretions*, W. C. Young, ed. Baltimore: Williams & Wilkins, 1961, p. 76.

Charniaux-Cotton, H., "Hormonal Control of Sex Differentiation in Invertebrates," in *Organogenesis*, R. L. DeHaan and H. Urpsrung, eds. New York: Holt, Rinehart and Winston, 1965, p. 701.

Crouse, H. V., "The Role of Ecdysone in DNA-Puff Formation and DNA Synthesis in the Polytene Chromosomes of *Sciara coprophila*," *Proceedings of the National Academy of Sciences*, vol. 61 (1968), p. 971.

Etkin, W., and L. I. Gilbert, eds., *Metamorphosis. A Problem in Developmental Biology*. New York: Appleton, 1968.

Hamilton, T. H., "Control by Estrogen of Genetic Transcription and Translation," *Science*, vol. 161 (1968), p. 649.

Hay, R. J., "Cell and Tissue Culture in Aging Research," in *Advances in Gerontological Research*, Vol. 2, B. L. Strehler, ed. New York: Academic Press, 1967, p. 121.

Jost, A., "Gonadal Hormones in the Sex Differentiation of the Mammalian Fetus," in *Organogenesis*, R. L. DeHaan and H. Ursprung eds. New York; Holt, Rinehart and Winston, 1965, p. 611.

Lillie, F. R., "The Theory of the Free-Martin." *Science*, vol. 43 (1916), p. 611; reprinted in *Foundations of Experimental Embryology*, B. H. Willier and J. M. Oppenheimer, eds. Englewood Cliffs, N.J.: Prentice-Hall, 1964, p. 137.

Saunders, J. W., Jr., "Death in Embryonic Systems." *Science*, vol. 154 (1966), p. 604.

# chapter 12

# Growth: Its Regulation in Relation to Differentiation

We began by saying that although differentiation, morphogenesis, and growth are separable conceptually—and to some degree, experimentally—we would stress their intercommunity. So far we have done just that; however, we have emphasized differentiation and morphogenesis. We should now redress that imbalance by considering some of the ideas on growth and the factors controlling it.

We said earlier that growth means permanent enlargement—that is, developmental increase in total mass. Growth is not differentiation, morphogenesis, or reproduction; the words are not synonymous. Nor is it cell division, although cell division may be one of its important ingredients. However, cell division can occur without growth; a prime example is the cleavage process, which results in a larger number of smaller cells.

Alternatively, growth can occur without cell division; that is, cells can grow larger without increasing their number. The growing oöcyte and the outgrowing neuron are good examples.

We have said what growth is not. Just what does growth involve? We get to know about growth by taking measurements at different times, comparing them, and noting a net gain by some criterion, usually mass or numbers. For example, we may measure wet or dry weights, total nitrogen, or total protein nitrogen content, cell number, or DNA. No single parameter is fully meaningful in itself. Wet weight may clearly be misleading; dry weight and nitrogen content, too, may include foodstuffs and products that fluctuate physiologically; and we have already pointed out the limitations of cell number or DNA. Thus, although our measurements must be chosen arbitrarily, we must define them carefully and insure that enough different kinds of measurements are taken that we understand the system. Therefore, we should ask in every measurement of growth not just for evidence of total increase, but for evidence showing whether the increase reflects increase in cell number or size.

The serial measurements thus taken define a *growth curve*. In Fig. 12-1 are depicted the growth of the chicken from embryo to maturity and the growth of a bacterial population. Both curves are *sigmoid* or S-shaped, as are most growth curves.

When a known bacterial population is inoculated into a given amount of medium in a series of Petri dishes or test tubes, incubated under constant conditions, and serial measurements of bacterial mass are taken, the following pattern is observed.

At first, little or no growth occurs. During this *lag* phase the bacteria adjust to their new environment and prepare for growth.

Now the cells begin to increase in number, during a brief period of increasing specific growth rate, leading to the second major phase, the *exponential* or *logarithmic* (in familiar laboratory jargon, "*log*") phase. This period is marked by a constant specific growth rate. During exponential growth in a constant environment, the average physiological properties of the cells are constant. This phase cannot continue in a finite volume of medium. The cells eventually pass through a brief period of decreasing growth rate, to enter the *stationary* phase, in which no growth occurs.

The S-shaped curve of growth applies to cultured cells of man, mouse, and chick; to cells of kidney and liver, as well as to *Escherichia coli*. It applies to the growth of carrot cells; it describes the growth of the brain, the chick embryo, and the human embryo.

Bear in mind that the curve describes the growth of the chick embryo brain and gonad, and of the embryo as a whole; yet the curves

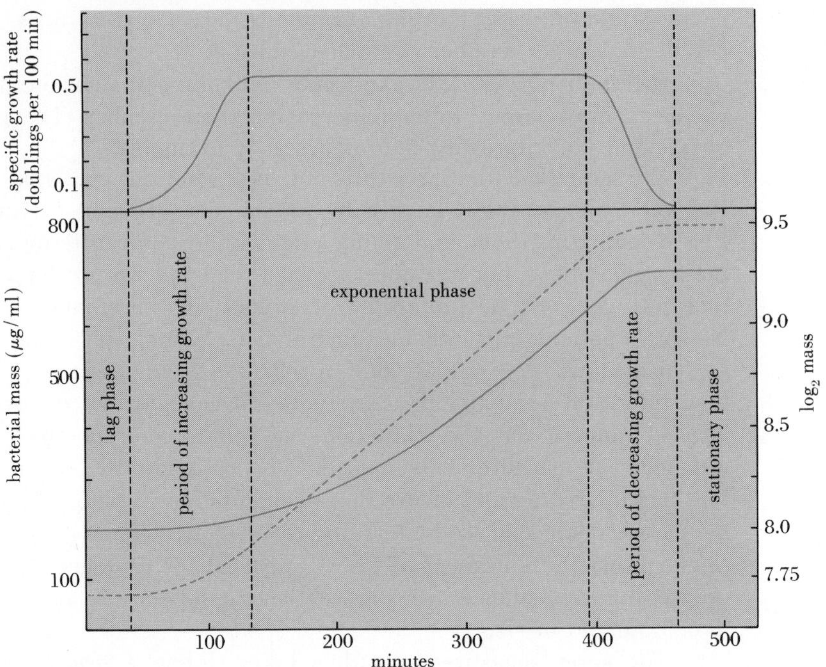


*Fig. 12-1 The bacterial growth curve. The lower part of the figure shows the increase in bacterial mass per milliliter (solid line) and the increase in the logarithm to the base 2 of the mass (broken line). The upper part shows the specific growth rate constant. (From W. R. Sistrom,* Microbial Life, *2d ed., Holt, Rinehart and Winston, Inc.)*

must not necessarily be superimposed in time. All the organs of an individual do not grow at the same rate. Nor, in fact, do parts of organs. Thus, growth must be regulated; there must again be interplay between cells and among the tissues of the body, insuring balanced growth and limiting size, for total mass remains relatively constant once maturity is reached.

## THE CELL CYCLE AND ITS CONTROLS

We have said that cell growth and cell division are not synonymous. Growth does occur by increase in the cells (and larger units) without cell division. However, cell division is such an integral part of many growing systems that we should first consider the factors controlling the cell cycle.

Apart from cleavage and other special circumstances, with each turn of the cell through its life cycle all of its elements—both nuclear and cytoplasmic—undergo a doubling. Not only are the chromosomes replicated, but mitochondria and ribosomes, lysosomes and membranes, as well. We know a little about many of the separate events: DNA is replicated semiconservatively—that is, half of each parent molecule is conserved in each daughter molecule, and thus the daughter molecule contains one old chain and one new chain. New mitochondria are produced by fission from the old, and so on. Yet how is the growth of each component adjusted to the pace of growth of the cell as a whole? Although we cannot give a comprehensive answer to this question, it is probably correct to say that cell growth is governed principally by the nuclear events of chromosome replication; moreover, we can say that chromosome replication itself is under genic control.

In *prokaryotes*, organisms without membrane-bound nuclei, DNA synthesis may be accomplished quickly. In *eukaryotes*, however, chromosomal replication may take several hours, different parts of chromosomes replicating asynchronously. Of course the chromosomes are replicated not at mitosis but between mitoses. In Fig. 12-2 we see examples of the duration of the cell life cycle. The $M$ (or $D$), mitotic (division) period is usually brief. The $S$ (or synthesis) period often takes about 6 to 8 hours. The $S$ period is bracketed by two periods, $G_1$ and $G_2$ ($G$ for "gap," an indication of how little we know). The lengths of these periods are highly variable. In rapidly growing bacteria, $G_1$ and $G_2$ are almost nonexistent.

The lengths of $S$, $G_2$, and $M$ tend to be constant in any given cell type under a variety of environmental conditions. However, $G_1$ may vary enormously. Changes in cell generation time are accomplished by expanding or contracting $G_1$. In one study of different epithelia of the gut, the average cell generation time varied between 17 hours (ileum) and 181 hours (esophagus), but $S + G_2 + \mathrm{M}$ remained constant at 10 hours. Thus $G_1$ ranged from 7 to 171 hours! During cleavage stages of embryogenesis, there is no measurable $G_1$. The fact that, following initial differentiative events, embryonic cells do have $G_1$ periods further suggests that the cycle itself differentiates under genetic control.

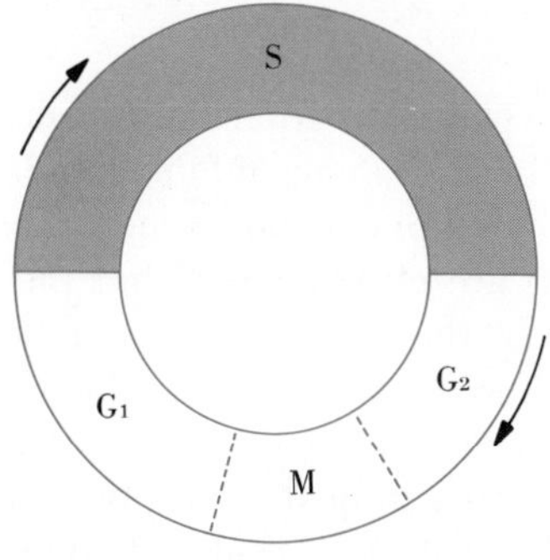


**Fig. 12-2** *Diagram of the mitotic cycle. The time occupied by mitosis is represented by M. Synthesis of DNA is represented by S and the periods before and after DNA synthesis are termed $G_1$ and $G_2$. In different cell populations the relative duration of each phase may be different. (Redrawn from D. E. Wimber, in* American Journal of Botany, *47:830, 1960.)*

It is during the $G_1$ period that the cell prepares for DNA synthesis. The requirements for DNA synthesis are known. In a general sense they include the precursors (deoxynucleoside triphosphates), the enzyme DNA polymerase and a DNA template primer. However, it is almost certain that the level of precursors does not usually regulate DNA synthesis. Their absence would prevent it, but their presence is not sufficient. The same can be said about DNA polymerase. Thus all of these essential "ingredients" may be present in $G_1$ even for long periods without the initiation of synthesis.

It is known that RNA and protein synthesis are required for a cell to pass from the $G_1$ into $S$ phase. The current hypothesis states that these syntheses are required in the formation of an "initiator" protein which somehow brings about the interaction between DNA and DNA polymerase. We have already mentioned that different parts of chromosomes replicate asynchronously, indicating that each chromosome may have a number of different replicating units or *replicons.* It has been estimated, for example, that one arm of a single X chromosome in the Chinese hamster may have 205 replicons. It is proposed, therefore, that there may be a specific initiator for each replicon. You may recall that we have already described one experiment suggesting the presence of such initiators in the cytoplasm of cleaving *Xenopus* eggs (page 187). Brain-cell nuclei transplanted to cleaving eggs resume DNA synthesis. We have also considered evidence favoring some form of the replicon hypothesis in *Xenopus*. In the multiple nucleoli formed in amphibian oöcytes, the genes for rRNA are replicated independently of other parts of the chromosome.

Certainly all phases of the cycle are equally "important." During the $S$ phase not only must DNA be replicated, but there must be a simultaneous cytoplasmic (and possibly nuclear) synthesis of histones and other proteins. The $G_2$ period is required for assembly of the mitotic apparatus. However, it seems likely that from the standpoint of greater understanding, more attention will have to be directed to the $G_1$ period and to the transition from $G_1$ to $S$.

## REGULATION OF ORGAN AND TISSUE GROWTH

In its development from the fertilized egg, the newborn human infant increases in weight about one billion times. The facts of growth are clear enough, but the underlying mechanisms have eluded us. In the discussion that follows, let us assume that the raw materials for growth are available. The nutrient supply is, of course, a limiting factor, but for the sake of our discussion, that supply—whether transshipped from the maternal circulation or available from the yolk—will be considered adequate.

What factors, then, govern size? Genetic constitution, of course. It is necessary only to mention that growth requires the synthesis of macromolecules and that the regulation of these processes involves the interplay of genes, the cytoplasmic machinery, and the microenvironment of the cell. However, this interplay is required also for differentiation and morphogenesis.

It is also said that the surface-to-volume ratio is important in determining size: When an organism grows, its volume increases faster than its surface area. The more it grows, the more food and oxygen it requires, and the more waste it will have to excrete. But since its surface area does not keep pace, adjustments have to be made. The methods evolved include using the food supply more efficiently and reducing the amount of toxic wastes produced; modifying the intake and outflow of food and wastes to make these processes more efficient; and, most pertinent here, changing shape to decrease the disparity between surface and volume. Yet if we reflect a moment we see that these adaptations themselves must require genic action.

It will be necessary, before continuing, to reenforce one of the points we made at the beginning of this chapter: organs may grow either by multiplying cells (or other functional units) or by enlarging existing units.

In mammals, for example, the blood may be increased indefinitely by increasing individual units, the blood cells. However certain fundamental units—neurons, muscle fibers, nephrons (the tubular unit of the kidney)—lose the ability to increase their numbers before maturity in birds and mammals, and can grow only by increasing the size of individual units. The ultimate dimensions to which a heart or skeletal muscle or kidney grows appears to be genetically determined. The mouse has about 12,500 nephrons; the elephant, 7,500,000. The fibers in adult hearts are alike in size in mammals over a wide range of sizes; this observation implies that the ultimate size of the heart is determined early in life when cardiac myoblasts stop dividing and start enlarging.

In contrast to the animals that attain a finite size, fishes continue to grow as long as they live. How do they enlarge their hearts and kidneys? Are they able to increase the number of functional units—nephrons, muscle fibers, and neurons? We don't know, but the answer is presumably in the affirmative.

Are there subtler specific controls of growth of organs and tissues? Let us take up the question by examining a series of experimental findings. If a kidney or part of the liver of a vertebrate is removed, the remaining kidney or residual liver mass begins to grow. We speak of this response as *compensatory hypertrophy*.

Consider the kidney. If one or even one-and-a-half kidneys are removed from a rat, the cells in the remaining portion begin to proliferate, a peak of mitotic activity being observed within 48 hours after surgery. However cellular proliferation results in tubular enlargement; it does not change the number of nephrons produced. Such experiments suggest that the size of the organ is somehow regulated by a "negative feedback"—in other words, that a product or activity of the organ operates in limiting its growth. Can such a factor be identified? There is evidence that it is humoral. If both kidneys, or one-and-a-half kidneys, are removed from a rat, which is then parabiosed to a normal rat, the cells of the kidney of the normal rat begin to proliferate. In short, these experiments show that some factor or factors do circulate and can regulate growth. Whether they are inhibitors derived from the kidneys themselves, or stimulators of extrarenal origin, is the point at issue.

There are clues to, and a more specific hypothesis about, the regulatory factor in liver regeneration. After partial hepatectomy, some of the remaining liver cells proliferate, and ultimately differentiate as the loss is repaired and the original mass restored. Thus a system exists that, as Glass put it, "informs the cells of the total mass to which they belong, maintains the mass in a state of equilibrium, and regulates the return of the mass to that state after any disturbance."

That the factor or factors involved are circulating was proved in the following ways: (1) results of experiments with parabionts are similar to those for kidney; (2) high concentrations of plasma from normal animals inhibit liver cells in culture; (3) when the concentration of plasma proteins is decreased in rats with intact livers, mitotic activity is increased; (4) when the concentration of plasma proteins is increased (by restricting fluid intake), mitotic activity in partially hepatectomized rats is reduced.

What is the inhibitory factor synthesized in liver? It appears to be one of the plasma proteins themselves. The plasma albumins and alpha and beta globulins are made in the liver, whereas gamma globulins are not. There is some evidence to suggest that the albumins may be one of the postulated inhibitory factors.

Albumins are synthesized on ribosomes; one can see how the concentration of albumin might regulate albumin synthesis, but how do you imagine the stimulation of synthesis of this protein triggers the production of the myriad products and structures that make up the liver?

## GROWTH CONTROL OF NERVE CELLS BY A PROTEIN FACTOR

It is the nature of neurons to establish morphological and functional connections with muscle fibers, gland cells, sensory receptors, and other end organs. We have already discussed the role of such end organs in the differentiation and maintenance of neurons, stressing their mutual dependency. We have described experiments in which Harrison, Detwiler, and Hamburger pioneered, wherein the peripheral field of a population of neurons was modified—either augmented or reduced—resulting in a corresponding increase or decrease in the nerve centers. To recall one of them, if the limb bud of the two-day chick embryo is extirpated, the ganglia destined to innervate the amputated limb show a decrease in mitotic activity. In addition, a number of ganglion cells suddenly die and, within a few hours, are removed by macrophages. This massive cell degeneration simulates the pattern of cell deaths that normally wipe out nerve-cell populations as the central nervous system is shaped.

It was a similar experiment that marked the beginning of one of the more remarkable trails of research of our time.

In an effort to determine whether motor and sensory fibers could innervate a fast-growing tissue, a fragment of rapidly growing tumor, a mouse sarcoma (number 180) was implanted into the body wall of a two-day chick embryo. The effect is a striking one: The tumor evokes an increase in volume of the sensory ganglia of the nerves invading it to two or three times normal size. The sympathetic ganglia adjacent to the tumor are also increased, to an even greater extent, being five to six times larger than controls.

The sensory neurons invading the tumor make no connections with the tumor cells, wandering around tortuously. The sympathetic fibers behave in even more bizarre fashion, sending fibers not only into the tumor but into visceral organs, which become flooded with fibers; all available space is thus filled up.

Levi-Montalcini proposed that the tumor contains a growth factor that selectively enhances sympathetic and sensory fiber outgrowth. Transplantation of tumor to the chorioallantoic membrane proved that the factor is released into the bloodstream, for the host organs are innervated profusely under these conditions. The turning point in the study was the demonstration that when ganglia and tumor fragments are cultured together *in vitro*, the ganglion is rapidly surrounded by a dense halo of fibers (Fig. 12-3). Now a method was available to assay the many fractions that would have to be studied if the factor were to be isolated.

A curious, accidental finding led to the next breakthrough. A nucleoprotein that was isolated from tumor was found to have some growth-promoting activity, and it was natural to inquire whether the activity lay in the nucleic acid or protein moiety. To degrade the nucleic acid, snake venom (containing the enzyme phosphodiesterase) was added to the factor, but in control experiments the venom itself was highly effective. What a curious and startling twist! The specific activity in the venon proved to be 1000 times higher than that of tumor homogenate. Chemically, it appeared to be a protein.

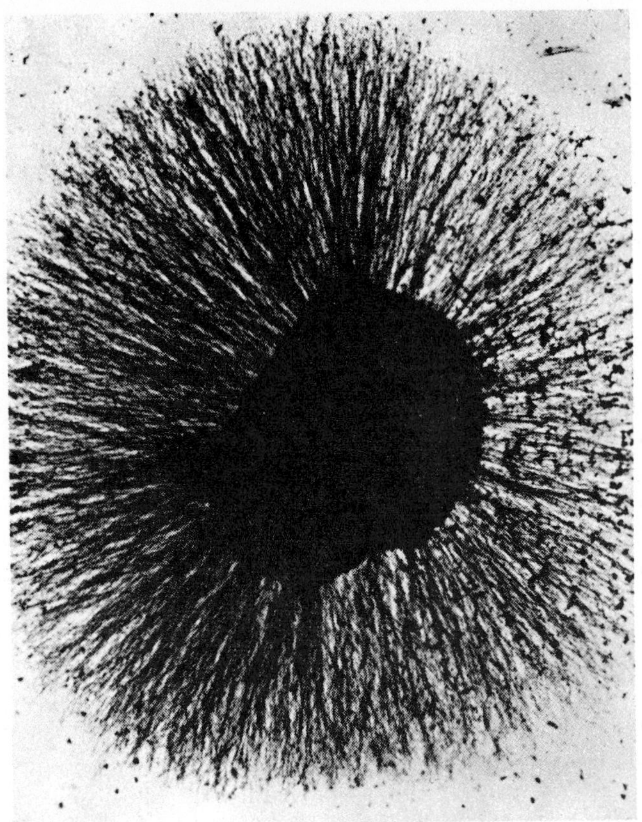

***Fig. 12-3*** *Microphotograph of a 7-day chick embryo sensory ganglion after 24 hours* in vitro *in a medium containing nerve growth factor. (From R. Levi-Montalcini and B. Booker, in* Proceedings of the National Academy of Sciences, *46, by courtesy of the Academy.)*

Snake venom is obtained from glands that are very similar to salivary glands. It was next found (this time, deliberately) that the submaxillary salivary glands of mice contain the factor. Its activity is 6000 to 10,000 times that of tumor factor and 10 times higher than that in snake venom.

This factor is so potent that a dose as small as 0.5 microgram per gram of body weight causes the developing sympathetic ganglia to increase fourfold to sixfold. Further tissue culture observations show that it affects ganglia of human fetuses as well as those of birds and rodents.

It is clear that sympathetic nerve cells are "receptive" to, and influenced by, growth-promoting protein.

But does the protein exist normally in the body, and does it play a role in the growth of the sympathetic ganglia? How would you go about proving the point?

Cohen and Levi-Montalcini prepared antibodies to the factor in rabbits. When this antibody is injected into newborn mice, rats, rabbits, and kittens, it prevents the growth of the sympathetic chain ganglia, which are virtually absent. The animals are "immunosympathectomized" (Fig. 12-4). This and other experiments give evidence that this protein may play an essential role in the growth and maintenance of sympathetic nerve cells. Might other nerve cells depend on specific factors for their growth?

Interest now centers on the nature of the factor and its mode of action. It is protein, but beyond that statement nothing "firm" can be adduced. The protein can be dissociated into subunits, but there is no agreement either on the protein's molecular weight, or on the number of subunits comprising it. Nor is it certainly known that all of the subunits have equal biological activity.

## DIFFERENTIATION AND GROWTH

It is so often said that differentiated cells do not divide that it is necessary to assert that many of them do! However the relations between differentiation and DNA replication and cell multiplication vary over a wide range. We have already referred to cell multiplication

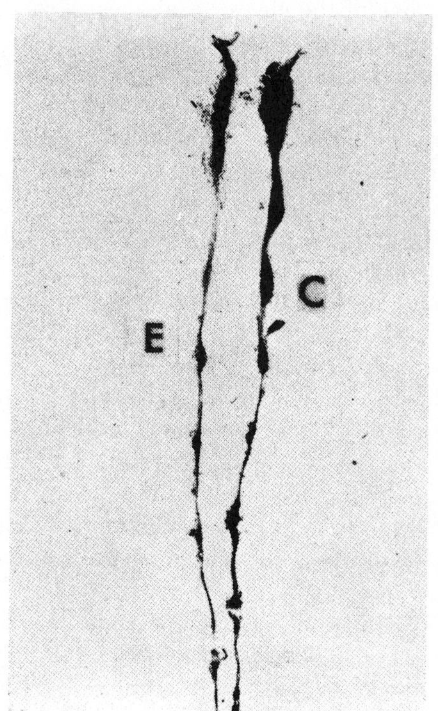


***Fig. 12-4*** *Immunosympathectomy. Effects of antiserum against nerve growth factor in a 3½-day-old rabit injected daily since birth. Stellate ganglion and sympathetic thoracic chains in control* (C) *and experimental* (E) *rabbit. (From R. Levi-Montalcini and B. Booker, in* Proceedings of the National Academy of Sciences, *46, by courtesy of the Academy.)*

in postnatal growth and reparative processes. Clonal cell cultures provide additional evidence that at least some specialized cells can divide.

Encouraged by the successful cloning of skeletal muscle, other workers have extended this approach to several other differentiated cell types of the chick embryo. Retinal pigment and cartilage cells have also been cloned and subcloned, showing that these differentiated phenotypes are stably inherited through many generations. Each of these cell types has clearly recognizable features that permit ready identification of differentiated clonal type. Moreover, these techniques have been extended to mammalian cells, rat liver cells having been cloned successfully.

Retinal pigment cells have been grown as clones from single cells while retaining their pigmentation and epithelial morphology. These cells have been subcloned six times, amounting to over 50 cell divisions, and have remained pigmented. Cartilage cells have been shown to retain their differentiated phenotype through at least 40 to 50 cell generations. Moreover, labeling studies show that the synthesis of DNA and of chondroitin sulfate (a distinctive product of the cartilage cell) occurs at the same time in the same cell.

In contrast to cartilage there are some cells in which there is a distinct discontinuity in time between DNA synthesis and differentiation, skeletal muscle being one of the clearer examples. Myoblasts divide repeatedly, but at fusion to form the myotube, DNA synthesis stops, not to be reactivated in the normal course of events (page 219). In several of the systems we have already examined (lens and pancreas, for example), a period of rapid cell division precedes specialization. It is possible that it has been the emphasis on these examples in which DNA synthesis and differentiation appear to be "mutually exclusive" that has so often resulted in the inaccurate generalization that differentiated cells cannot divide.

Instead of emphasizing "mutual exclusivity," however, we would stress the possibility that, in some cells at least, new transcription may *depend upon* an immediately preceding DNA replication. Among the several lines of evidence that might be brought to bear on this problem, none is more pertinent than that being derived from studies of cell transformations induced by viruses. This evidence can be summarized very briefly once the background of the work is made clear. Although the pioneering studies of Peyton Rous on the chicken tumor that bears his name (Rous sarcoma) established the role of a virus in causing a tumor, the full impact of the finding was felt only within the last decade when increasing numbers of virus-induced tumors came to light. During the same period, it was established that viruses could transform cells in culture. For example, if a piece of human skin is cultured, fibroblasts derived from it outgrow all other cells and soon one has a homogeneous fibroblastic culture. These cells, like most freshly isolated cells in culture form monolayers, one cell thick, with little tendency to overlap. The formation of such sheets involves a number of interactions. One type of interaction is reflected in *contact inhibition of movement*, first extensively described by Abercrombie and Heaysman. At points of contact between cultured cells, their active ruffled border is immobilized, and cells cease to move over each other. They still move on the free area of the surface, however, until a monolayer is formed. Many tumor cells are subject to contact inhibition of movement when in contact with normal cells, but not with one another. Tumor cells show an increased tendency to overlap.

A second phenomenon is sometimes loosely referred to as *contact inhibition of growth*. Cells frequently cease to grow when they reach a critical *saturation density*. This phenomenon also appears to be a factor in monolayering. However, rapid replacement of the medium will allow cells to "escape" from growth inhibition, and multilayering results. When cell cultures are inoculated with a tumor virus, morphologically transformed cells appear. The growth properties of these cells also are greatly altered. They now proliferate rapidly, with multilayering and loss of orientation being common. When transformed cells are returned to animals, tumors are produced. Normal fibroblastic cells synthesize abundant collagen; transformed cells make very little of it. (It should be added that failure to synthesize the normal products of a cell is not an inevitable consequence of tumorigenesis; the cells of some tumors of endocrine glands produce large quantities of normal hormones). Frequently, new tumor-specific antigens are made in transformed cells; their synthesis is permanent. These products are specific for the virus (they are the same molecules regardless of species of the host), but they are not components of the viral particle. Nongrowing cells are not generally susceptible to virally induced tumor formation, whereas rapidly growing cells and young animals are more susceptible.

It is now clear that one of the first critical steps in the production of such a tumor is the synthesis of cellular DNA. Both DNA viruses and RNA viruses, as exemplified by polyoma virus and Rous sarcoma virus, respectively, induce DNA synthesis, and the DNA produced is clearly identifiable as cellular, not viral.

If we are to continue using viral transformation of animal cells as a model, it is necessary to ask what is required for viral transformation and for maintenance of the transformed state? Is it essential that the viral genome be retained? Must viral message be produced continuously? Is a tumor virus integrated into the genome or does it persist in the cytoplasm as an *episome*, an automomous particle? The technique of somatic hybridization (page 188) has been used to study these questions. When cell hybrids are made between mouse and human cells, there is a rapid and extensive loss of the human chromosomes with no concomitant loss of mouse chromosomes. It was reasoned that if hybrids were made between normal mouse cells and human cells transformed by a tumor virus (simian virus 40), and if all human chromosomes were lost, then if the viral genome were integrated into a human chromosome, it too would disappear. Such experiments have been performed; they show that a viral antigen (T) is absent only from hybrid cells that have lost most or all of their human chromosomes, providing suggestive evidence that the viral genome is integrated into the chromosomes of transformed cells.

The full impact of this kind of study is yet to be felt, but these facts further strenghten the argument that for a major new differentiation or transformation, the cellular genome must be replicated, in whole or in part. In the process, new genes must be activated, or old ones reactivated allowing the change in direction.

Possibly the most compelling evidence for our argument has emerged from studies of *transdetermination* in *Drosophila.* By transdetermination we mean a process whereby cells determined or programmed for a given role are changed, so they differentiate in another direction.

To appreciate the importance of the findings, we need to review some of the main features of the development of *Drosophila. Drosophila* undergoes a complete metamorphosis (page 254), its life cycle having four distinct stages: embryo, larva, pupa, and adult or *imago.* Many of the adult organs are not derived from homologous larval organs but are formed from special cells contained in larval structures called *imaginal disks* (referring to the imago). The imaginal disk cells proliferate during larval life but do not differentiate detectably. After pupation cell proliferation ceases in the disks and differentiation begins, leading to the formation from the disk cells of specialized adult organs, such as wings, eyes, antennae, legs, and genitalia.

A mature larva contains several different imaginal disks, each a separate structure precisely located in the larva. A single disk may contain from 10,000 to 20,000 cytologically indistinguishable cells. Despite this cytological uniformity, a single disk gives rise to several distinct structures in the adult. The wing which is formed from the cells of the wing disk contains bristles (composed of four different cell types), trichomes, sensillae (at least three different types), and cuticle. Thus, a disk for a single adult organ can specify at least ten different cell types.

Although the cells in the different disks *appear* uniform, they are already determined.

When an imaginal disk is removed from one larva and transplanted into the abdominal region of other larva, both the transplanted disk and the host develop normally. The resulting adult has a fully differentiated eye or leg, for example, within its abdominal cavity. Thus an imaginal disk is capable of undergoing its characteristic sequence of differentiation autonomously, independent of the location it normally occupies in the larva and pupa.

However, when in place of a larval host, an adult host serves as the recipient of a transplanted disk, the results are strikingly different. When a disk is inserted into the abdominal cavity of a young adult and recovered about three weeks later (just before the end of the adult's normal life-span) it is found that the disk cells have *proliferated*, but there is no evidence of differentiation. The transplant may now be divided into several pieces and each may be separately injected into a fresh adult host. This technique, developed by Hadorn, establishes long-term "cultures *in vivo*" of disk cells.

The cultured disk cells have the following characteristics. First, they do not differentiate, retaining the cytological appearance of the original imaginal cells. Their chromosome number and morphology remain unaltered. Second, although the cultured cells retain their original determination when they are first tested by transplantation into a larval host, the determination of cultured cells eventually may change. These exceptions are significant. Determination may be lost, resulting in a subline of cultured cells that are cytologically indistinguishable from the original cells and continue to proliferate in adults, but do not differentiate at all when tranplanted into a larval host. Or part of the determination specificity in some sublines of the cultured cells may be lost and the number of different structures that are formed at differentiation is reduced. Finally, some cells exhibit a *new* pattern of differentiation. They have been *transdetermined.*

For example, in a culture started from an antenna disk there can occur cells that differentiate into wing structures which normally are formed only from cells of the wing disk. Transdetermination can originate in a single cell, which subsequently gives rise to a clone of uniformly transdetermined cells. The adult structures produced by transdetermined cells after transplantation into a larval host are indistinguishable from the structures normally produced by the imaginal disk for those structures.

Cells of different disks transdetermine in different and predictable ways. For example, palpus disk cells transdetermine preferentially to head, wing, and antenna cells, whereas antenna disk cells transdetermine preferentially to wing and tarsus cells.

Therefore the palpus cells, which undergo transdetermination, cannot be identical with the antenna cells, which undergo transdetermination. This observation is especially important because it eliminates the possibility that transdetermination affects only undetermined embryonic cells, which might conceivably be present in all cultures of disk cells. Thus, it appears that a transdetermination event occurs in a determined cell, eliminating its preexistent determination specificity and concomitantly establishing a new specificity.

The mechanism of transdetermination is not known. It is known, however, that the changes are dependent on the proliferating activity of the cells, the rate of transdetermination being closely correlated with the rate of proliferation (Fig. 12-5).

We have then evidence from several directions forcing us to reshape our ideas of differential gene expression. Earlier we encountered the differential replication of part of the genome when we described the phenomenon of specific gene amplification (page 179).

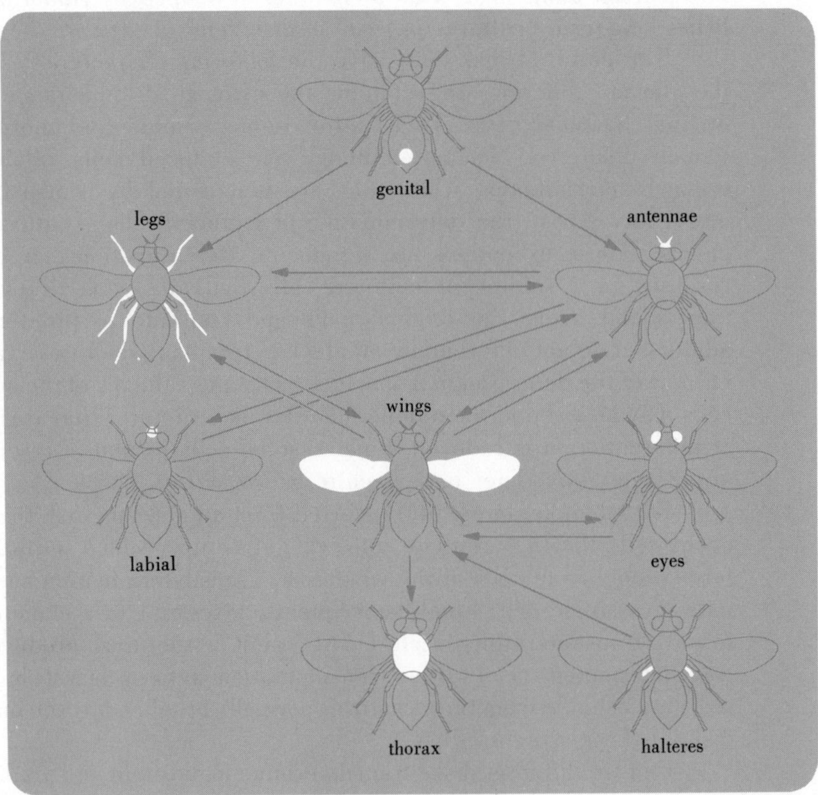


**Fig. 12-5** *Transdetermination Sequence undergone by seven kinds of imaginal-disk cells is shown by arrows. Genital cells may change into leg or antenna cells, whereas leg and antenna cells may become labial or wing cells. In most instances the final transdetermination is from wing cell to thorax cell; the change to thorax appears to be irreversible. (From "Transdetermination in cells," by E. Hadorn. Copyright © 1968 by Scientific American, Inc. All rights reserved.)*

Why a requirement for DNA synthesis preceeding differentiation? In specific gene amplification, a selective replication of genes provides a mechanism for insuring the accumulation and preservation of genes required for synthesis of critical products during a limited period of development.

However, the requirement for DNA synthesis preceding other differentiative events is not so easily explained. The concept of differential gene expression, as we have come to know it, calls for transcription resulting from the sequential activation of specific genes (with the concomitant inactivation of others), one after another. Nothing about the concept gives a clue as to why an intervening activation of DNA synthesis might allow a new direction in differentiation. However, the concept of differential gene expression fails to take into account one set of existing facts about development. Development is not a smooth continuum; as we have learned, it proceeds in "blocks." Rather than a sequence of expressions of individual genes, perhaps we should envision the activation of blocks or groups of related genes: those active in cleavage, those during gastrulation, and so on (page 158). DNA synthesis might be required to switch from one major block of activities to the next. Within a given sequence of a related group of genes, new transcriptions might occur without DNA synthesis; but to turn off a major block and activate another, DNA synthesis would be needed.

We do not know enough about the nature of the factors that maintain genetic silence and regulate transcription to permit a truly satisfactory formulation of the problem. We have already summarized the earlier arguments for the role of histones and have shown that the range of messenger RNAs transcribed in mammalian organs is restricted by the masking of much of the genome, possibly by histones and other proteins (page 185).

Perhaps in that earlier section lies a helpful clue, in the idea that there might be only one time in the life of the cell when a gene could be freed of its existing repressors: during S phase, when both DNA synthesis and histone synthesis occur.

## *STABILITY AND IRREVERSIBILITY ARE NOT SYNONYMOUS*

Earlier we expressed doubt that differentiation involves a stable, irreversible change in DNA; one reason given for that doubt is that in many animals the limited repertoire of a specialized cell may abruptly be increased. Here we refer to the changes observed during regeneration of lost parts in the adult. A second reason is that at least some highly specialized plant cells are capable of a truly remarkable, renewed proliferation and differentiation. Mature cells taken from a differentiated tissue can be made to behave in the same way as the fertilized egg: proliferating and passing through stages of development that closely resemble those of the normal embryo, and producing a full plant. These cells appear to have intrinsic properties that, in the normal plant, are held in check (Chapter 5).

### *The Regeneration of Body Parts in Animals*

Although constraints, once imposed, are not readily overcome, the genome in many cells remains capable of reactivation. One example is the regeneration of body parts in the salamanders and newts.

When a limb or tail of a salamander is amputated, the wound is closed promptly, in a matter of hours, by the movement of the bordering epidermal cells. The thin epidermal layer thickens subsequently; for a few days there are the usual inflammatory processes in the site, and histolysis of adjacent tissue.

After the phase of wound healing there occurs a more intense breakdown of tissues bordering the original wound. Muscle cells especially are affected, along with connective tissue, cartilage, and bone.

At the same time, cells resembling primitive embryonic mesenchyme cells emerge. Although the source of these cells was a matter of long-standing dispute, the evidence now available suggests that they arise from the old tissues by a process of morphologic dedifferentiation (Fig. 12-6). Freed from restraints during histolysis, these cells may multiply and contribute to the blastema of regeneration—a cellular mound of mesenchyme cells, beneath the wound epithelium, that give rise to the new part. Gradually the blastema enlarges, forming an obvious outgrowth, a bud; in the case of the limb, by the end of three or four weeks a new structure is emerging, growth is slowed, and differentiation and morphogenesis are initiated. By the fifth week after amputation, digits appear and movements of the limb begin.

Knowing that neurons grow out from the spinal cord, you would immediately surmise that after they were severed at amputation, they, too, would begin invading the stump. They do. Two or three days after amputation, during the time of wound healing, regenerating nerve fibers are found among the wound tissues. Are they necessary for regeneration? They are; if the regenerate is denervated during the early phases of dedifferentiation and enlargement, growth ceases and the blastema becomes scar tissue. There can be no question that nerves are necessary. It may be added that no single type of nerve—motor, sensory, or sympathetic—is critical; however, for any given limb, a *threshold volume of neuroplasm is essential.* It was formerly thought

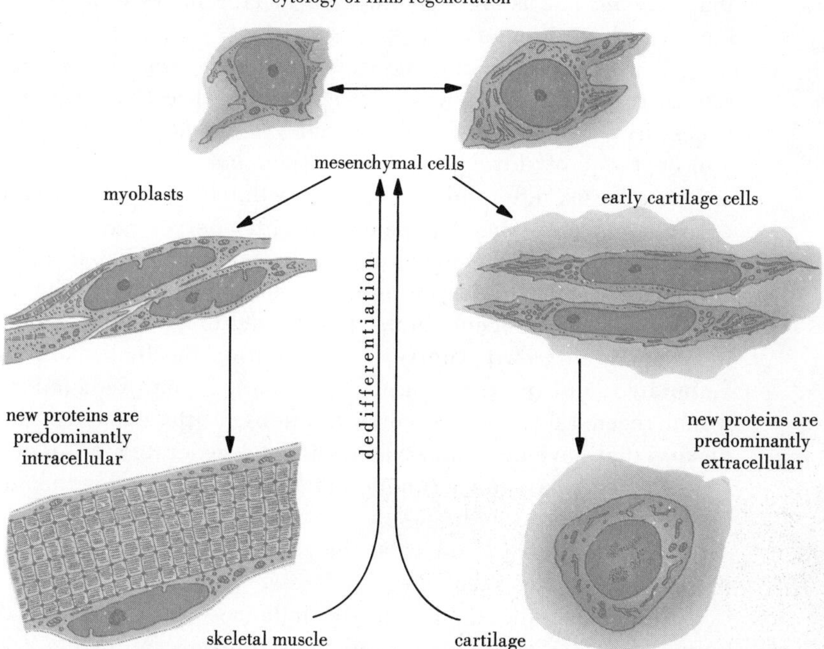


*Fig. 12-6 Diagram summarizing the changes in fine structure that occur in the mesenchymal cells of the blastema as they differentiate into myoblasts and muscle cells, on the one hand, and early cartilage cells and cartilage, on the other. The formed cell types apparently dedifferentiate (arrows, center of figure) into mesenchymal cells when the limb is amputated. (After E. D. Hay, "Cytological Studies of Dedifferentiation and Differentiation in Regenerating Amphibian Limbs," in* Regeneration, *D. Rudnick, ed. Copyright 1962, The Ronald Press Co.)*

that it was the number of fibers that was important. The evidence was the following: Normally, sensory nerves are most involved, for there are more sensory than motor fibers in the limb, and the number of motor fibers is too small to support regeneration. However, if the sensory supply is kept out, motor nerves bifurcate and increase the number of fibers to such an extent that they are effective.

Moreover, in species like the frog—in which, in contrast to the salamander, the capacity for regeneration is lost at metamorphosis—regeneration may be stimulated by increasing the nerve supply to the amputation site.

However, some species having few fibers are capable of regeneration—*Xenopus laevis*, for example. *Xenopus* and the mouse each have few fibers; *Xenopus* regenerates a limb, the mouse does not. However, the *Xenopus* fibers are large in diameter, whereas those of the mouse are small. When one calculates the volume of neuroplasm, he finds that *Xenopus* has as much neuroplasm in a few fibers as other regenerating forms have in many.

Thus there is abundant evidence that nerves play an important role in regeneration. However, we should realize that nerves are not necessary for the original embryonic development of the limb. If nerves are kept out of developing limbs, such *aneurogenic* limbs undergo morphogenesis, differentiation, and growth, although of course they do not function properly. It appears that after nerves have been present in a limb, the limb becomes dependent on, or "addicted to," nerves. If this is so, what would you predict to be the outcome of the following experiment? Aneurogenic limbs are allowed to develop in salamanders by keeping outgrowing nerves from entering the limb. Now they are amputated (and the nerve supply kept out). If you predicted that they would regenerate, you are correct. Limbs, of the same age and complexity, that have been innervated cannot regenerate after denervation.

Do nerves assume a function that is accomplished in their absence before they enter a limb?

We still do not know what the nerves do in regeneration. Do they produce a "trophic agent"?

It appears likely that nerves influence the structures they innervate in ways other than by impulse transmission or the release of acetylcholine. Hypothetical "trophic agents" have been postulated to account for the influences of nerve on several end organs. There is evidence that highly specific attributes of nerve determine the metabolic and synthetic patterns of the muscles they innervate. Yet we do not know the nature of the postulated trophic agents.

## FURTHER READING

Abercrombie, M., "Contact Inhibition: the Phenomenon and Its Biological Implications," *National Cancer Institute Monograph*, no. 26 (1967), p. 249.

Goss, R. J., *Adaptive Growth.* New York: Academic Press, 1965.

______, "Kinetics of Compensatory Growth," *Quarterly Review of Biology*, vol. 40 (1965), p. 123.

______, *Principles of Regeneration.* New York: Academic Press, 1969.

Hadorn, E., "Transdetermination in Cells," *Scientific American*, vol. 129 (1968), p. 110.

Hay, E. D., *Regeneration.* New York: Holt, Rinehart and Winston, 1966.

Levi-Montalcini, R., "Growth Control of Nerve Cells by a Protein Factor and Its Antiserum" *Science*, vol. 143 (1964), p. 105.

Sachs, L., "An Analysis of the Mechanism of Neoplastic Cell Transformation by Polyoma Virus, Hydrocarbons, and X-irradiation," in *Current Topics in Developmental Biology*, Vol. 3, A. Monroy and A. A. Moscona, eds.

New York: Academic Press, 1967, p. 129.
Stoker, M., "Contact and Short-Range Interactions Affecting Growth of Animal Cells in Culture," *ibid.*, p. 107.

chapter **13**

# *Meristems and the Control of Postembryonic Plant Development*

In Chapter 12 we saw that it is not unusual for differentiated cells to divide. In fact, some cells continue division throughout the entire life-span in most organisms. The continued production and growth of new cells should result in continued enlargement of the organism, but in vertebrate animals, with the exception of fishes and a few others, the adult reaches a genetically determined size, which is then maintained. How is adult size stability achieved in the face of continued cell production? The answer lies in a balance between cell formation and cell elimination. While new cells are being added to the organism, senescent cells are being removed. Cell replacement takes place within the framework of existing tissues, so although the cells might be gradually replaced, the tissue maintains an unchanging size and appearance. This mode of selective cell replacement is possible in animals because the cells are mobile and are capable of changing their shape and contacts with adjacent cells.

In plants, however, cell walls are rigid and firmly cemented together, and selective cell replacement cannot occur. Senescent and dead plant cells are either removed totally, as in shedding of leaves and bark, or are retained in the body of the organism throughout its life-span, as are the dead cells in wood. Nor are new cells formed within the framework of the existing tissues. They are added in an accretionary manner in localized regions called *meristems*. This results in the continued enlargement of the plant throughout its life-span. The most important of the meristems are the *apical meristems*, situated at the tip of each shoot and root. They consist of cells partially differentiated from the embryonic condition, which retain the capacity for mitosis and produce a succession of tissues and organs.

The apical meristems are delimited in the embryo but do not contribute significantly to its development. They become active in postembryonic development, starting with germination, and most of the vegetative organs and all of the reproductive organs are formed by their activity during this period. The continued initiation of organs, which has no counterpart in animal development, has led to the meristems being called regions of "permanent embryogeny." In some plants other meristems are initiated during post embryonic development. These are lateral in position, sheathing each stem and root in zones of dividing cells. There are two *lateral meristems*: the vascular cambium, which produces wood and phloem, and the cork cambium, which produces the corky bark. Neither of these meristems produces organs, and they are usually not present in herbaceous plants. Because they are not essential for primary plant development we shall pay little attention to them in this chapter.

## THE CONTROL OF MERISTEM ACTIVITY

Apical meristems are continuously active in annual plants, and are still functional when the plant dies. Are the meristems perhaps capable of growing for unlimited periods of time under suitable conditions? This idea received striking confirmation when root tips were excised from a tomato plant and grown in a sterile nutrient medium. The root tips were isolated in 1933 and have since been subcultured at weekly intervals by transferring tips to fresh culture medium. They are still growing actively some 39 years later; this is a far longer life-span than is normal for a tomato plant.

In perennial plants, periods of dormancy intervene between periods of active meristem growth. The kinds of organs produced by the meristems may vary. The leaves of juvenile plants are often different from those of the adult; bud scales may replace vegetative leaves in advance of periods of dormancy; and reproductive organs ultimately replace vegetative ones. All of these changes suggest that the meristems are subject to developmental regulation in their growth. How is their activity regulated? Does each meristem function as a self-regulating cell-producing region, or is this activity regulated by stimuli transmitted to the meristem from other parts of the plant? Does a meristem organize the tissues and appendages it produces, or is the differentiation of tissues and organs controlled by interactions with older parts of the plant or the external environment?

These are the questions we shall examine in this chapter. The answers to them will determine the importance to be attached to the meristems, whether they are to be considered as autonomous developmental organizers or as plastic regions that are molded in response to instructions received from other parts of the plant.

## MERISTEMS AS CENTERS OF CELL PRODUCTION

The organization and activity of the apical meristems can best be understood if we first examine the range of cellular structure and then ask questions that can be tested experimentally. The structure of the root tip is simpler than that of the shoot tip because lateral roots are not formed close to the meristem, so we will begin by examining the root.

In longitudinal sections of the root tip of tobacco, files of cells converge to a region just behind the tip and terminate in three transversely arranged cell layers (Fig. 13-1). These three cell layers and their immediate derivatives are the apical meristem. Each layer in the meristem is continuous with differentiating cells in the older parts of the root. Derivative cells of the anterior meristem layer differentiate as the root epidermis and the root cap. This latter is a forward derivative of the meristem, so the apical meristem is actually in a subterminal position in the root and is entirely surrounded by differentiating cells. The middle layer of the meristem produces cells that differentiate as the cortex; derivatives of the inner layer differentiate as procambium and subsequently as the vascular system of the root. Thus the same three tissue systems that were initiated in the embryo continue to be differentiated immediately behind the meristem.

There is no sharp boundary between the meristem and cells that are beginning to differentiate. How do we distinguish the meristem

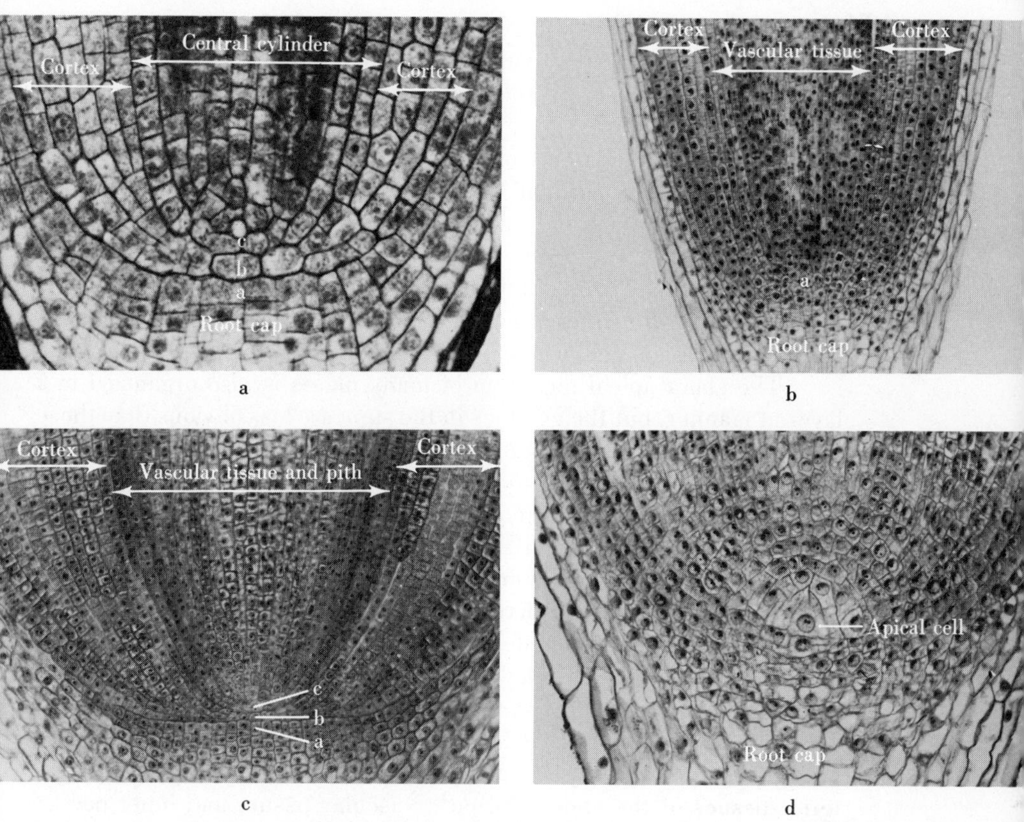


***Fig. 13-1*** *Longitudinal sections of the root tip of different plants showing the terminal meristem and the differentiating tissues.* (a) *Root of tobacco* (Nicotiana tabacum). *The three meristem layers are labeled a, b, and c, and the derivative tissues of the root are identified.* (b) *Root of pea* (Pisum sativum). *The meristem contains a single cell layer, labeled a, which gives rise to all the tissues of the root.* (c) *Root of corn* (Zea mays). *There are three meristem layers, labeled a, b, and c, but layer c derivatives differentiate only as root cap. (In tobacco this cell layer initiated both the root cap and the epidermis.)* (d) *Root of the fern* Botrychium. *A single enlarged apical cell can be seen in the meristem, and this divides in planes parallel to each side to initiate all the tissues of the root.* (a) *×350;* (b–d) *×75. (a from K. Esau,* Plant Anatomy, *1953, by courtesy of the author and John Wiley and Sons.)*

cells from the derivative cells? Mitosis is not limited to the meristem cells, but occurs also in differentiating cells for about a millimeter behind the root tip. However, meristem cells retain the capacity for continued mitosis, whereas the differentiating cells undergo only a limited number of divisions. This difference in behavior is related to the positions occupied by the meristem cells and the derivative cells. If a root is decapitated so that the meristem is removed, cells that had started to differentiate will regenerate a new apical meristem and resume mitosis. Using the criterion of capacity for continued division, the meristem consists of about 1000 cells and is 0.1 mm or less in diameter.

The apical meristems of roots of other species may differ from that of tobacco. In pea and in pine roots there is a single layer of meristem cells from which all the cells in the root differentiate (Fig. 13-1). In corn there are three meristem layers, but the root cap and the epidermis originate from different meristem layers (Fig. 13-1). The roots of many ferns are differently organized, the tip being occupied by a single, large apical cell, which is the initial of all the root cells (Fig. 13-1).

The shoot apical meristem of many plants is also organized in a layered manner, but the cell files in the stem are less obvious than those in the root because their regularity is soon disturbed by leaves, which develop very close to the apical meristem. In fact, the positions of the youngest leaf primordia are convenient markers by which we can define the boundary between meristem and differentiating cells. Defined in this way, the apical meristem of a tobacco shoot is a hemispherical mound about 0.25 mm in diameter (Fig. 13-2). The superficial cells in the meristem are arranged in two distinctive layers, called *tunica* layers. These layers result, like similar layers in the root meristem, from restriction of cell division to a plane perpendicular to the surface. The surface layer of tunica cells differentiates only as epidermis, and the underlying layer gives rise to most of the cortex in tobacco. The interior tissues of the stem—the pith, vascular tissue, and inner part of the cortex—are derived from meristem cells situated under the tunica and dividing in various planes. This part of the meristem is called the *corpus*. The tunica–corpus pattern of meristem organization is charac-

**Fig. 13-2** *Longitudinal sections of the shoot tip of different plants showing the terminal meristem, young leaf primordia, and differentiating tissues in the stem.* (a) *Shoot of tobacco* (Nicotiana tabacum). *The terminal meristem has a two layered tunica (T) and beneath this is the corpus (C). Below the corpus cells are beginning to differentiate as pith (P). On the right side of the meristem is a newly emergent leaf primordium (LP) in which a procambial strand is visible (PS). Below the leaf the stem tissues, cortex (Cx) and epidermis (E) can be seen.* (b) *Shoot of castor oil plant* (Ricinus communis). *The terminal meristem consists of a superficial tunica and underlying corpus, but these are further subdivided into zones which can be observed in this figure. The central zone cells (CZ) are large and lightly stained. Peripheral zone cells (PZ) are smaller, and are more intensely stained. Under the central zone is a zone of file meristem (FZ) that contains cells which divide in a horizontal plane. These will differentiate as pith. (Labeling as in part a.)* (c) *Shoot of the maidenhair fern* (Adiantum). *The center of the meristem has a single enlarged apical cell (AC) that initiates all the tissues of the shoot by divisions parallel to its inclined inner walls. (Labeling as in part a.) All ×75.*

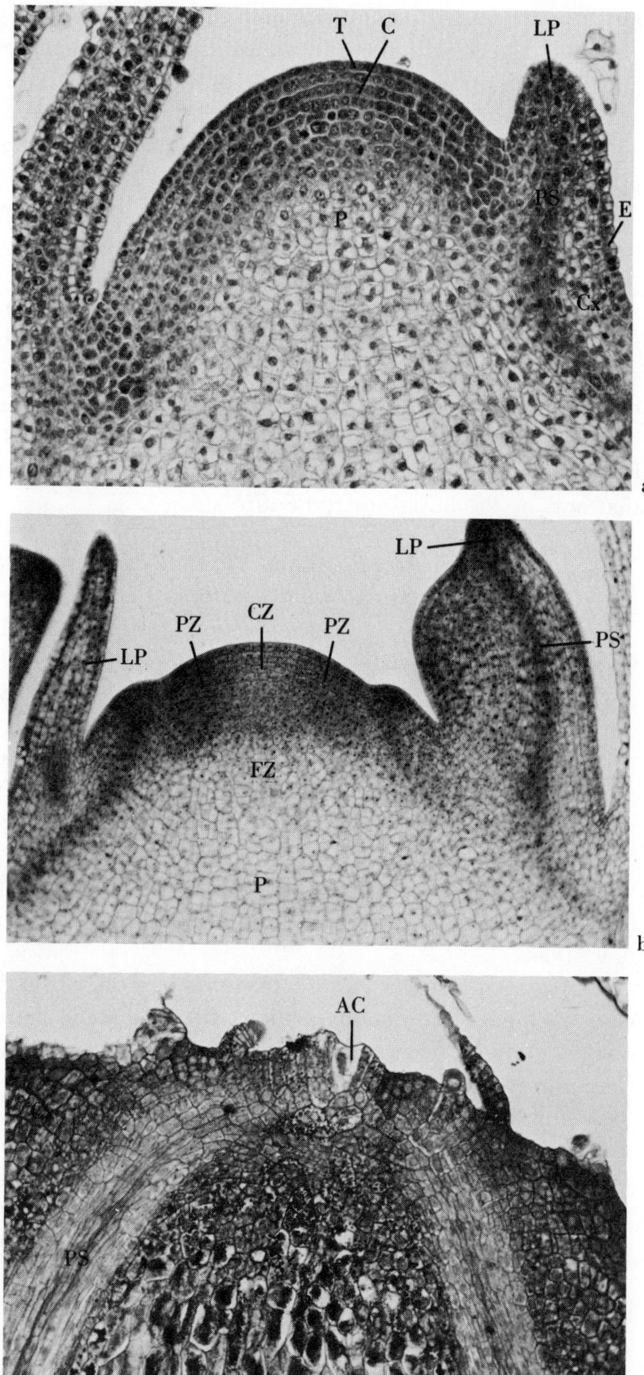
T
C
LP
PS
E
P
Cx
a
LP
PZ
CZ
PZ
LP
PS
FZ
P
b
AC
PS
c

teristic of flowering plants, although there are differences in the number of tunica layers in different species and at different times of development within a plant.

Cells within different parts of the apical meristem of a tobacco shoot have a uniform cytological appearance, but in some flowering plants the meristem is subdivided into distinctive *cytohistological zones*. The central cells of the tunica layers and the corpus are vacuolated, and divide infrequently. This central zone is surrounded peripherally by cells that are small, divide frequently, and are highly stratified, and it overlies a zone in which cells are arranged in vertical files (Fig. 13-2). The apical meristems of most conifer shoots possess a similar zonation pattern, except that there are no tunica layers. The superficial cells divide in irregular planes and function as initials of all the cells of the shoot. In most ferns the meristem forms a single superficial layer of enlarged cells within which is a central larger apical cell that acts as the initial for the entire shoot (Fig. 13-2).

### The Role of Meristem Layers

What is the developmental significance of the various patterns of apical meristem organization that we have just examined? One of the most consistent features of meristems is the layered appearance. Each of the meristem layers is continuous with certain of the differentiated cells of the mature organ and appears to produce only these cells during development. Do the meristems, therefore, consist of cell layers that are already determined to form only those tissues with which they are in continuity? Or does the meristem consist of cells that all retain the potential to differentiate as any cell type of the mature organ, even though they will normally form only one or a few cell types because of the position they occupy within the total mass of meristem cells?

These questions cannot be answered by observing microscope slide sections of meristems, which present a static picture fixed at one instant in time. What is needed is a method that can take the dynamic, time-related aspects of growth into account. We require a technique by which the developmental fate of cells differentiating from any of the meristem layers can be followed. The method of dye marking used so successfully to follow animal cells does not work in plants, but others are available. The best of these involves inducing a mutation in one of the meristem layers and following the distribution of derivative mutant cells in the differentiating organ. Apical meristem mutations result from treatment of germinating seeds or seedlings with solutions of the alkaloid colchicine. Mutant cells are polyploid and can be distinguished in the plant after a period of growth by their greater size, increased number of nucleoli, enlarged nuclei, and by the increased chromosome number in dividing cells.

When seeds of *Datura stramonium* were treated with colchicine, plants were found in which either of the two tunica layers or the corpus of the shoot apical meristem were separately converted to the polyploid state. Such mutationally layered plants are called *periclinal chimeras*, and some of the meristem types obtained are illustrated in Fig. 13-3. The plants most useful for our present purpose are those in which the second tunica layer is polyploid and the outer tunica layer and corpus are diploid. The second tunica layer would normally be expected to give rise to most of the cortex, and cross sections of the stem reveal the exclusive presence of mutant cells in the outer part of the cortex (Fig. 13-4). However, the inner boundary between mutant cells derived from the second tunica layer and normal diploid cells derived from the corpus is extremely irregular. In some places diploid cells extend far out into the cortex; in other places the vascular tissue and even some of the pith is composed of mutant cells. This result is not consistent with the idea that the apical meristem layers are already determined to produce only specific derivative tissues, and suggests that the cells in the meristem layers are able to produce a greater array of cell types than they give rise to during normal development.

## Distribution of Cell Division in the Meristem

In the meristems illustrated in Fig. 13-3, all the cells of each layer behave uniformly. If the layer is polyploid, all the cells are polyploid. The most likely explanation is that a centrally placed cell in the layer mutated in response to colchi-

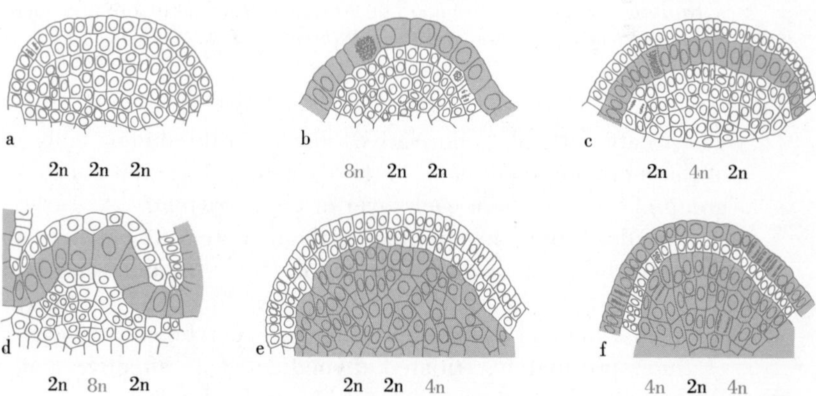


**Fig. 13-3** *The shoot apical meristem of normal* (a) *and periclinal chimeras* (b–f) *of* Datura. *In each apex the ploidy level of the nuclei in each of the three meristem layers is shown, the first number being for the surface layer of the tunica and the last number for the corpus. Polyploid cells are larger than diploid cells in the same meristem and contain larger nuclei. (From A. G. Avery et al.,* Blakeslee—The genus Datura, *1959, by courtesy of the Ronald Press.)*

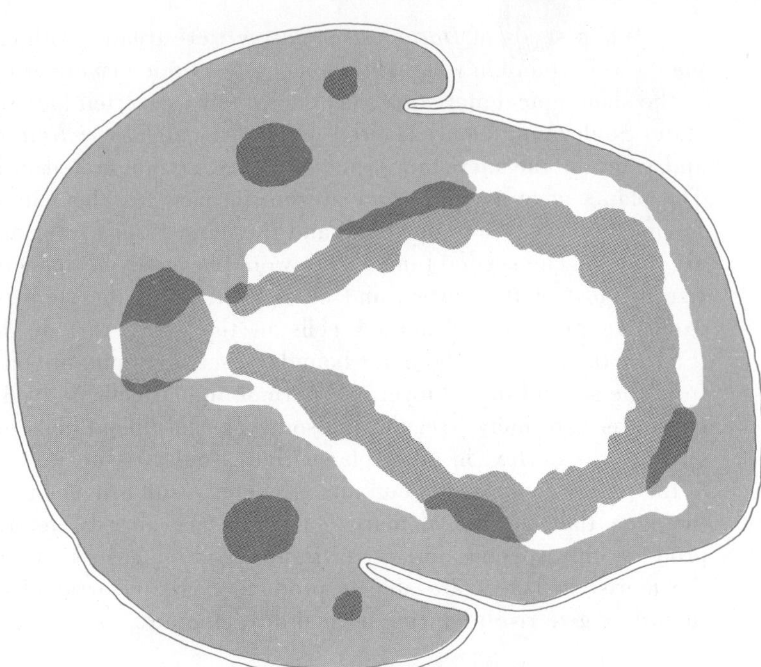

*Fig. 13-4 Cross section of the stem of a periclinal chimeral peach plant in which the constitution of the three apical meristem layers was 2n, 4n, 2n. In this stem section the tissue derived from the tetraploid second layer of the tunica is shown in color. The entire epidermis, which is derived from the outer tunica layer, consists of diploid cells. The outer part of the cortex is composed entirely of tetraploid cells, and tetraploid cells extend into deeper tissues in several places so that some of the inner cortex and some of the vascular tissue is tetraploid. (From H. Derman,* The American Journal of Botany, *40:154, 1953, by courtesy of the author and the Botanical Society of America.)*

cine treatment and its derivatives displaced the diploid cells as growth of the meristem continued. Is there, then, a single initial cell or a small group of initial cells in each layer of the meristem?

In describing the structure of the meristems of both root and shoot we used the terms *initial cells* and *apical cells*, but what evidence is there that these cells perform an initiating function? To determine this we need a method by which dividing cells can be marked over a period of time; fortunately, tritiated thymidine is as effective a marker of DNA synthesis in plant cells as it is in animal cells.

Solutions of tritiated thymidine have been applied directly to the apical meristems of vegetative shoots of lupin and several other species of flowering plants by Buvat and his associates, and to the root meristem of corn, mustard, and other species by Clowes. The results of both these experiments were in surprising disagreement with what might have been expected. In all cases the cells located in the central part of the meristem either failed to incorporate thymidine, and were, therefore, not synthesizing DNA preparatory to division, or they incorporated thymidine much more slowly than the surrounding meristem cells (Fig. 13-5). The function of the central mitotically *quiescent zone* is still being actively studied and is still not thoroughly understood. By extending observations on its behavior to other developmental stages it has been found that there is no quiescent zone in the meristems of the embryo or seedling. As the root and shoot apical meristems enlarge in juvenile growth some of the central cells stop dividing or divide more slowly, and the quiescent zone enlarges until it may comprise 200 to 500 cells in later stages of vegetative growth. In the shoot apex, divisions are resumed in this zone when the meristem begins to develop as a flower and much of the floral meristem comes from these cells.

The centrally placed cells in the meristem divide actively during seedling stages of growth and may continue to divide slowly during vegetative growth, so they do function as the ultimate source of the cells of the shoot or root, and in this sense contain the initial cells. However, they may function in other ways. The quiescent zone coincides with the central zone of vacuolated cells in those shoot meristems which are cytohistologically zoned. Perhaps this group of distinctive cells performs some function other than active cell division. If these cells perform no function it should be possible to destroy them without affecting growth of the meristem. This can be done by exposing the apical meristem while viewing it under a binocular dissecting microscope, and then puncturing the center of the meristem with a fine glass or steel needle.

The experiment has profound effects on development. Growth stops and is resumed in the peripheral region, which becomes organized as one or more apical meristems. In control experiments, where punctures were made in the peripheral part of the meristem, development was not affected. The conclusion from these experiments is that the central part of the meristem functions to regulate and integrate mitotic activity in the rapidly dividing peripheral regions. The mechanism by which this interaction is achieved is at present unknown. Possibly the slow rate of division in the central cells permits the synthesis of a factor essential for integrated meristem function.

## *The Regulation of Meristem Growth*

So far in our investigation we have examined the regulation of mitotic activity within the apical meristems. We must now ask whether continued mitotic activity is an autonomous function of the meristems, or whether mitosis occurs only in response to directing stimuli transmitted from other parts of the plant. The most direct test of these alternative hypotheses is to excise meristems and grow them in sterile culture. If they grow, they must be autonomous; if not, they must be dependent on exogenous stimuli. However, there are certain qualifications we must first make.

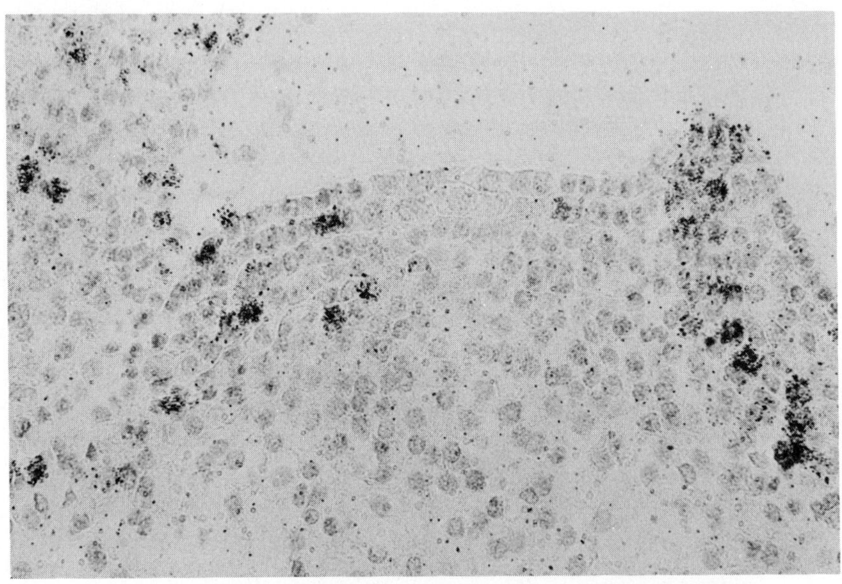
a

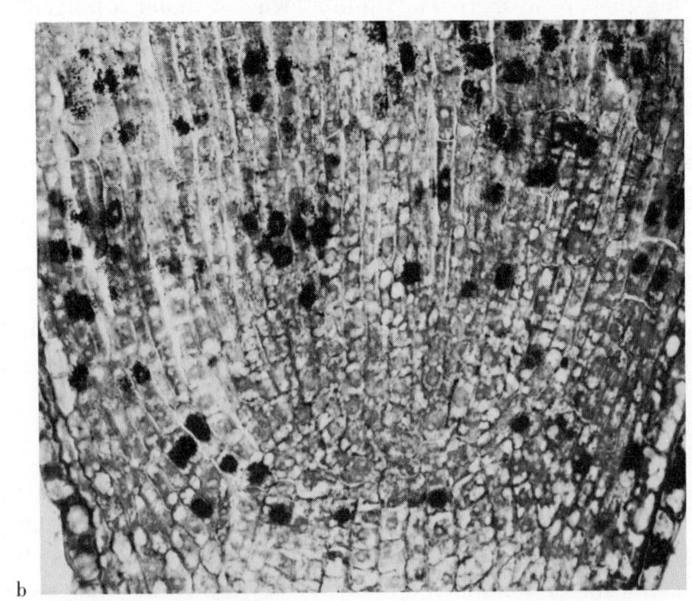
b

Without an energy source, roots will certainly be unable to grow and since energy-containing carbohydrates are normally synthesized in the photosynthetic shoot and not in the root, we should supply these carbohydrates in the culture medium. Similarly, vitamins are synthesized in leaves and we should supply these. Given these limitations, the question then is: Will root and shoot apical meristems excised from the plant without any surrounding organs and as few derivative cells as possible continue mitotic activity and organized development on a simple nutrient culture medium? This experiment has been performed on root tips of pea and in shoot meristems of several ferns including *Adiantum*. In all of these mitotic activity continued in an essentially normal manner. Shoot meristems of *Adiantum* grew successfully on culture media containing only sugar and inorganic salts, indicating a high level of apical autonomy (Fig. 13-6).

Results of similar experiments on the shoot meristems of flowering plants have produced conflicting results. Sunflower, orchid, and potato meristems will grow successfully on very simple culture media producing shoots with normal tissues and leaves. However, lupin meristems have so far not been grown indefinitely. After forming several leaves they stop mitotic activity, and no further addition of vitamins, amino acids, or hormones will cause them to resume development. Whether this indicates a requirement for some stimulus from other parts of the plant or whether it simply indicates inadequate culture conditions is not known. However, the success in growing detached

**Fig. 13-5** *Longitudinal sections of terminal meristems after labelling with tritiated thymidine.* (a) *Shoot apex of sunflower* (Helianthus annuus). *The meristem had previously been excised from the plant and grown in sterile culture on nutrient medium. After 48 hours in sterile culture tritiated thymidine was added to the medium, and the meristem was allowed to incorporate thymidine into DNA for 24 hours. The shoot tip was then fixed, sectioned, and attached to microscope slides. The slides were then dipped into liquid photographic emulsion, and the emulsion-coated slides were stored in the dark. They were subsequently developed in photographic developer and stained to show the cells. In the figure accumulations of silver grains in the emulsion show as dark spots and indicate nuclei that were synthesizing DNA during the experimental period. It can be seen that although several of the meristem cells did synthesize DNA none of these was in the central part of the meristem. Thus this part is the quiescent center.* (b) *Root apex of leafy spurge* (Euphorbia esula). *The meristem was immersed in a solution of tritiated thymidine while still attached to the plant and then processed as the sunflower shoot tip described above. Here also it can be seen that none of the cells in the central part of the meristem has incorporated thymidine into DNA and this region is, therefore, the quiescent center. (*a *from T. A. Steeves et al.,* Canadian Journal of Botany, *47:1367, 1969;* b *from M. V. S. Raju et al.,* Canadian Journal of Botany, *42:1615, 1964, by courtesy of the authors and reproduced by permission of the The National Research Council of Canada from The Canadian Journal of Botany.)*

a

b

**Fig. 13-6** *Growth in sterile culture of terminal meristems excised from* Adiantum *shoots.* (a) *Excised meristems were grown in a culture medium containing only inorganic salts and sugar. Each meristem has given rise to a complete plant consisting of stem, leaves, and roots.* (b) *Improved growth of excised meristems which resulted from enriching the culture medium with yeast extract, a source of vitamins and amino acids. (From R. H. Wetmore,* Brookhaven Symposia in Biology, *6:22, 1954, by courtesy of the author and the Brookhaven National Laboratory.)*

meristems of several plants is sufficient proof that the meristem is essentially independent of the rest of the plant, and is autonomous in its mitotic activity.

## MERISTEMS AS DEVELOPMENTAL ORGANIZERS

Cells produced by the meristems become integrated into tissues and organs. What are the stimuli that act on a particular group of differentiating cells, causing them to become a part of the vascular system, or a leaf, or a lateral shoot? Where do these stimuli arise? Two ideas have influenced scientific investigation of these questions. The first is that the apical meristem is the organizer of development of the cells which it produces. Not only does it produce the cells, but it also determines their fate, and through its activity molds them into the tissues and organs of the plant. On the other hand, some of the evidence of plant development has been interpreted as indicating that new tissues and organs differentiate in response to stimuli coming from other parts of the plant. These exogenous stimuli are thought to determine the pattern of development of the cells produced by the meristem. In this view the meristem is simply a cell-producing region and is otherwise unimportant in development. These questions have been investigated extensively using surgical and sterile culture techniques that permit developing regions of the plant to be isolated from one another.

### Meristem–Leaf Interaction

A leaf is initiated at the margin of the shoot apical meristem by local intensification of cell division in several of the superficial meristem layers. Enlargement of the newly formed cells causes the leaf to emerge as a hemispherical, radially symmetrical primordium (Fig. 13-7). As the leaf primordium continues to enlarge, the side facing toward the shoot apical meristem becomes progressively flattened because the rate of cell growth in this region is less than in other regions, and the leaf (which is then about a millimeter tall) resembles a cone with one flattened side. During this time the leaf has been growing in height through mitotic activity of a terminal cell group, but these cells cease mitosis and further elongation results from divisions and cell expansion throughout the leaf. The dorsoventral leaf blade develops through the activity of a marginal meristem, which is continuous along the edge of the primordium. In plants with compound leaves the marginal meristem is not continuous, but is formed in an intermittent manner along the margin, initiating local outgrowths that develop into the individual leaflets.

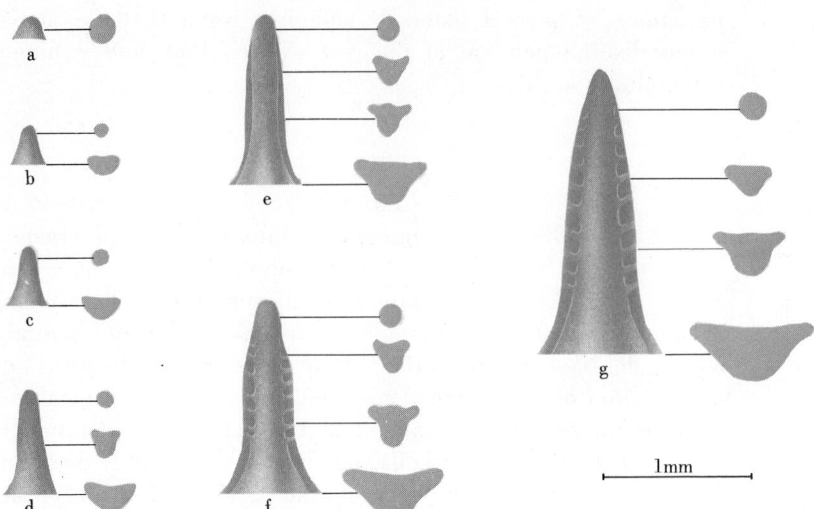


*Fig. 13-7 Diagrams of early stages of leaf development in tobacco* (Nicotiana tabacum). (a) *The newly emergent leaf primordium is hemispherical. As it elongates its upper surface becomes progressively flattened so that the leaf is bilaterally symmetrical* (b–d). *Subsequently the leaf blade develops by the activity of a marginal meristem* (e–g). *(From G. S. Avery,* The American Journal of Botany, *20:566, 1933, by courtesy of the author and the Botanical Society of America.)*

One of the most interesting aspects of leaf development is the change from radial to bilateral symmetry, which occurs in the earliest stages of leaf development. The apparent cause of this change is the reduced growth rate in cells facing the shoot apical meristem. Wardlaw suggested that these leaf cells were inhibited in their development by action of the apical meristem. To investigate this possibility he made a small surgical incision between a radially symmetrical leaf primordium, or the site at which a leaf primordium would soon emerge, and the shoot apical meristem in the fern *Dryopteris* (Fig. 13-8). The partially isolated primordium continued to develop. However, it developed not as a dorsoventral leaf, but as a radially symmetrical shoot with its own apical meristem around which leaf primordia developed. When the incision isolating the primordium from the apical meristem was delayed until the leaf had become recognizably bilateral, its further development was not affected and it matured as a typical leaf. These results were in agreement with Wardlaw's hypothesis, and they indicated further that the apical meristem produces only one kind of lateral appendage. The subsequent differentiation of the primordium as a leaf or a lateral shoot depends on whether development occurs in a region of interaction with the shoot apical meristem when it develops

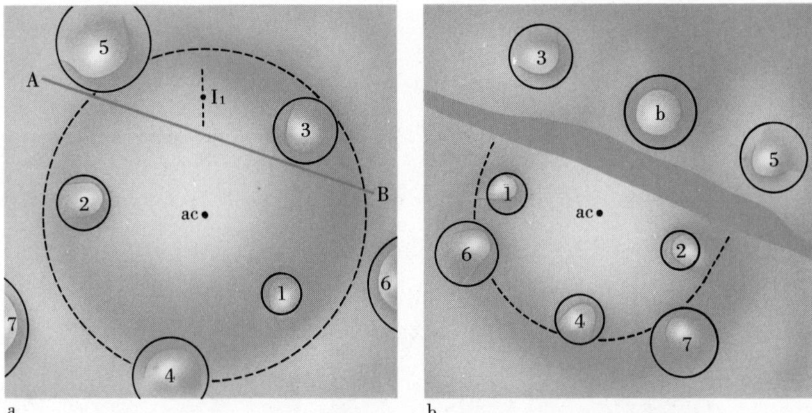


*Fig. 13-8 Surgical isolation of a prospective leaf site from the shoot apical meristem in the fern* Dryopteris. (a) *The operation is shown diagrammatically. An incision A–B is made with a microscalpel between the site at which the next leaf ($I_1$) will emerge and the meristem, the center of which can be identified by the enlarged apical cell (ac). In the diagram, leaves are numbered 1 to 7 in order of increasing age.* (b) *A shoot some time after the operation. The incision now appears as a wide scar. In the position of $I_1$ a bud with radial symmetry (b) has developed. (Adapted from C. W. Wardlaw,* Growth *(supplement), 13:93, 1949, by courtesy of the author and the editors.)*

as a leaf, or in the absence of interaction when it develops as a shoot.

The action of the apical meristem on the lateral primordium is that of a trigger, setting development onto a new course, and prolonged interaction between the meristem and the lateral primordium is not necessary. When bilateral leaf primordia of *Osmunda* were excised from the plant and grown separately in sterile culture they continued to develop and formed typical, although small, dorsoventral leaves (Fig. 13-9). Thus, once a primordium is determined as a leaf it does not require further morphogenetic stimuli for its development. However, excised leaves growing in sterile culture can still be influenced in their development by external stimuli. The level of nutrients in the culture medium alters the pattern of development to produce either simple juvenile type leaves or larger adult leaves (Fig. 13-9), but the basic determination as a leaf is not modifiable.

### Leaf–Leaf Interaction

The leaves formed around the apical meristem are regularly spaced. In most plants leaf primordia are initiated singly and the successive leaves are arranged in a helical array around the stem with a constant angle of divergence. Fig. 13-10 is a diagrammatic repre-

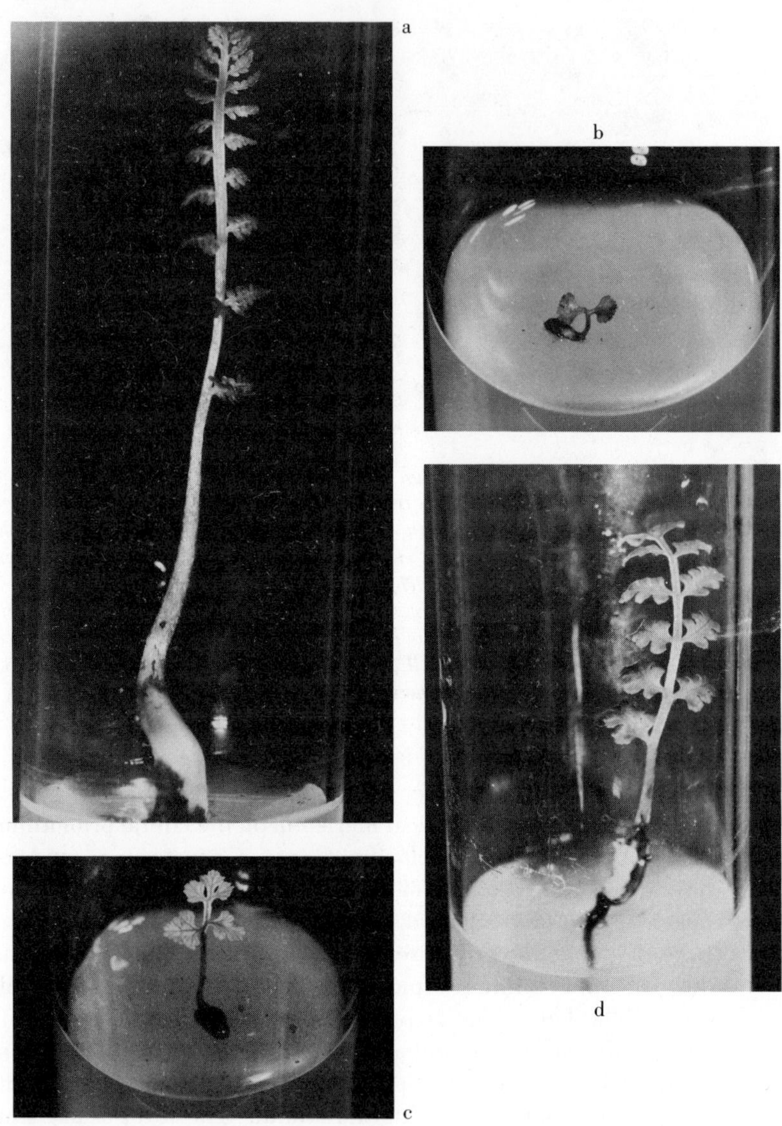


**Fig. 13-9** *Leaf cultures of the cinnamon fern* (Osmunda cinnamomea). (a) *An excised leaf which has developed to maturity in culture.* (b–d) *The effect of nutrients on development of excised leaves in culture. In* (b) *the culture medium contained 0.006 percent sucrose. The leaf has matured as a small, juvenile type leaf. In* (c) *the sugar concentration in the medium was 0.025 percent and in* (d) *it was 1.0 percent. It can be seen that raising the sugar concentration increases the amount of growth and the developmental complexity of the leaf.* ×2.5. *(From I. M. Sussex and M. E. Clutter,* Phytomorphology, *10:87, 1960, by courtesy of the International Society of Plant Morphologists.*

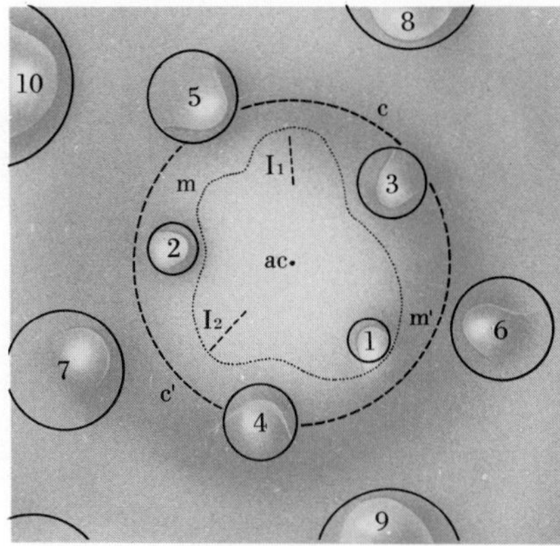


***Fig. 13-10*** *Diagrammatic representation of the shoot apex of the fern* Dryopteris. *Leaves surrounding the terminal meristem are numbered 1 to 10 in order of increasing age. The positions at which the next two leaves ($I_1$ the next leaf, and $I_2$ the next but one) will emerge are indicated. The center of the apex is occupied by an enlarged apical cell (ac), and the margin of the meristem is indicated in the diagram by the dashed line $m-m^1$. The shoot tip is conical, and the base of the cone is shown by the heavy dashed line $c-c^1$. (Adapted from C. W. Wardlaw,* Growth *(supplement), 13:93, 1949, by courtesy of the author and the editors.)*

sentation of the apical meristem and young leaf primordia of *Dryopteris*. The youngest leaf primordium, $P_1$ in the diagram, is situated about 137° around the circumference of the stem from the next older primordium, $P_2$, and occupies a position between two older primordia, $P_4$ and $P_6$. With this information it is possible to predict the positions at which the next leaf primordia will emerge. Thus, the next leaf will lie between $P_3$ and $P_5$, and its presumed position is shown in the diagram as $I_1$. The leaf which will arise after $I_1$ will lie between $P_2$ and $P_4$ and its position is shown as $I_2$ in the diagram.

What causes the regularity in leaf position? Wardlaw has examined this question, again using surgical methods. He predicted that the position occupied by a leaf is in some way established by the older leaves which border it. To test this hypothesis he destroyed the $I_1$ site in *Dryopteris* shoots by a fine needle puncture, and examined the apex as it continued to grow after the operation to see what effect the absence of a leaf in the $I_1$ position would have on later formed leaves. His prediction of the result, shown in Fig. 13-11A, was that the next two leaves, $I_2$ and $I_3$, would not be affected, but that the following leaf, $I_4$, would

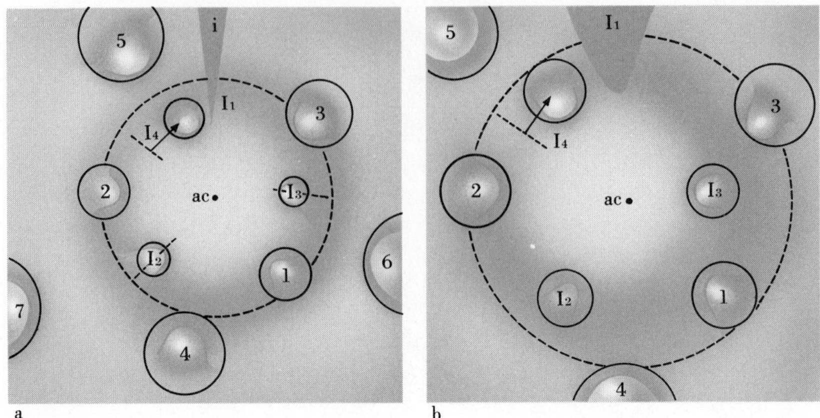


**Fig. 13-11** *The effect of destroying a leaf on the positions of subsequent leaves in* Dryopteris. (a) *A diagram predicting the effect of an incision (i) made in the $I_1$ site thus preventing the formation of a leaf there. It is anticipated that the next two leaves to emerge ($I_2$ and $I_3$) will occupy their normal positions, but that leaf $I_4$, which arises next after $I_3$, will emerge closer to the operated $I_1$ position than is normal.* (b) *The result of an experiment is shown in which the prediction is seen to have been correct. (Adapted from C. W. Wardlaw,* Growth *(supplement), 13:93, 1949, by courtesy of the author and the editors.)*

emerge closer to the operation site in $I_1$ than would be the case in unoperated controls. The result of such an experiment is shown in Fig. 13-11B, which is seen to support his prediction. Wardlaw postulated that each emergent leaf primordium and the apical meristem of the shoot produces around itself a *physiological field* which inhibits the initiation of new leaves within its boundary. Thus, in the apex shown in Fig. 13-11 the absence of the leaf and its accompanying physiological field at $I_1$ permitted the leaf $I_4$ to develop in a region of the apex that would normally not be available to it.

The physiological field theory proposes that a new leaf will emerge when two older leaves become sufficiently spaced from one another and from the apical meristem by growth that their effective fields do not overlap (Fig. 13-12). Because of the small size of the postulated fields it has not been possible to identify their nature. Thus, it is not known whether they result from substances produced by the developing leaf primordia and the apical meristem that diffuse outward from them, or whether the fields represent regions drained of some essential metabolite by the developing organ. But the experimental results clearly indicate that the positions occupied by new leaves are determined both by the apical meristem and by older leaves.

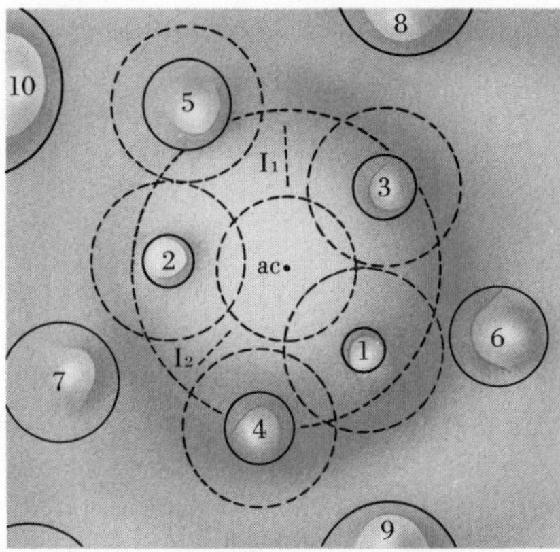


*Fig. 13-12* *A diagram of the shoot apex of the fern* Dryopteris *showing the distribution of physiological fields around the center of the meristem and the young leaves. It is seen that the position $I_1$ is already outside the boundary of the postulated fields and it is predicted that a leaf will soon emerge at this site. The $I_2$ position has just become cleared of physiological fields associated with the other leaves and a leaf may be expected to emerge there some time later. (From C. W. Wardlaw,* Growth *(supplement), 13:93, 1949, by courtesy of the author and the editors.)*

**Meristem–Vascular Tissue Interaction**

The cells produced by the apical meristems differentiate in continuity with older tissues of similar types, and it seems possible that differentiation of new cells may occur by induction, with the inductive stimuli coming from the older cells. When apical meristems of roots or shoots are excised and grown in sterile culture, as we saw previously, they produce tissues that differentiate normally and occupy normal positions in the organ. Although this might suggest that differentiation of tissues is regulated by the meristem and not by the previously differentiated tissues, the possibility must be considered that the older tissues had already induced a physiological "prepattern" in the meristem prior to excision and that tissue differentiation then proceeds in accordance with this patterning. Although it might appear difficult to design experiments to test these alternatives some very ingenious experiments carried out by Torrey on pea roots have provided conclusive answers.

Half-millimeter-long pea root tips, which contained little more than the root cap and the apical meristem, were excised and grown in sterile culture. When the vascular tissue which had differentiated in the roots was studied it was found to be distributed differently from that in a normal pea root. Pea roots usually contain three strands of xylem, but the roots grown from the excised tips sometimes contained only two or one xylem strand (Fig. 13-13). There was a gradual return to the three-stranded condition with further growth. The results indicate that if there is any prepattern it does not function in excised root tips, and that variation in the number of strands seems to be related to variation in meristem growth.

A second experiment by Torrey showed even more conclusively that there is no influence of the older preformed tissues on the pattern of differentiation of new cells. He took the bases of the pea roots from which the tips had been excised and grew them in culture. They regenerated new apical meristems from the cut surfaces and these continued the growth of the root. When the distribution of vascular tissue in the newly formed part of the root was studied it was found to bear no relation to the three xylem strands in the basal region. Depending on the composition of the culture medium there were up to six xylem strands, which originated independently of the preexisting xylem strands. Even in those regenerated roots that contained three xylem strands, the strands did not always originate in continuity with older xylem strands.

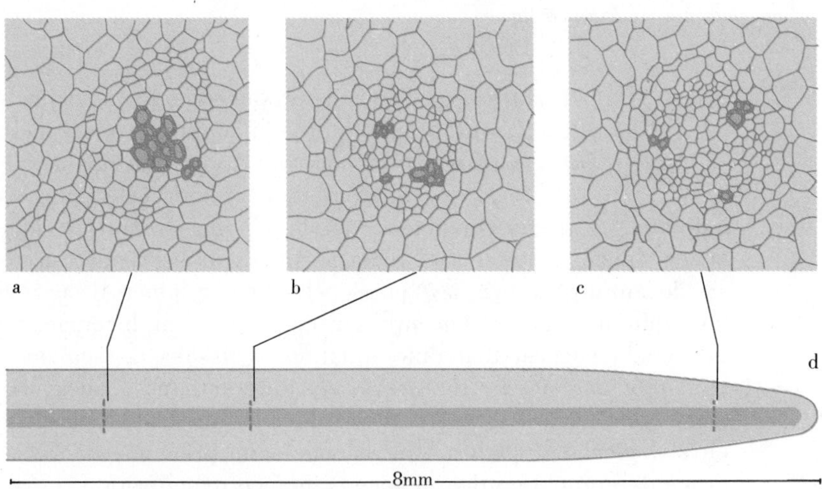


**Fig. 13-13** *Changes in the distribution of vascular tissue in excised pea roots during growth in sterile culture.* (a) *The base of the root which was formed soon after the root has been excised and started to grow in sterile culture. There is a single strand of xylem in the vascular tissue at this level in the root. In* (b) *taken about 1.4 mm further from the base than* (a) *there are two large xylem strands and the first cell of a third. In* (c) *taken 4.6 mm closer to the tip than* (b) *there are three evenly spaced xylem strands.* (d) *The positions from which the cross sections* (a–c) *were taken.* (a–c *from J. G. Torrey,* The American Journal of Botany, *42:183, 1955, by courtesy of the author and the Botanical Society of America.)*

These experiments demonstrate clearly that the differentiation of the vascular tissue of an organ is under the control of the apical meristem.

### *Leaf–Vascular Tissue Interaction*

In the shoot, as in the root, the initial differentiation of vascular tissue occurs in relation to the activity of the apical meristem, but its terminal phase of differentiation to form mature cells appears to be related more closely to the developing leaves. Procambial strands extend from the stem into each leaf from the time the leaf first emerges on the apical margin (Fig. 13-14), and the maturation of procambium as xylem and phloem occurs in a system of interconnecting vascular strands which are continuous with the vascular tissue of the root.

The influence of the leaf on differentiation in its associated procambial strand has been studied experimentally. If an expanding leaf of *Coleus* is cut off, the number of xylem cells that differentiate in the stem below the leaf is reduced. It seems that the leaf supplies something required for differentiation of procambial cells as xylem cells. Jacobs has studied the interaction between the developing leaf and vascular tissue differentiating in the stem, and has obtained information indicating that xylem differentiation is chemically controlled. He found that if one of the vascular strands in the stem is cut, new xylem cells are regenerated by redifferentiation of pith cells, and vascular continuity is restored around the wound (Fig.13-15). Removal of young expanding leaves above the cut reduces the number of xylem cells that are regenerated. The number of xylem cells that differentiate can be increased in defoliated plants if a concentration of the plant hormone indoleacetic acid equivalent to that obtained from leaves is supplied to the stem through the base of one of the removed leaves.

Indoleacetic acid is known to be synthesized in leaves and to move in a predominantly basipetal direction in the stem. Since the regeneration of xylem around the wound also followed a predominantly basipetal path, Jacobs concluded that differentiation of xylem is controlled by leaf-produced indoleacetic acid.

The evidence we have examined in the past few pages points strongly to the conclusion that apical meristems function as autonomous developmental organizers. Although young leaves may influence the terminal stages of vascular tissue differentiation and the positions that new leaves occupy, it is the apical meristems that regulate the initial differentiation in the derivative tissues, as well as the initial symmetry of the lateral appendages. However, the development of plants is re-

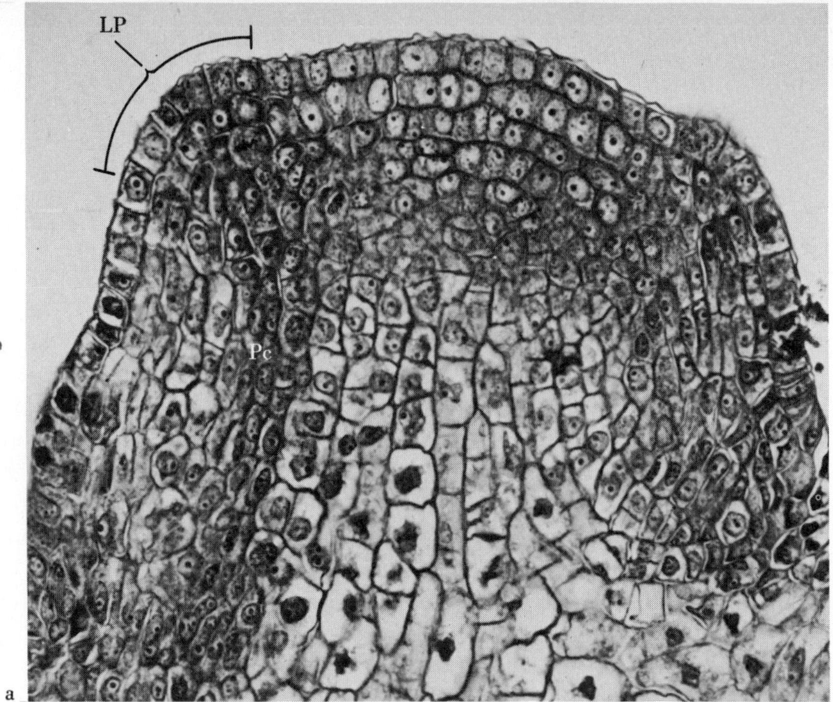


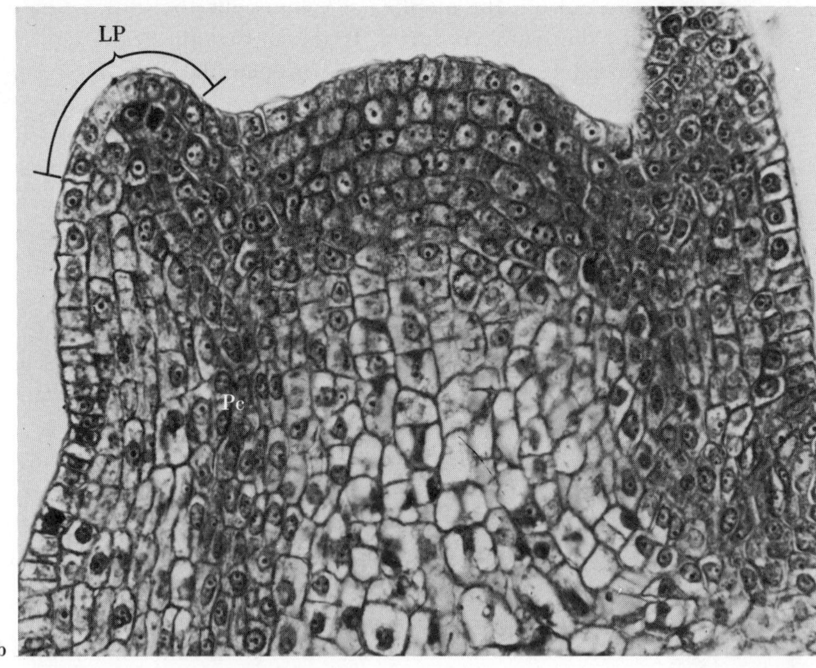


***Fig. 13-14*** *Longitudinal sections of the shoot apex of potato* (Solanum tuberosum) *showing differentiation of the leaf trace procambium (Pc) into a leaf primordium (LP). In* (a) *the leaf is just emerging on the left side of the meristem but its procambial strand is already differentiating. Shown in* (b) *is a slightly later stage of development. On the right side of the meristem a marginal portion of an older leaf is visible. (From I. M. Sussex,* Phytomorphology, *5:253, 1955, by courtesy of the International Society of Plant Morphologists.)*

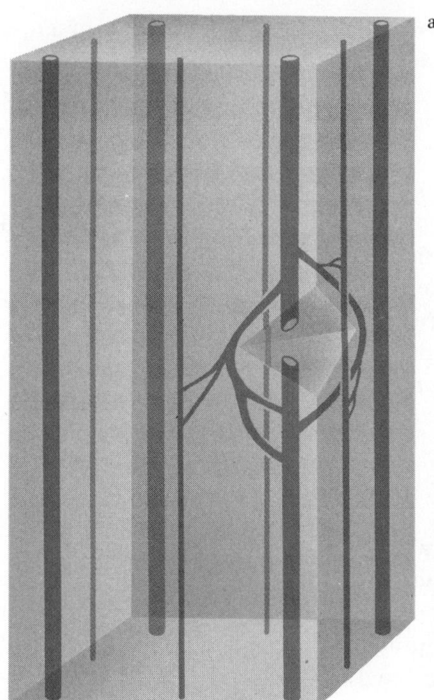

***Fig. 13-15*** *Regeneration of pith cells in the stem of* Coleus *as xylem after a vascular strand has been severed by a cut.* (a) *A diagram showing the cut in the corner of the stem which severed a major vascular strand. Newly regenerated strands of xylem cells are shown which reconnect the two parts of the severed strand. Some of the regenerated strands have established connections with other vascular strands in the stem.* (b) *A photograph in the zone of regeneration showing the severed end of the vascular strand on the right and some of the regenerated xylem on the left. (*b *from L. W. Roberts and D. E. Fosket,* The Botanical Gazette, *123:247, 1962, by courtesy of the authors and the University of Chicago Press.)*

markably responsive to environmental change, and in the next section we shall consider whether some aspects of meristem function are subject to direct or indirect regulation by the environment.

## INTERACTIONS OF THE PLANT AND THE ENVIRONMENT

The way in which plants respond to environmental change represents one of the major differences between them and animals. Because of their mobility, animals can respond by a change in their behavior pattern. Thus, they tend to move from an unfavorable to a more favorable environment as in seasonal migrations, and when disturbed with respect to illumination or gravity, for example, they may simply move so as to resume optimal orientation.

For plants, which are fixed in position, this mode of response is not possible; instead, they respond to an environmental stimulus by a change in their developmental pattern. Some plant growth responses are quite rapid. When a plant is laid horizontally so that its orientation in the gravity field is altered, or when it is illuminated from one side only, the plant alters its pattern of growth in such a way as to resume its normal orientation in respect to the stimulus, and this takes only a few hours. These responses are caused by differential growth of cells differentiating behind the meristems, and the meristems themselves are not affected. The physiological mechanisms underlying these responses are excellently analyzed in several chapters in *The Living Plant*, in this series.

In contrast, the responses of plants to some other environmental stimuli take weeks or months for completion. These involve the formation of new kinds of lateral organs and considerable changes in meristem growth. Thus, to contend with unfavorable climatic conditions meristems may produce protective bud scales in place of the usual vegetative leaves, or they may produce flowers in response to seasonal changes of temperature or day length. The physiological mechanisms of these developmental responses are also discussed in *The Living Plant* and we need not review them extensively here. Instead, we will concentrate on those aspects of the plant–environment interaction that bear on the question of meristem autonomy.

### Low-Temperature Effects

Biennial plants must be exposed to low temperature before they will produce flowers. Under usual conditions a biennial plant grows vegetatively during the first season, then becomes dormant and is subjected to low temperature for several weeks or months during winter. When growth is resumed the following season the shoot vegetative meristem is rapidly converted to a floral meristem and flowers are produced. Some interesting experiments have suggested that it is the meristem that perceives the low temperature. Cabbage plants kept under uniform high temperature will maintain continuous vegetative growth and will reach heights of several feet. They flower only after the plant has been exposed to low temperature. However, flowering occurs if the shoot tip alone is chilled while the rest of the plant is kept at the high temperature. This experiment suggests that the growing region of the plant responds directly to temperature, but experiments with a variety of rye are more conclusive.

Plants of the Petkus variety of rye do not flower until they have been exposed to low temperature. In usual horticultural practice, seeds are planted in the late fall or winter. They do not germinate and are exposed to cold during winter. This process is known as *vernalization.* Germination in the following spring is soon followed by flower formation. If Petkus rye seeds are planted in the spring without having been exposed to low temperatures the plants continue vegetative growth throughout the entire season, or flower only after prolonged vegetative growth, thus making them an uneconomical crop. The seeds can be experimentally vernalized if they are first allowed to imbibe water—this step is essential—and then maintained at 5°C for 14 weeks. Progressively shorter exposures to this temperature results in progressively more vegetative growth being made before flowers are formed.

Embryos excised from the rye seeds can be vernalized in sterile culture, and isolated shoot tips consisting of the apical meristem and a few leaf primordia can also be vernalized in culture and will initiate flowers on further growth. Thus, the low temperature seems to be perceived directly by the meristem, which then responds by an altered pattern of growth.

In some biennial plants, though not Petkus rye, the requirement of low temperature for flower initiation can be replaced by treatment with gibberellic acid, one of the plant hormones. This suggests that in normal development a change in the hormonal balance in the meristem causes it to be converted to the floral condition. Here, then, is one case in which the pattern of meristem growth is controlled by a stimulus from an external source, and in biennial plants we must modify our concept of meristem autonomy to include these facts.

### Photoperiodic Effects

In other plants, flowering follows exposure to days of appropriate length; this phenomenon is known as *photoperiodism.* Plants fall into three general classes in their flowering response to photoperiod. Some are *short-day plants* which initiate flowers only when the light period of the alternating light-dark cycles is less than a certain critical length. Short-day plants include *Chrysanthemum*, *Poinsettia*, soybean, and *Xanthium.* In the second class are *long-day plants*, which flower only when the critical day length is exceeded. Examples of long-day plants are lettuce, radish, spinach, and many varieties of tobacco. Other plants—such as dandelion, sunflower, and tomato—flower regardless of the day length and are referred to as *day-neutral plants.* Some plants require only a single inductive photoperiod to cause flowering, whereas others require a succession of inductive photocycles.

The photoperiodic stimulus is perceived in mature leaves, and the meristem itself is insensitive to direct stimulation by the photoperiod. In the leaves the photoreceptor is *phytochrome*, a chromoprotein, which exists in interconvertible active and inactive forms. The particular balance between the two is determined by the length of day and the spectral quality of the light. Phytochrome has absorption maxima at 660nm[1] and 730nm, and when red light is absorbed the active form of the pigment is formed. In some manner, which has not yet been fully elucidated, this causes a shift in the hormonal balance of the plant, and vegetative meristems then develop as floral meristems.

The change in the pattern of meristem growth is extreme. Cells in the central mitotically quiescent zone resume division and synthesize increased amounts of RNA and protein. The floral meristem first enlarges, then becomes progressively smaller, as lateral organs are produced. Finally all of its cells differentiate as floral parts, and meristem function ceases. The floral organs—the sepals, petals, stamens, and carpels—are initiated sequentially, and the meristem appears to pass through a succession of developmental stages in each of which a new kind of lateral organ is initiated (Fig. 13-16).

Is the floral meristem determined at one step by the stimulus received from the leaves, or does the succession of different lateral organs result from the arrival in the meristem of a succession of specific organ-forming substances? Answers to these questions have come by combining surgical and sterile culture techniques. If a floral meristem of tobacco is excised from the plant when it is just starting to initiate sepals, the first of the floral organs to be formed, it will continue to form normal floral organs in the normal sequence when grown in a very simple culture medium. It seems clear that the floral meristem is determined at a single step by the flowering stimulus coming from the leaves, but how is the successional formation of floral appendages regulated?

If, when the floral meristem is excised from the plant it is also bisected vertically into two halves, each will form a flower when grown in sterile culture. But the structure of the flower differs depending on the developmental stage the flower had reached when it was bisected. What we are interested in is the kind of organs formed along the side of the floral meristem next to the bisecting incision. It is found that the organs formed here are always the same kind as those being formed on the intact side of the meristem (Fig. 13-17). In meristems bisected when all the sepals had just been formed the first organs appearing next to the bisection are petals. If the meristem is bisected after petals and stamens have been initiated, a carpel is the organ formed next to the bisection. Thus, in the bisected meristems, reorganization does not involve a return to the initial condition but a continua-

[1] 1 nanometer (nm) equals 1 millionth of a meter.

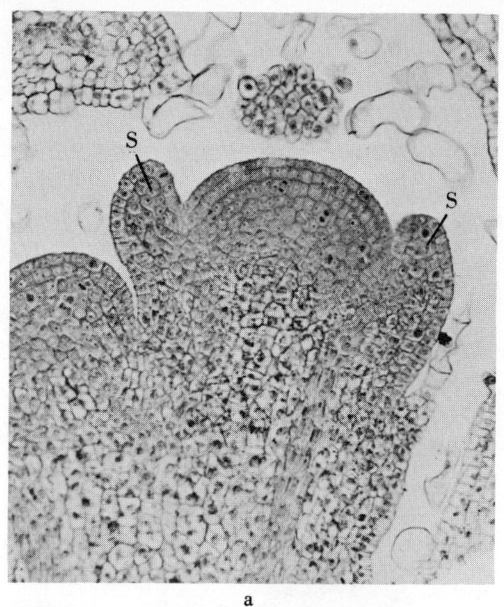


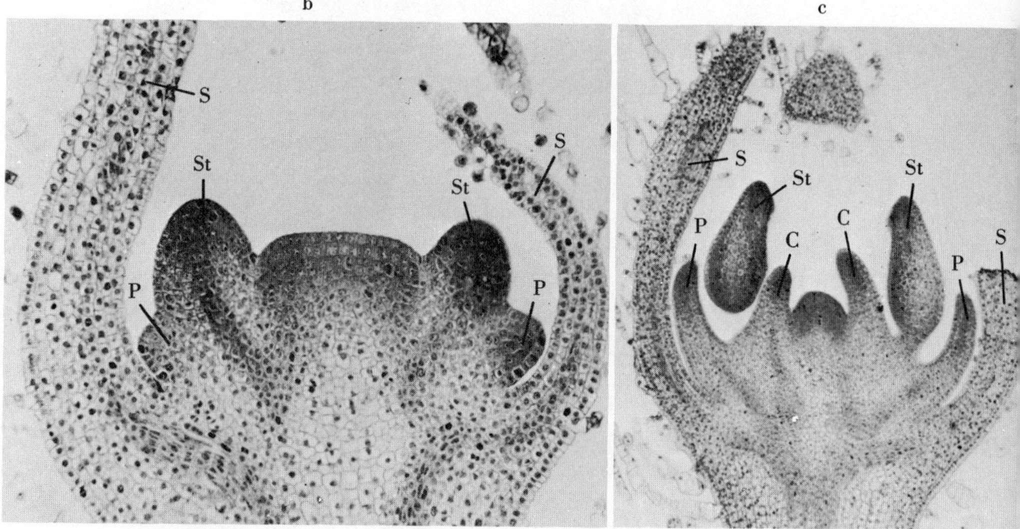


***Fig. 13-16*** *Stages in the development of a floral meristem of tobacco* (Nicotiana tabacum). (a) *The primordial sepal stage. The primordia (S) on either side of the meristem are two of the five sepals, the only floral organs which have yet differentiated.* (b) *The primordial stamen stage. The youngest primordia to have differentiated are stamens (St). Parts of two petals (P) appear below the stamens, and two of the elongating sepals (S) are below these. The meristem has become flattened prior to carpel initiation.* (c) *A flower in which all of the organs have been initiated. The youngest primordia are carpels (C). It can be seen that each kind of lateral organ formed in the flower has a distinctive morphology.* a–b × *88,* c × *70. (courtesy G. S. Hicks)*

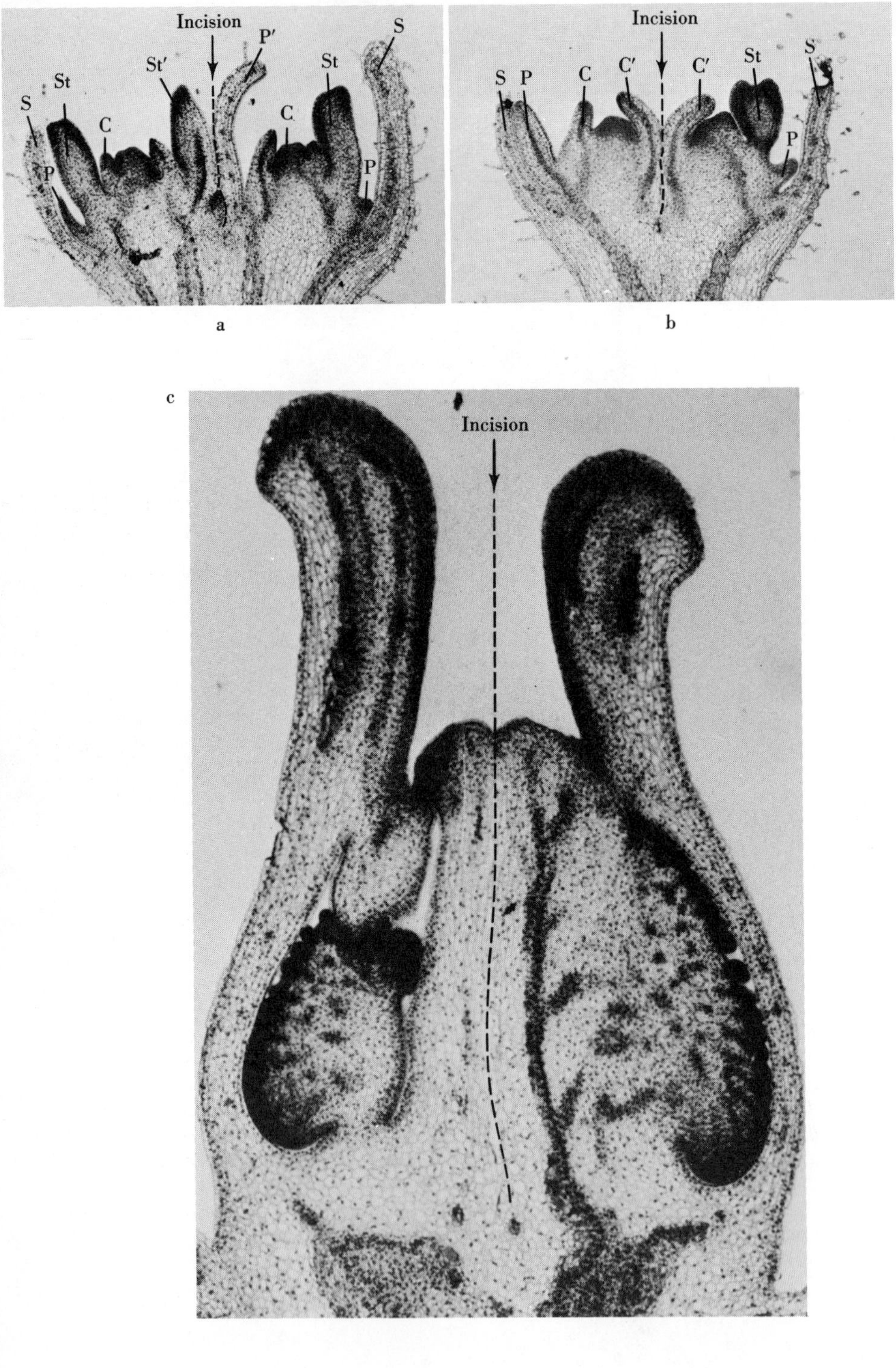

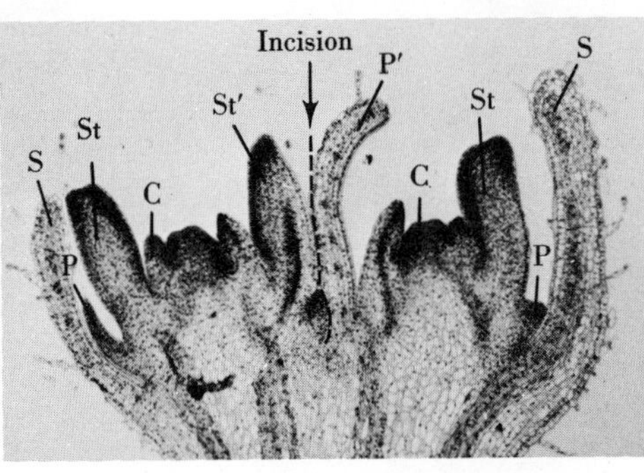


a

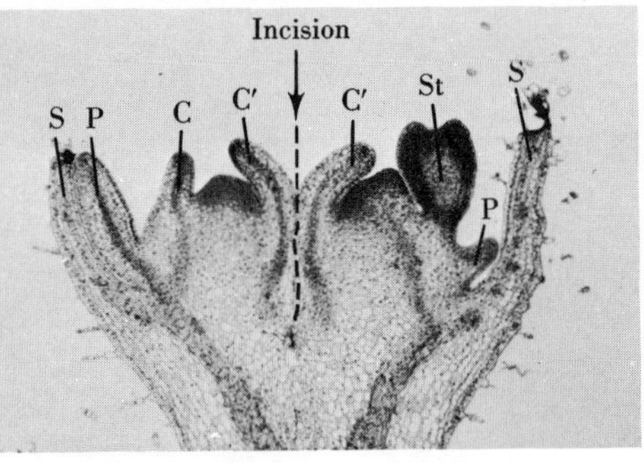


b

c

tion of those developmental processes that have already been set in motion by the triggering action of the floral stimulus.

## THE MERISTEMS AND THE PLANT

We have seen in this chapter that the meristems possess a high level of developmental independence. Given a supply of basal nutrients the apical meristems of both shoot and root are capable of normal development when grown in isolation from the rest of the plant, and are essentially autonomous. The evidence for this is stronger in the ferns than in the flowering plants, but it seems unlikely that the meristems in different plant groups are fundamentally different in their organization. Thus, the conclusion that meristems are developmental organizers seems to be well-founded, and it can be said that the meristems make the plants rather than the other way round.

The way in which the meristem regulates its own mitotic activity and the development of tissues and organs to which it gives rise is still not thoroughly understood. Because of the small sizes of the interacting parts, it has not been possible in most cases to determine whether the interaction is chemical or has some other cause. However, the evidence so far suggests that induction as it occurs in animal-cell differentiation does not play a major role in plant development. The chemicals that control plant development are more likely to be hormones, with molecular weights of a few hundreds, than to be macromolecules. Whether this is another consequence of the cell wall, which increases

***Fig. 13-17*** *Regenerative capacity of floral meristems of tobacco* (Nicotiana tabacum) *at different developmental stages. In each of the figures the floral meristem was bisected, and the two halves were excised from the plant and grown together in tubes of sterile nutrient medium. Development of the outer side of the flower was normal in each case, but regeneration of organs along the inner side of the flower next to the bisecting incision varied in relation to the stage of development reached by the flower when the cut was made.* (a) *Meristem bisected before sepal primordia were initiated. Stamens ($St^1$) and petals ($P^1$) were regenerated on the inner side of each flower and were similar to stamens (St) and petals (P) formed normally on the outer side.* (b) *The result obtained when the meristem at the primordial sepal or primordial petal stage is bisected. The capacity to regenerate stamens or petals on the inner side of the meristem has been lost, but carpel regeneration ($C^1$) still occurs.* (c) *Meristem bisected at the primordial carpel stage (a stage slightly younger than that reached by the flower in Fig. 13-16c). All regenerative capacity has been lost. In this figure only the inner part of the flower is shown. All × 35. (From G. S. Hicks,* Canadian Journal of Botany, *48, 1970, by courtesy of the author and reproduced by permission of the National Research Council of Canada from the Canadian Journal of Botany.)*

the space between plant protoplasts and which may act as a molecular sieve to restrict the movement of large molecules is not known, but it provides an interesting basis on which to speculate about the different control mechanisms in the two groups of organisms.

## *FURTHER READING*

Allsopp, A., "Shoot Morphogenesis," *Annual Review of Plant Physiology*, vol. 15 (1964), p. 225.

Ball, E., "Sterile Culture of the Shoot Apex of *Lupinus albus*," *Growth*, vol. 24 (1960), p. 91.

Clowes, F. A. L., *Apical Meristems*. Oxford: Blackwell, 1961.

Cusick, F., "Studies of Floral Morphogenesis. I. Median Bisections of Flower Primorida in *Primula bulleyana* Forest," *Transactions of the Royal Society of Edinburgh*, vol. 63 (1956), p. 153.

Cutter, E. G., "Recent Experimental Studies of the Shoot Apex and Shoot Morphogenesis," *Botanical Reviews*, vol. 31 (1965), p. 7.

Gifford, E. M., "Developmental Studies of Vegetative and Floral Meristems," *Brookhaven Symposium in Biology*, vol. 16 (1964), p. 126.

Jacobs, W. P., "The Role of Auxin in Differentiation of Xylem around a Wound," *The American Journal of Botany*, vol. 39 (1952), p. 301.

Laetsch, W. M. and R. A. Cleland, eds., *Papers on Plant Growth and Development*. Boston: Little, Brown, 1967.

Steeves, T. A., "On the Determination of Leaf Primordia in Ferns," in *Trends in Plant Morphogenesis*, E. G. Cutter, ed. London: Longmans, 1966.

Sussex, I. M., "The Permanence of Meristems: Developmental Organizers or Reactors to Exogenous Stimuli?" *Brookhaven Symposium in Biology*, vol. 16 (1964), p. 1.

———, and Clutter, M. E., "A Study of the Effect of Externally Supplied Sucrose on the Morphology of Excised Fern Leaves *in vitro*," *Phytomorphology*, vol. 10 (1960), p. 87.

Torrey. J. G., "Auxin Control of Vascular Pattern Formation in Regenerating Pea Root Meristems Grown *in vitro*," *The American Journal of Botany*, vol. 44 (1957), p. 859.

Wardlaw, C. W., "Experiments on Organogenesis in Ferns," *Growth*, vol. 13 suppl. (1949), p. 93.

Wetmore, R. H., "The Use of *in vitro* Cultures in the Investigation of Growth and Differentiation in Vascular Plants," *Brookhaven Symposium in Biology*, vol. 6 (1954), p. 22.

*chapter* **14**

# *The Biological Basis of Individuality*

From the fertilized human egg, about 0.14 mm in diameter, only barely visible to the naked eye (Fig. 14-1), we emerge as individuals with combinations of behavioral, chemical, and structural properties that stamp us as unique, except for those few of us who have an identical twin. We believe that this concept of the uniqueness of the individual applies equally to man and mouse, to fish and fowl. Whether each cockroach and each jellyfish, or each oak and each linden, is unique is less certain, largely because the techniques on which our ideas of individuality are based cannot yet be applied equally well to all living forms. We will not be far wrong, however, if we assume at the outset that, apart from those creatures that in nature or by

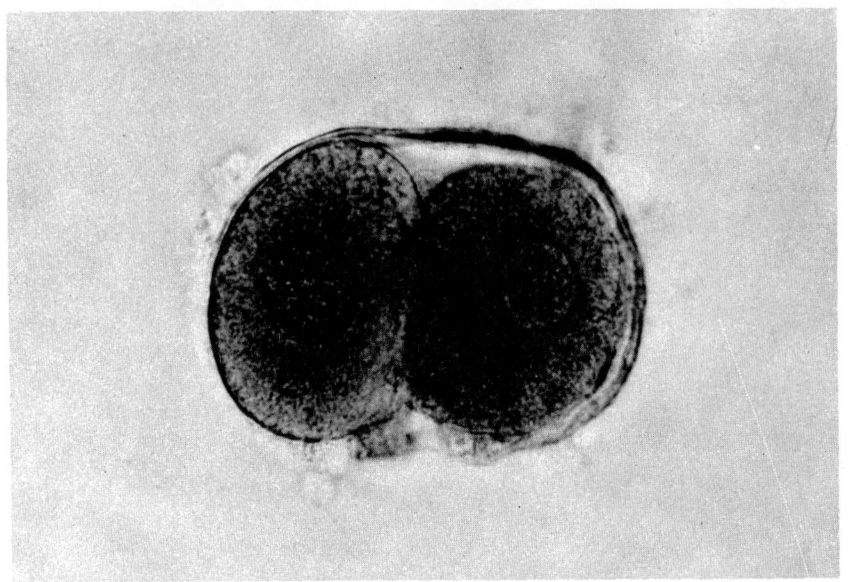

***Fig. 14-1*** *Photomicrograph of 2-cell human ovum.* × *500. (From A. T. Hertig, J. Rock, E. C. Adams, and W. J. Mulligan, in* Contributions to Embryology, *35, by courtesy of Carnegie Institution of Washington.)*

the artifice of the experimenter, or animal or plant breeder, have identical genes, each living being is unique.

Let us begin by asking, "What is the evidence for the uniqueness of the individual?" Not all the questions we have asked about development can be answered as directly as this one. We can tell, at least in the vertebrates, whether two animals are identical or whether they are different simply on the basis of whether grafts of their skin can be exchanged successfully. The failure of *homografts*, as grafts between nonidentical members of the same species are called, leads to one of the key generalizations of our time. This fundamental argument, established by several generations of biologists and surgeons, states that the response of an animal to such a graft is an immunological reaction, related to allergy and delayed hypersensitivity. The response is similar in all the vertebrates from fishes to man. At first a skin homograft heals; its epithelial cells begin to divide. For a time, then, such grafts behave like *autografts*—that is, grafts of the individual's own skin, transplanted from one part of the body to another, which persist for the life of the animal. But, in contrast to autografts, homografts are eventually thrown off following the breakdown of their blood vessels. The time required for their destruction varies in different species and, in the cold-blooded animals, also depends on the temperature; in many mammals the process is completed within 10 to 12 days.

We have already alluded to the fundamental factor determining whether a graft will be accepted or destroyed: the genetic relation between donor and host. The "rule" is well known: if a graft carries what are called "histocompatibility" genes (even one) that are absent in the host, it may evoke an antagonistic response. The absence in the donor of such genes as may be present in the host is not significant. The remarkable, subtle nature of these differences is shown best in certain highly inbred lines of mice: grafts made from females to males succeed, whereas those made from males to females fail. Otherwise genetically identical to the females, the males differ by containing a Y chromosome and the products of the genes contained in it, which are responsible for the incompatibility.

Thus we know that the cells of each individual—not just skin cells or muscle cells or kidney cells, but all the cells of the body—contain genetically controlled specific molecules or groups of molecules that stamp it as unique. And, as we shall learn, during development specific lines of cells emerge with the capacity of reorganizing these subtle differences.

Perhaps this last assertion will begin to answer some of the questions that have already occurred to you in reading this introduction. We began with a discussion of the uniqueness of the individual, advancing as evidence for it the fact that grafts between adult individuals are destroyed except in animals having identical genotypes. In nearly every other chapter of this book, however, you were confronted with evidence that appears to provide an exception to the rule: successful grafts between embryos, even of different species; chimeric tissues composed of embryonic chick and mouse cells; and numerous other examples. The fact is that homografts, even heterografts, are accepted and maintained throughout at least part of embryonic life. The time at which an animal acquires the capacity to recognize, and to react to, an antigenic stimulus, be it bacteria, virus or graft, depends on two factors: the nature of the stimulus and the species. This observation evokes two fundamental, related questions: (1) At what stage during development of an embryo does the set of properties that mark it an individual, capable of evoking an immune response, appear? (2) What factors underlie the gradual development of the capacity to respond?

We shall be concerned chiefly with the second question, for, as will be apparent shortly, explorations of the first are just beginning. However, we can hardly come to grips with the development of immunological mechanisms until we have taken a closer look at them

## *MECHANISMS OF TRANSPLANTATION IMMUNITY*

How are homografts destroyed? What is the evidence that the homograft reaction is an immunological reaction? Does it differ significantly from other immunological reactions? As we usually think of them, immunological reactions have two principal components: (1) an antigenic stimulus and (2) an organism, which responds to the stimulus by producing and releasing a specific antibody.

### *Nature of the Stimulus*

Antigens are molecules that possess the ability to evoke the production of a specific antibody or other response when they are administered to a suitable animal. Polysaccharides, proteins, and lipoproteins are antigenic; nucleic acids are poor antigens. Bacteria and viruses are antigenic by virtue of antigens contained within them or on their surface.

But what are the transplantation antigens—the antigens emitted by a graft? These must have *individual specificity*; that is, they must be alike for all members of the same genotype—hence the name *isoantigens.* They are not confined to one organ or to a group of organs; they are found in all the cells of the body of a given individual.

Our information about them is inadequate. Because they are genetically controlled, and because all nucleated cells contain them, it was once thought that the DNA (or DNA combined with protein) of an individual might itself be isoantigenic. However, the evidence thus far has not supported this hypothesis. At present, attention is centered on two large classes of molecules, the lipoproteins and mucoproteins, in which proteins are combined with lipids and polysaccharides, respectively.

### *Nature of the Response*

The immunological nature of the homograft reaction is suggested by a simple experiment. It is a commonplace experience in the field of immunology to observe that an animal that has been immunized to a given antigen reacts more quickly when the same antigen is injected later on. The elapsed time between the end of the first immunization and the second challenge can be weeks, months, or even years. How many of us have had chicken pox or measles a second time? For reasons we do not yet understand, we have an "immunologic memory." So it is with the homograft reaction. On the average, a mouse of one strain (for example, CBA) will destroy a skin graft from a mouse of another strain (A) in 10 to 11 days. If a second graft is made to the first recipient from a second A-strain mouse, even to another site on the body, it will be destroyed more rapidly, usually in about six days. From this experiment we learn that (1) the reaction is systemic, rather than local, and (2) that it behaves like an antigen–antibody reaction, displaying the property of "memory" or recall.

We now enter an area in which fact and hypothesis are difficult to distinguish. What is the nature of the difficulty? Until recently, all of the evidence for the immunologic nature of the reaction has been circumstantial. It is as tight a chain of circumstantial evidence as any ever forged. However, there has been no direct evidence proving that the reaction is effected by an antibody. We should review at least one experiment pointing up the problem. We know that antibodies may be transferred *passively*. All of us are familiar with administration of antibody (gamma globulin) to protect the unborn child of a woman who has been exposed to rubella (German measles) early in pregnancy. If antibodies are formed in response to a homograft stimulus, it should be possible to transfer them passively. The following kind of experiment has been tried repeatedly (Fig. 14-2).

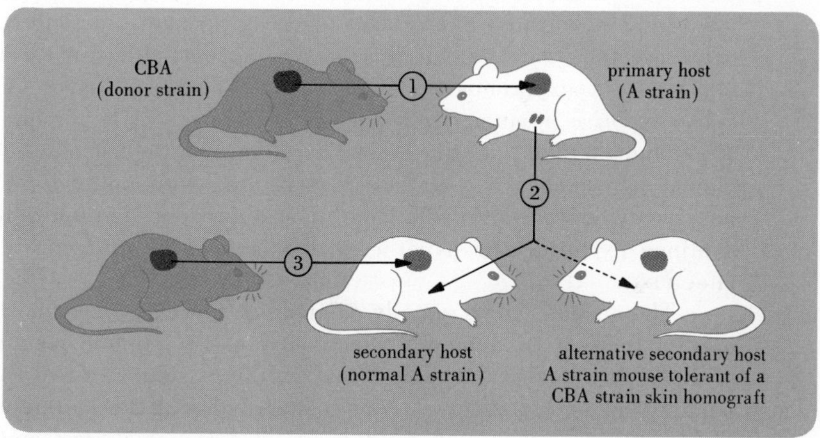


**Fig. 14-2** *Adoptive immunization. Transfer of transplantation immunity by means of sensitized lymph node cells.* (1) *A normal A-strain mouse (primary host) is sensitized by grafting it with homologous skin from a* CBA*-strain donor. After rejection of the homograft, the regional lymph nodes are excised from the sensitized animal and a cell suspension prepared from them.* (2) *These cells are then injected into a second* A*-strain mouse (secondary host). The latter animal may be either a normal* A*-strain mouse that is challenged with a homograft of* CBA*-strain skin* (3) *or an* A*-strain mouse that has previously been rendered specifically tolerant of a* CBA*-strain homograft. Accelerated rejection of the test graft in the case of the first secondary host, or prompt abrogation of tolerance in the case of the alternative secondary host, are indicative of the transfer of immunity. (From Fig. 1, R. E. Billingham and W. K. Silvers, in* Transplantation of Tissues and Cells, *p. 108, by courtesy of the Wistar Institute.)*

A skin graft is made from a CBA to an A-strain mouse; at the time it is being destroyed, at 10 to 11 days, the serum containing gamma globulin is transfused to a second A-strain mouse. The second mouse is now challenged with a CBA graft. If antibody has been transferred, this graft should be destroyed in less than 10 to 11 days. The median survival time should approach six days, the usual period in which a second graft is destroyed ("second-set time"). Almost invariably the result is 10 to 11 days, indicating that antibody is not involved. However, if the experiment is modified only slightly, it succeeds. If instead of transfusing serum, living lymphocytes are transfused, the graft is destroyed in second-set time.

Thus it appears that the homograft reaction requires not just antibodies, but living lymphocytes. If lymphocytes are required, what do they do? Two lines of thought have to be pursued.

The first holds that the homograft reaction does not differ fundamentally from any antigen–antibody reaction. Antibody is produced in response to the isoantigenic stimulus. However, the stimulus may be slight and the amount of antibody may be small, perhaps insufficient to be concentrated at the reaction site by transfusing serum. Antibody must be concentrated at the right place at the right time and lymphocytes are the "carriers." If this idea is correct, either of the following kinds of experiments should succeed: (1) transfusion of living cells (*adoptive* immunity); (2) passive transfusion of larger quantities of gamma globulin than those used heretofore, or the use of other techniques by which the graft site is saturated with antibody. Enough positive evidence is available from these approaches to require that we continue exploring the hypothesis that *cell-bound antibodies* are the effectors.

The second line is equally compelling, however. It holds that not only is there a fundamental dichotomy in the immune mechanisms of animals, but also that it is based on the existence of two separate populations of lymphocytes, one concerned with the production of antibodies, the other with fundamentally different allergic and graft-rejection reactions. This concept is based on evidence from two disparate lines: immunologic deficiency diseases of man and the development of antibody-forming mechanisms. We shall emphasize the latter.

## *DEVELOPMENT OF ANTIBODY-FORMING MECHANISMS*

When do animals develop the capacity to respond? At birth? Not necessarily. Maturity in transplantation immunity like other physiological properties, does not necessarily correspond with birth. It has been said that birth is a "moveable feast" in the calendar of development. If an investigator is ingenious and skillful enough, he can show, by making grafts to animals in the uterus, that fetuses of some species, like the sheep, can reject homografts at least 54 days before birth. The gestation period of the sheep is about 150 days. Moreover, the sheep fetus makes antibodies to some purified antigens when it is only 40 days old—that is, 40 days *in utero*. These experiments provide evidence on two points: (1) When the homograft reaction is first detected, it is complete. Fetal homograft reactions follow the same course and take the same time as those in the adult. (2) The sheep fetus begins to respond to different antigens at different times—to bacteriophage at 40 days (or earlier), to ferritin at 66 days, to ovalbumin at 125 days. This poses a problem: Why not all at once? It seems unlikely that once immunologically competent cells are present, they would not respond to all antigens. We can provide an answer to our question. It may not be entirely correct, but it probably will point us in the right direction. However, first we need to provide additional information.

At least in some species, and for some antigens, the formation of antibody may require the cooperation of at least two kinds of cells, one type that "recognizes" the antigen and "processes" it, and a second that responds by synthesizing globulins. We have already identified the responding cell as the *lymphocyte*. The other cell of the pair is the *macrophage*. Macrophages and related cell types engulf antigens, and the enzymes contained in their lysosomes, the cell particulates specialized in catabolic processes, metabolize them. How do macrophages then interact with lymphocytes?

The next step is more than mere conjecture, but it is not solidly established, either. One kind of mechanism that is being explored is described in the followng experiment. An antigen is mixed with a population of macrophages; after a suitable interval, the macrophages are homogenized, and ultimately their RNA is extracted. This RNA is added to a population of lymphocytes growing in culture. Ordinarily lymphocytes do not make antibody *in vitro* unless they have been primed with antigen before they were explanted. Since these lymphocytes are not so primed, the control cultures produce no antibody. But the lymphocytes to which this RNA is added make antibody specific for the antigen that was added earlier to the macrophages.

For a time it was believed that the macrophage was induced to synthesize an "informational" RNA, specific for the antigen—that is, that the antigen in some manner acted as an inducer or derepressor. Then this specific RNA would be transferred to the lymphocyte, there to direct antibody synthesis on polyribosomes. However, this idea appears to be an oversimplification. It has not been possible to prove that the RNA extracted is free of antigen. In fact, it is the combination of antigen with RNA that appears to be effective. Traces of antigen, so small as to be ineffective alone, become effectively stimulatory when combined with RNA. This complex of antigen with RNA has been called a "superantigen"; however, calling it that doesn't tell us how it operates.

In the chick and mouse it appears that the immune mechanisms start operating about the time of hatching or parturition, although the evidence thus far is piecemeal. It would not be surprising to find that the sequence is much like that in sheep. At the other end of the time scale, homografts can be exchanged between rats during the first two weeks after birth.

Thus at birth—or before or after birth, depending on the species—there must exist in each vertebrate populations of cells capable of responding to a stimulus, a stimulus which we may designate simply as "foreignness." Clearly, these cells already differ from their fellows, but the difference is magnified as a result of the stimulus. What are the steps by which such cells differentiate? This is another example of cellular differentiation in which the initial stage is the consequence of purely intrinsic reactions, and the final one is the consequence of an extrinsic stimulus.

## *THE TWO-COMPONENT CONCEPT OF THE LYMPHOID SYSTEM*

Our problem then has become one of cellular differentiation. Macrophages take up antigens, and are modified by them. Lymphocytes respond to antigenic stimuli, some, if not all, of which are mediated by macrophages. Once signaled, some lymphocytes undergo a series of changes as they elaborate the machinery for protein synthesis, being transformed into *plasma cells*, specialized for the synthesis and secretion of antibody. Other lymphocytes become specialized for homograft immunity.

How do we know that such a division of labor exists, and what is its cellular basis? The first clues came from studies of deficiency diseases in man. As first described by Bruton, and studied extensively by Good and co-workers, some *agammaglobulinemic* patients lack the ability to make antibodies. These patients do not have plasma cells, and the lymphoid tissues associated with the gut (tonsils, intestinal "Peyer's patches") are deficient. On the other hand they reject homografts quite well, and may develop allergies. They have a well-developed thymus.

In other deficiency diseases, although the evidence is less complete, the situation may be exactly reversed; that is, patients with plasma cells, making antibodies, lack the thymus. Such studies suggest that quite possibly antibody formation requires lymphocytes somehow derived from or dependent upon the lymphoid structure of the gut, whereas homograft reactions require thymus-dependent lymphocytes.

These organs have long been enigmatic. Consider the thymus: it is rich in lymphocytes, yet it is not truly effective in antibody production; it is maximal in size in young animals, when immunological capacity is minimal, and generally decreases in size as the animal grows older.

Experiments have now shown that the thymus plays a key role in the development of the lymphoid system. If the thymus is removed from an older animal, no effects of consequence are observed. However, if the newborn animal is thymectomized, there is a diminution in lymphocytes. Moreover, the capacity to mount a homograft response is eliminated or drastically reduced. It has been found that when rudiments of the spleen of the 13- or 14-day mouse embryo are cultured *in vitro* they fail to become lymphoid unless embryonic thymic rudiments are cultured with them. What is the explanation? Two possibilities have been envisioned, and are being tested. The first is that the thymus is the source of lymphoid cells responsible for the homograft reactions. By this we mean that the normal origin of these lymphoid cells in the spleen and lymph nodes is by immigration of the cells from the thymus. The second possibility states that the thymus produces a diffusible factor necessary for lymphoid differentiation. Thus far there is evidence in favor of both roles; we may find that the thymus plays a dual role. Moreover, the humoral factor may be required either for stimulating the proliferation of immunologically competent cells or, as seems more likely, for inducing the differentiation of lymphoid precursors to lymphocytes.

In birds, the division of labor is clear: cells destined to play a role in the homograft reaction are derived from (controlled by) the thymus, whereas those destined to make circulating antibodies owe their origin to another organ, the bursa of Fabricius, a cloacal lymphoid structure whose role is analogous to that of the tonsil and Peyer's patches in mammals (Fig. 14-3).

We can carry our search for the origin of these cells a step or two further. We must consider the possibility that the thymus and bursa (or tonsil) are only "way stations" in the movement of these white blood cells from some primordial site to their definitive positions in bone marrow, lymph nodes, and spleen. There is ample evidence in the mammal that the production of red blood cells begins in the blood islands of the yolk sac. In time, the site shifts, first to body mesenchyme, then to the liver; until, finally, erythrocytes are produced in the bone marrow.

The analogy to the red-cell system is not based entirely on conjecture. When embryos are parabiosed by fusion of their yolk sacs

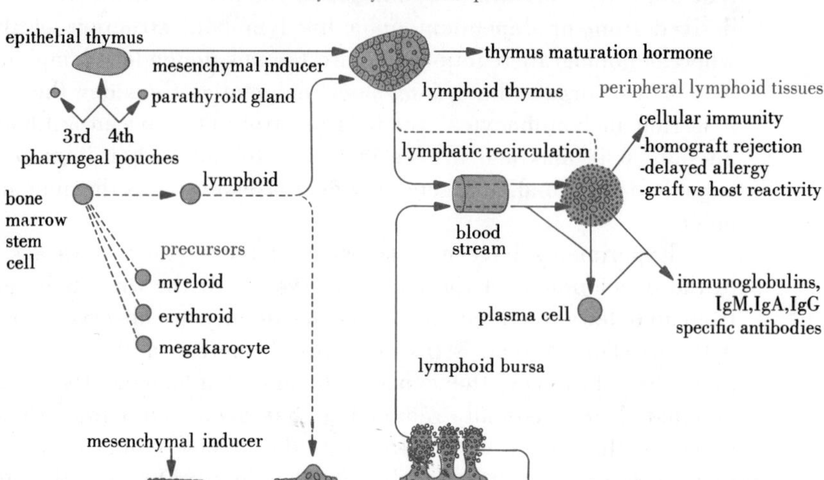


***Fig. 14-3*** *Differentiation of the bone marrow stem cell. There are two lines of differentiation of the lymphoid stem cell: a* thymus-dependent *population, which gives rise to lymphocytes responsible for the cell-mediated immunological responses; and a thymus-independent/*bursal-dependent *population, which is ultimately responsible for plasma cell development and immunoglobulin synthesis. (From* Immunologic Diseases in Man, *R. A. Good, ed., National Foundation Press.)*

before the thymus has formed, the cells of the thymus are found to be chimeric, that is derived from both yolk sacs, indicating migration into the thymus as it is formed. What other cells migrate from the yolk sac early in development (see page 93)?

## *THE CLONAL SELECTION THEORY OF ANTIBODY PRODUCTION*

Thus far we have been considering major classes of immunologically competent cells. In doing so we have spoken of "antibodies," much as we have earlier spoken of "hemoglobins" or "crystallins." Although there are clearly several hemoglobins and crystallins, the range of antibody diversity is enormous; that is, the number of different antibody molecules which one animal can synthesize must be very large. An early estimate of one million may be orders of magnitude too low. How is this diversity generated? What is the role of an antigen? Does it somehow "instruct" the responding cell to make specific antibody, or does it "select" a cell already determined to make specific antibody and stimulate its further response? To put it in the terms we have used for other systems, is a given cell determined to make antibodies, generally, with the specificity imposed by antigen, or is each cell dernined to make specific antibody with the antigen's role being to magnify the response? We believe that the evidence favors the selective hypothesis, as advanced first by Burnet in 1959.

It is necessary to emphasize at the outset that the specificy of an antibody resides in its primary structure, that is in its amino acid sequence.

All of three main classes of immunoglobins, γ-G, γ-A and γ-M are composed of two types of polypeptide chains, heavy and light, which are linked by disulfide bonds. Each light chain, for example, contains a constant half and a variable half. Possibilities for diversity therefore must occur in the variable parts of the light and heavy chains. Different antibodies have different amino acid residues at the variable positions, and identical residues in other stretches that are conserved. Thus, at some point in the determination process the enormous potential for diversity must be generated. The mechanism must be one that permits partial rearrangement and partial conservation of the gene product. None of the levels of control we discussed in Chapter 8 appears to provide exactly the kind of mechanism required for antibody production.

There exists, then, in the bone marrow of the adult, a population of blood stem cells, descendants of cells that had their origin in the yolk sac. The evidence currently available suggests that these stem cells are initially determined only to be blood cells. They may give rise to erythrocytes, granulocytes, or lymphocytes. However, when one of these stem cells is stimulated (triggered, for example, by thymic hormone) to become a lymphocyte, a decisive step must occur, committing that cell and its progeny to the production of antibody molecules of one particular specification. The nature of that decisive step is unknown; some believe it may require a kind of "scrambling" mechanism to bring about a reassortment in nucleotide sequences before transcription; others speculate about the generation of diversity at the translational level.

Whatever the mechanism, the cell now makes its specific antibody, the synthesis of heavy and light chains proceeding on separate polyribosomes. Although there is not complete agreement on the subject, it appears probable that a single cell ordinarily produces only one class of light chain and one class of heavy chain.

But we have discussed antibody formation without mentioning antigen. It appears that once a lymphocyte is determined to make a specific antibody, it proceeds to do so, but only in very small quantities. The antibodies so produced are held in the lymphocyte, where they serve as *antigen receptors.* Somehow the combination of the receptor with antigen stimulates the proliferation of this specific line of cells, and antibody is now produced in bulk.

### IMMUNOLOGICAL SUPPRESSION

Clearly, the capacity for recognizing individual or species differences arises during development; perhaps you already see a corollary to this statement. If an embryo acquires the capacity to distingush "self" from "nonself" at some critical point in its life history, then it should be possible to confuse the embryo by exposing it to foreign cells before that critical stage is reached; such cells should then persist, behaving as if they were the animal's own cells. Genetically, they should remain foreign; that is, they should retain the specific complement of genes with which they were endowed; their cell products should be typical of their origin. But these products will not be identified as foreign because the mechanism for receiving the signal, "foreign," will not respond to this stimulus. Consider the following now-classic experiment by Medawar and his associates: several hundred thousand living spleen cells of an adult strain-A mouse are injected into a 17-day-old mouse fetus of strain CBA. The injected mouse is born a few days later and is reared until it is six months old. At that time a skin graft from a mouse of strain A is made to it. In the normal, untreated CBA mouse, such a graft would be destroyed within 10 to 12 days, but the treated mouse is *tolerant*, having been exposed to A cells before they could be recognized as foreign. Such a graft may persist for the life of the animal. The graft retains its own genetic identity, but the host's reaction mechanisms for this specific strain have been suppressed. To prove that the host, not the graft, has been altered, the following experiment is performed. Six months after a successful graft has been made, a few million adult lymph node or spleen cells from another normal CBA mouse are injected into a tolerant CBA mouse bearing such a graft. Now the graft is destroyed; tolerance is abolished. This experiment proves a fundamental point: in tolerance, the graft has not been modified; under the influence of its genes, it continues to produce and release into the circulation its individual recognition signals (Figs. 14-4 and 14-5).

We now believe that actively acquired tolerance, as the phenomenon was first called, is but one example of a broader spectrum of *immunological paralysis* or suppression. It has been found in adult animals that although there is a wide range of concentration of antigen that elicits antibodies, above a limiting concentration (which varies with the antigen) the same antigen is inhibitory. When the antibody-

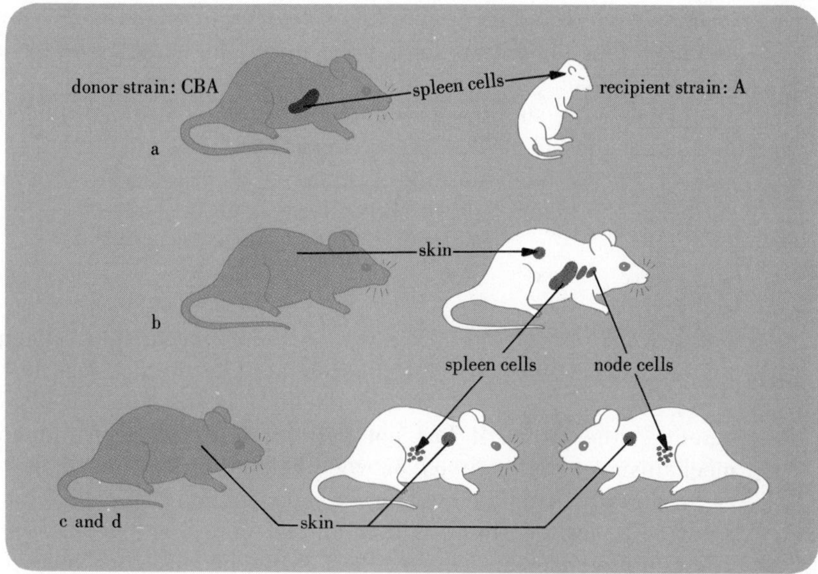


**Fig. 14-4** *Induction of tolerance, and the test for persistence of donor strain cells in the organs of tolerant animals.* (a) *Injection of newborn* A-*strain mice with CBA spleen cells.* (b) *Displaying tolerance by grafting with CBA skin.* (c) *Removal of host organs, preparation of cell suspensions and their injection into normal A-strain "secondary hosts."* (d) *Grafting of the secondary hosts with CBA skin. (After R. Billingham and L. Brent, in* Philosophical Transactions of Royal Society of London, *242, by courtesy of the Society.)*

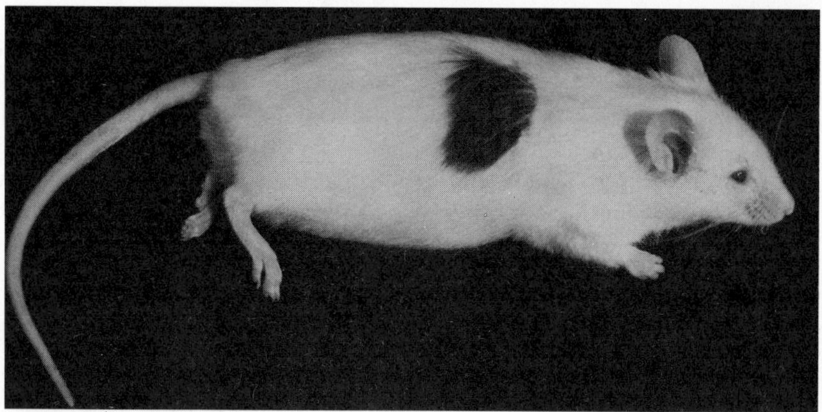

**Fig. 14-5** *Acquired tolerance in mice. Photograph shows a CBA skin homograft that had been transplanted 50 days previously to a specifically tolerant A-strain host. The graft has regenerated a dense crop of hairs. (From Fig. 3, R. E. Billingham and W. K. Silvers, in* Transplantation of Tissues and Cells, *p. 11, by courtesy of the Wistar Institute.)*

forming system is just beginning to develop in embryos and newborn animals, the amount of antigen required to inhibit will be correspondingly small.

How is the suppression effected? We don't know. Although it appears that the effect of high concentrations of antigen (relative to the size of the antibody-forming system) is to repress the *synthesis* of antibody, the manner in which the repression is achieved is unknown.

### GRAFT-VERSUS-HOST REACTIONS

We may examine one further consequence of experiments like the one just described. When the host is immunologically incompetent—either when it has not yet developed its own immunologic mechanisms, as in embryos, or when its lymphoid system is destroyed, as in heavily irradiated adult animals—immunologically competent adult cells grafted from another animal may react *against the host.* The consequences are severe. Pathologic changes are observed in the liver, the thymus, and especially the spleen, which is markedly enlarged. Animals with *runt disease* (Fig. 14-6), as it is sometimes called, are stunted and unhealthy in appearance, and, especially in the case of the chick, frequently die before birth. The reaction appears to be more complex than originally thought. It is basically an immunological reaction, as suggested by several kinds of evidence, including studies in which inbred lines of animals are used. However, the initial proposition that the enlarged spleen resulted from a colonization of the spleen by the donor cells, which then proliferated, appears to be only partly correct. More rigorous studies, in which the donor cells have been followed cytologically (using the sex chromosomes of the chicken as a marker) show that not more than a third of the cells in the enlarged spleen are from the donor. The cells of the host (in the chicken, the *granulocytes*, blood cells unusually rich in lysosomes) proliferate rapidly, accounting for the bulk of the enlargement. We do not know the stimulus for the proliferation of host cells.

### ORGAN TRANSPLANTATION

Recent experience in transplanting hearts and kidneys is based in large part on the basic discoveries we have just described. The proposition may be stated directly in a series of questions and answers. Is it possible to suppress the immunological response in man completely? It is, by irradiation, drugs, or antibodies made against lymphocytes. However, the patient cannot be kept in this state in-

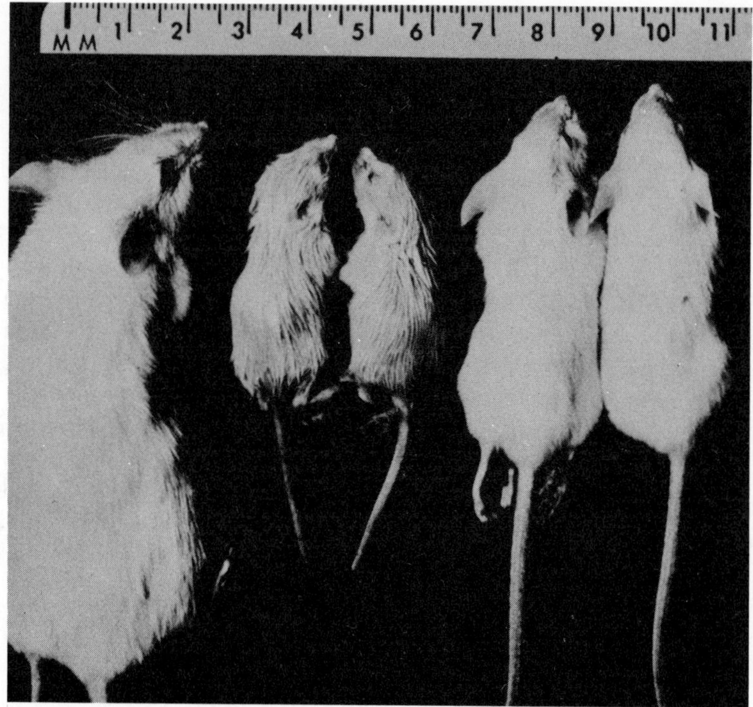


***Fig. 14-6** Runt disease in mice, produced by intraperitoneal injection of spleen cells from an unrelated strain. The photograph shows a mother and litter of runted and normal offspring. Note the characteristic "oily" coats of the two animals, in which runting was induced with homologous immune cells. (From J. Jutila and R. Weiser,* Journal of Immunology, *88, by courtesy of The Williams & Wilkins Company.)*

definitely, unless he is to be sustained in a germ-free environment. Eventually the immunologic mechanism will regenerate.

Since the administration of an antigen or a graft to an unresponsive embryonic host results in suppression or tolerance as the immunological system develops, might it not also happen that a graft made to a host rendered immunologically ineffective by irradiation, would also specifically suppress the rejection mechanism as it regenerated? The answer is yes.

Obviously the prognosis in any such graft is difficult for several reasons. Basically, we fail to understand the underlying mechanism of immunological suppression. On the practical side, we have to contend with both host-versus-graft and graft-versus-host reactions, with the host being essentially a "battleground" of two conflicting cell populations. It is a question of delicate balance: totally suppressing both host-versus-graft and graft-versus-host reactions until the grafted organ is firmly entrenched; as the immune mechanisms are restored, both components will remain in balance.

## *FETUS AS HOMOGRAFT*

The mammalian fetus is, by our definition, a homograft (or allograft). It contains genes that are absent from its host, the mother. Early in development, the trophoblast becomes intimately associated with the uterine epithelium. Thus it would appear that there would be ample opportunity for rejection. Why is the fetus not rejected? Are the embryo and associated trophoblast nonantigenic? Or is the fetus somehow protected from the maternal response?

A number of experiments have shown that the embryo and trophoblast are, in fact, antigenic. Moreover evidence has accumulated to show that maternal lymphocytes are potentially capable of destroying the fetus. Thus the fetus must be protected from the immunological hazards of pregnancy by some kind of barrier. The nature of that barrier is not yet known with certainty, although evidence points to the existence of a protective layer of mucoprotein between the fetal and maternal components.

## *AUTOIMMUNITY*

We have seen the severe consequences of immunological reactions in both the "conventional" host-versus-graft and rarer graft-versus-host combinations. Does an animal ever develop immunity to its own antigens? The answer is yes; the problem is a medically important one for man himself. We can do little more than introduce the concept. A number of diseases in man are thought to be the consequence of autoimmunization. How is it possible? Let us look at one experiment. If one testis of an adult guinea pig is removed, emulsified in oil and mixed with an *adjuvant* (killed bacteria or any substance that increases the immunological response) and the mixture is then reinjected into the animal from which the testis was removed, after a few weeks the remaining testis will become *aspermatogenic*; that is, the spermatogenic elements will be completely destroyed (Fig. 14-7). All of the evidence points to the conclusion that the reaction is immunological; it is related, if not identical, to the homograft reaction. Why should sperm be autoantigens? Possibly because they are produced relatively late in life, after the immunological mechanisms have matured. They have been isolated, having developed in a "physiological quarantine." When suddenly

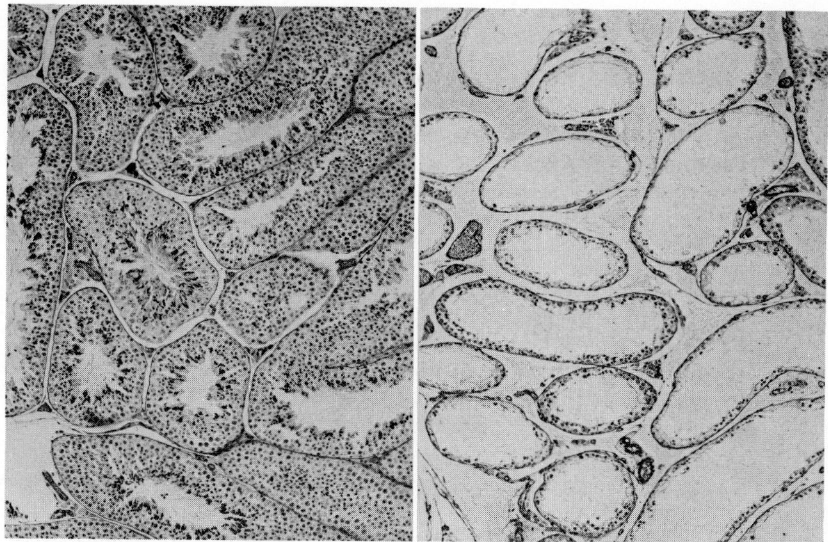

***Fig. 14-7*** *Induced aspermatogenesis in the guinea pig.* (Left) *Normal testis of adult guinea pig;* (Right) *Aspermatogenic testis of guinea pig two months following intramuscular injection of an extract of adult guinea pig testis with adjuvant. (By courtesy of D. W. Bishop.)*

these antigens enter the bloodstream they are recognized as foreign. Other tissues that appear to be autoantigenic because they have been in physiological quarantine include brain, lens, and thyroid. It is therefore not only theoretically possible, but in fact demonstrable, that disease and death may result from autoimmunization. As a result of a genetic defect, or a defect produced by infection or some other stress, tissue antigens may enter the circulation. The body reacts to them producing a response similar to a homograft reaction. The reaction extends to the organ itself, resulting in damage or destruction. At a recent conference at least eleven such diseases were discussed—and the list is growing. We see here another example of the way in which basic and clinical problems are interwoven. And an advance from either direction has an impact on the field as a whole.

To all investigators in this field the true excitement lies in unraveling the mechanisms through which we acquire the marvelous capacity not only of recognizing an antigen and reacting to it in a highly specific fashion, but also of "remembering" what we have "learned." What is the molecular basis of immunologic memory? Possibly our first breakthrough, our first glimmer of understanding of memory, will come in studies not of the brain but of the lymphocyte.

## FURTHER READING

Auerbach, R., "The Development of Immunocompetent Cells," in *Control Mechanisms in Developmental Processes*, M. Locke, ed. New York: Academic Press, 1967, p. 254.

Billingham, R. E., and W. K. Silvers, "Skin Transplants and the Hamster," *Scientific American*, January 1963, p. 118.

Burnet, F. M., "The Mechanism of Immunity," *Scientific American*, January 1961, p. 58.

———, "The Thymus Gland," *Scientific American*, November 1962, p. 50.

———, *The Integrity of the Body*. Cambridge, Mass.: Harvard University Press, 1962.

DeLanney, L. E., and J. D. Ebert, "On the Chick Spleen: Origin, Patterns of Normal Development and their Experimental Modification," Carnegie Institution of Washington publication 621, *Contributions to Embryology*, vol. 37 (1962), p. 57.

Good, R. A., *Immunologic Deficiency Diseases in Man*. New York: The National Foundation, 1968.

Jerne, N. K., "Summary: Waiting for the End," *Antibodies; Cold Spring Harbor Symposia on Quantitative Biology*, vol. 32 (1967), p. 591.

Kabat, E., *Structural Concepts in Immunology and Immunochemistry*. New York: Holt, Rinehart and Winston, 1968.

Medawar, P. B., *The Uniqueness of the Individual*. New York: Basic Books, 1957, p. 80.

Miller, J. F. A. P., "The Thymus and the Development of Immunologic Responsiveness," *Science*, vol. 144 (1964), p. 1544.

Silverstein, A., "Ontogeny of the Immune Response," *Science*, vol. 144 (1964), p. 1423.

Smith, R. T., R. A. Good, and P. A. Miescher, *Ontogeny of Immunity*. Gainesville, Fla.: University of Florida Press, 1967.

Sterzl, J., and A. Silverstein, "Developmental Aspects of Immunity," *Advances in Immunology*, vol. 6 (1967), p. 337.

# Index

## D

W



X



Y



Z